Tarahumara Medicine

Tarahumara Medicine

*Ethnobotany and Healing
among the Rarámuri of Mexico*

Fructuoso Irigoyen-Rascón
with Alfonso Paredes

University of Oklahoma Press : Norman

This book is published as part of the Recovering Languages and Literacies of the Americas initiative. Recovering Languages and Literacies is generously supported by the Andrew W. Mellon Foundation.

Also by Fructuoso Irigoyen-Rascón
Cerocahuí: Una Comunidad en la Tarahumara (Mexico City, 1974)
(and Jesús Manuel Palma Batista) *Rarajípari: La Carrera de Bola Tarahumara*
  (Chihuahua, Mexico, 1994)

Library of Congress Cataloging-in-Publication Data
Irigoyen Rascón, Fructuoso.
Tarahumara medicine : ethnobotany and healing among the Rarámuri
of Mexico / Fructuoso Irigoyen-Rascón with Alfonso Paredes.
pages cm
Includes bibliographical references and index.
ISBN 978-0-8061-4828-1 (hardcover : alk. paper) —
ISBN 978-0-8061-4362-0 (pbk. : alk. paper)
1. Tarahumara Indians—Medicine. 2. Tarahumara Indians—Ethnobotany.
3. Tarahumara Indians—Rites and ceremonies. 4. Traditional medicine—
Mexico—Tarahumara Mountains. I. Paredes, Alfonso. II. Title. III. Title:
Ethnobotany and healing among the Rarámuri of Mexico.
F1221.T25I76 2015
972.16004′974546—dc23
                                    2015015807

The paper in this book meets the guidelines for permanence and durability of the Committee on Production Guidelines for Book Longevity of the Council on Library Resources, Inc. ∞

Copyright © 2015 by the University of Oklahoma Press, Norman, Publishing Division of the University. Paperback published 2016. Manufactured in the U.S.A.

All rights reserved. No part of this publication may be reproduced, stored in a retrieval system, or transmitted, in any form or by any means, electronic, mechanical, photocopying, recording, or otherwise—except as permitted under Section 107 or 108 of the United States Copyright Act—without the prior written permission of the University of Oklahoma Press. To request permission to reproduce selections from this book, write to Permissions, University of Oklahoma Press, 2800 Venture Drive, Norman OK 73069, or email rights.oupress@ou.edu.

To my wife, Josefina,
 my daughter, Josefina,
  and my sons, Tocho and Saul

Contents

List of Illustrations · ix
List of Tables · xi
Acknowledgments · xiii

Introduction · 3
1. The Tarahumara Ecological Habitat · 15
2. A Historical Review of the Tarahumara People · 36
3. Rarámuri, the People and Their Culture · 58
4. Affiliative Social Activities of the Tarahumara People · 88
5. Great Life Occasions and Ceremonies: Birth and Death among the Tarahumaras · 94
6. Major Festivities of the Tarahumaras · 102
7. Loss-of-Health Conceptual Schemes of the Tarahumaras · 113
8. Rarámuri Healers · 135
9. The Jíkuri Ceremonial Complex · 140
10. Compendium of Tarahumara Herbal Remedies and Healing Practices · 155
11. The Tarahumaras: A Conventional Medical Perspective · 267

Notes · 285
Bibliography · 357
Index · 383

Illustrations

All photographs are by the author.

Figures

Urique Canyon · 21
Mixed pine-oak forest in Tarahumara Country · 28
Tarahumara man smiling for the camera · 60
Tarahumara men listening to a *nawésari* · 61
Blond Tarahumara children from Kwechi · 62
Tarahumara girls playing · 62
Tarahumara women sunbathing · 63
Tarahumara girl's face · 63
Young Tarahumara woman holding her child · 64
Nawésari ceremony · 80
Group of *pintos* · 104
Group of *pariseo* (pharisee) dancers · 105
Tenanches with the saints of the pueblo · 107
Matachín dancers · 112
Jíkuri (peyote) ceremony · 146
Jíkuri ceremony—peyote is placed in a hole · 147
Jíkuri ceremony—the *sipáame* speaks to the peyote · 148
Blooming field of *Stevia serrata* (*ronínowa*) in Tarahumara Country · 241
Tarahumara girl suffering from kwashiorkor-type malnutrition · 273
Same Tarahumara girl after treatment · 273

Radiographic image of a Tarahumara child with severe gastroenteritis · 275
New patient clinic visits for acute infectious gastroenteritis · 276

Maps

Tarahumara Country physiographic areas · 16
Tarahumara Country population centers · 19
Tarahumara Country isotherms · 26

Tables

Annual temperatures and rainfall in Tarahumara Country ᛫ 23–25
Acute illnesses in rainy season and in winter ᛫ 277
Chronic illnesses ᛫ 278
Volume of clinical work at Clínica San Carlos, Norogachi, Chihuahua ᛫ 283

Acknowledgments

I would like to acknowledge the trust and help that I received from the several Rarámuri men and women who opened to me the gates of a centuries-old culture that evolved on the American continent long before the arrival of Europeans. In particular, I appreciate the help of Erasmo Palma Tuchéachi, a native Rarámuri scholar and keen observer of nature. He assisted me a great deal in the process of learning the Rarámuri language. He helped to draft *Chá-okó*, a manual of healing terms in Rarámuri. He shared with me information on many different herbs and remedies from a group of more than one hundred collected by me in Norogachi. Palma volunteered information, such as the synonyms for names of plants and their uses, and he even provided numerous specimens.

I should also like to mention Luis Moreno Bakasórare and Guadalupe Córdova de Gardea. They helped to clarify the obscure steps incorporated in some of the healing procedures. Gardea, a mestizo woman married to a Rarámuri, provided information independently from that obtained from Tarahumara informants, including a list of Mexican folk remedies also present in the Tarahumara list of remedies.

Many other resource persons could be mentioned, such as those who assisted me in collecting items such as a tangle of *matarí* roots, a *ru'síwari* sample, a fragment of obsidian (*rawí*), and baskets full of *a'rí* and *batagá*.

Diana Schofield-Meléndez deserves a special mention; through some sort of magic and infinite patience, she transformed my very disconnected original manuscript into a readable account.

Special thanks also go to the late Campbell W. Pennington, a former professor from Texas A&M University, a gifted geographer, and author of the seminal book *The Tarahumar of Mexico*. Also deserving acknowledgment is David

Brambila, S.J., a remarkable linguist with a profound knowledge of the Tarahumaras. Carlos Díaz Infante, S.J., and Luis González-Rodríguez also contributed importantly to help me gain knowledge of the Tarahumaras. They provided me with advice, copies of their own work, and the significant work of others. Similar appreciation is deserved by Dennis and Dorothy Mull, William Breen Murray, Enrique Salmón, and Beatriz Uribe Moctezuma.

Alfonso Paredes, M.D., the editor of this book, should be acknowledged. Dr. Paredes was instrumental in creating the present structure of the book, strengthening the thematic focus, and applying academic referencing conventions. He also assisted in incorporating the suggestions by the peer reviewers the University of Oklahoma Press solicited and responding to the copyediting. Dr. Paredes and I met several years ago during his tenure as professor at the University of Oklahoma Medical School. He visited the Rarámuri land several times to conduct field research on the Tarahumaras and their environment. On one of his trips, he arrived at the Rarámuri land with a group of scientists from the OU Department of Medicine to conduct an ecological study, which included psychobiological, cardiovascular, nutritional, and work capacity studies on the Tarahumaras. This material was later published in several scientific journals. It was on one of these occasions that I joined his group as a field assistant and interpreter and as a careful observer of the technologies that they brought with them. Incidentally, recently I had the opportunity to write with Dr. Paredes a book chapter on Jíkuri, the Tarahumara peyote cult, and the ceremonial activities associated with this interesting plant.

It is very important to express my appreciation for the assistance from the staff of the Archivo General de Indias in Seville, Spain, and the Bancroft Library at the University of California at Berkeley. They allowed me access to the invaluable collections they have under their custody.

I should also mention all those professionals who worked with me in the Mexican Health departments, Instituto Nacional Indigenista (INI) and Secretaría de Salubridad y Asistencia (SSA).

I naturally should acknowledge the help of my professors from the Universidad Autónoma de Chihuahua, particularly Octavio Corral Romero, M.D., and Miguel Aranda, M.D., and most especially the help of my friends and colleagues, Rodolfo Reyes Grajeda, M.D., and Jorge Avila Hernandez, M.D.

Last but not least, I am especially indebted to the nurses who supported my medical work in Tarahumara Country through the years, most especially the

Sisters of Charity of Saint Charles Borromeaus, Sister Mary Elise Andris, and Sister Mary Rosalía Mendoza.

In a very special manner I wish to express my gratitude to my wife, Josefina, for her patience and support during the preparation of this book. I also appreciate the support of my daughter, Josefina, and my sons, Tocho and Saul. Throughout the several years drafting the book, I stole much time and attention from all my family; they however graciously realized the importance of the task.

Tarahumara Medicine

Introduction

The main objective of this book is to describe the healing practices of the Tarahumara Indians, their herbal cures, rituals, and conceptions of disease. A comprehensive description of their traditional herbal remedies is presented, including their botanical characteristics, properties, attributed effects, and uses. Their curing rituals and ceremonies are carefully outlined. The healing practices of the Tarahumaras appear to be a blending of religious lore, magic, and careful observations of nature. This aspect of their culture also has socially agglutinating functions that has contributed to give them identity as a group.

In examining the evolution of their culture, a most significant event is considered: the strategic change that took place when the Tarahumaras abandoned belligerence and subscribed to passive resistance in confronting the Spanish intrusion and that of other alien groups. It is through this passive resistance that their identity has evolved and strategies reinforcing tribal cohesion came to be. This issue remains relevant in our increasingly smaller world.

Regarding the topic of evolution of traditional medicine, the following quote from Sejourné deserves attention. She noted: "The error seems to exist in assuming science to have evolved out of magic and religion, when in fact it is an entirely different phenomenon. The problem of domination over the external world is thus confused with the philosophical problem of existence, which is a problem foreign to science. If our knowledge of magical and religious thought in the Pre-Columbian world is not to be hopelessly vague, we must try to understand the concepts of life such modes of thought imply."[1]

It is important to continue examining some of the relationships between the Tarahumara healing system and conventional medicine. The term "traditional medicine" has been used within social science contexts as a descriptive term contrasting it with scientific medicine. In this sense the term is roughly a synonym for what was called "folk medicine" in classic American functionalist anthropology. In a strictly medical environment, the term seems to have negative implications; that is, being opposed to the innovative, to the modern and changing. Gamio expressed concern as early as the 1920s about the attitude of the medical profession toward native healers and herbalists who were frequently demeaned and even considered dangerous.[2] A more benign appreciation was adopted by Naranjo, who saw these empirical folk healers as depositories of popular knowledge, their practices a remnant of common therapies of years bygone.[3] Some of the views just stated might have fostered attempts to label traditional medicine with other terms: Indian medicine, aboriginal medicine, and parallel medicine. The approach called "parallel medicine" even implied that the standard medical system could use the particular traditional system of the area to render care to those who did not have access or found cultural barriers to conventional medical care. Some even believe that these approaches could be integrated into the delivery of primary care health services.[4]

Some investigators have examined traditional medicine to obtain information about substances, drugs, or techniques that could be used in conventional biomedicine. As a matter of fact, a great deal of interaction and interdependence has often occurred between traditional medical systems and science. The contribution of empirical discoveries from native cultural groups to the body of knowledge on therapeutic agents in scientific medicine always has been present.[5] This may be illustrated by many examples but more recently we could mention the case of the southern Mexican wild yam, *cabeza de negro* (*Dioscorea* sp.), a medicinal plant used by the ancient Aztecs;[6] this plant has rendered the raw material from which sapogenins[7] were chemically converted into progesterone—first by chemist Russell E. Marker and later by George Rosenkranz—and developed into a pharmacological agent, the contraceptive pill which has changed moral, social, and cultural values in the contemporary world.[8] The synthesis of steroids from the same raw materials has resulted in a constellation of useful drugs. Another example is curare and its alkaloid, tubocurarine—with its muscle-paralyzing effects—obtained from several plants of the Amazonic jungle (*Strychnos* sp. and *Chondodendron* sp.) that revolutionized some areas of modern surgery.[9]

In studying the healing practices of the Tarahumaras, I have tried to examine them from two perspectives, the "etic" and "emic." According to the etic perspective, the concepts and categories identified in a culture should have a meaning and even utility for the outside observer from the dominant culture. On the other hand, according to the emic perspective, the cultural distinctions that are observed should have a meaning for the members of the indigenous society.

The outside observer should consider the concepts of disease that the culture has and, in many instances, make the inferences of efficacy of remedies and rituals based on the assumptions and expectations of that culture. This encompasses the study not only of the so-called culture-bound syndromes but also of conditions, which may have a correspondence in standard medical nosology that should also be seen from the point of view of the sufferer within his or her own culture. Students of traditional medicines must be aware of stumbling blocks that may hinder the evaluation of the observed facts. These include lack of knowledge of the language or awareness of the particular psychological characteristics of the cultural group. Also, some ethnocentric biases, such as folklorism, belief in an assumed uniformity of the group, and assumptions of infantilism, may veil racial prejudices.[10] As Lookout states, "Unfortunately those with the closest contact with Indians are often the least scientific in their judgments while those who occasionally set up a rigorous study design have insufficient knowledge of Indians and their cultural heritage."[11]

More importantly, profound differences may exist between conceptions that view the universe as divided into the material and spiritual while the Indians may view in the universe an integrated continuity between directly understandable phenomena and non-directly comprehensible events. Many of the daily Indian actions may have a magical-religious meaning that could easily be overlooked by to the non-Indian.

The Tarahumaras, or Rarámuri, the Great Runners

The Tarahumaras are one of the oldest aboriginal groups of North America; their progenitors probably were members of the Clovis culture that populated North America 11,050 to 10,800 years before the present. They are characterized by their graceful and dignified demeanor and their athletic build. The men's attire consists of a loincloth, shirt, cotton girdle, bead collar, and headband or straw hat. Their women wear long pleated skirts, loose cotton blouses, and belts. This attire however varies with the region. Their population is estimated at 75,000 to 100,000. Although the migratory movement to Chihuahua City and other cities is constantly increasing, both in intensity and permanence,[12] most of them live in communities scattered through the rough mountains and deep canyons of the Sierra Madre Occidental located in the north of Mexico, also known as Sierra Tarahumara. The Tarahumaras live in small, widely dispersed hamlets called *rancherías,* alongside streams that enable them to cultivate small parcels of land.

These men and women are striking for their kindness and peaceful ways. Many of their customs and traditions date from pre-Hispanic days. They speak their own language and call themselves Rarámuri; that means "nimble feet."[13] This is alluding to their remarkable agility and endurance demonstrated during the Rarajípari, a rather physically demanding kickball race that sometimes covers a distance of more than 150 kilometers. In earlier times these people were known for their stamina as running military couriers. Tradition states that they could chase deer until the animals fell exhausted. Their considerable geographic and cultural isolation has been partially overcome only recently by modern methods of communication.

Considerations in the Study of Tarahumara Healing Practices

In my work, I have relied on informants as well as participant observation. In addition, careful attention was given to the anthropological literature.

In evaluating materials obtained from informants, I have kept in mind possible sources of bias and followed a process of cross-validation detailed below. Some informants consider themselves proud connoisseurs and are interested in communicating their own interpretations of facts to the researcher, adding liberally their personal views as if these were those of the culture at large. Other informants are sensitive about issues of "tribal confidentiality" and may attempt to disclose only information that they believe is "not confidential" or safe to share; it is therefore important to attempt to validate the information, comparing it with data from various sources.

Participant observation also has some shortcomings. Like in physics' Heisenberg's uncertainty principle, the observer may influence the observed events by his or her very presence in the setting.[14] As an example, one may find striking differences between the behavior of participants in a *bajíachi* celebration when nuns or priests are present and behavior displayed when only individuals of their own group are in attendance. During the bajíachi, corn beer (*batari*) is consumed and disruptive actions, exceedingly joyous displays, and sexual innuendoes are common. The amount of batari consumed may be the same but the behavior when persons alien to the culture are present may differ.

The process of data collection has been long and laborious. I began gathering information about the concepts of illness and the therapeutic methods used by the Tarahumaras in 1972 during my required social service year when I was assigned to a medical clinic in Cerocahui, a small community in the Tarahumara heartland. At this point, most of the information available to me was through the exchanges with patients during medical consultations. This information was collected with a mixture of amazement and amusement. Impressed by this experience, I decided to write the thesis required for graduation from medical school, giving it an anthropological slant, including descriptions of the health concepts and remedies used by the Indians. This was an alternative to writing a report summarizing conventional health data on prevalence and outcomes of the medical disorders encountered. The plan was submitted to the medical school thesis committee. Surprisingly, it was approved. Through the process of preparing the thesis, it became necessary to do a comprehensive review of the available anthropological literature from authors such as Lumholtz, Bennett and Zingg, Brambila, Kennedy, Artaud, and others.

Once the thesis was written, wishing to share the information with a broader audience, my next task was to find a publisher. This was not easy. Contact was made with experts from the National Autonomous University of Mexico, such as Fernando Beltrán Hernández, M.D., and at the National Indigenist Institute (INI), Professor Alfonso Villa Rojas and Juan Rulfo, an accomplished writer.[15] Their advice and criticisms were in general positive and useful, but their comments ranged from "This is very important information of anthropological significance but without the scientific rigor expected in anthropological writings" to "The material is more like fiction than scholarly work."

The thesis was published as a book in Spanish under the sponsorship of the Department of Human Ecology of the School of Medicine of the National Autonomous University of Mexico under the title *Cerocahui: Una Comunidad en la Sierra Tarahumara*. The book has been reprinted several times within a span of forty years. An English translation is however not yet available.

After graduation from medical school, I returned to the Tarahumara region as a physician of the Clínica San Carlos,[16] a medical mission in Norogachi, a very traditional Tarahumara settlement, where I spent five years. During this period, I continued to gather information from patients and their relatives, enriching my knowledge of their concepts of health and disease, use of herbal remedies, and familiarity with their rituals.

By this time I had acquired an extensive knowledge of the Tarahumara language, a very important instrument of inquiry. As I gained familiarity as well as an in-depth knowledge of the language, it became apparent to me that the Tarahumaras not only spoke differently, they appeared to *think* differently.

I developed thus a close rapport and gained the trust of the Indians. This gave me the opportunity to observe and participate in their ceremonies, including those involved in Jíkuri, the peyote cult. Other important events followed: in 1974, I joined a research group from the University of Oklahoma as a field aide and translator; the team included Alfonso Paredes, M.D., Art Zeiner, Ph.D., and Lawrence Cawden, a media expert.[17] The group was in Norogachi to conduct a study on the physiology of ethanol and patterns of consumption of alcohol in healthy Tarahumaras. This participation exposed me to issues of scientific methodology and gave me access to a network of contacts and opportunities to present my observations on the Tarahumara peyote cult at the University of Oklahoma.

During this period, various missionary activities from religious groups were taking place in the area. Some were very old, such as the missions of the Jesuits present in the area since the seventeenth century; others were very recent and driven by a self-critical and effervescent intellectual reframing of the 1970s. In particular, a team of Catholic Marist Brothers was involved in a project in Kwechi, a hamlet in the region. As part of the project, they were gathering systematic information about the Tarahumara culture while trying to become incorporated in the everyday life and activities of the Indians.[18] The place had been chosen under the assumption that the population was relatively free from western influences. The approach of the Brothers was a form of participant observation. The Marists were influenced by theoretical views of Lévi-Strauss, liberation theology, and to a lesser extent Marxist analysis, then in vogue. The group intended to prepare a report that could serve as a framework for future educational projects among the Tarahumaras. I was invited to participate in the activities of the group, probably because they recognized my familiarity with the Tarahumaras and their language. The project lasted three years, giving me abundant opportunities to observe and collect information. Some of the data collected by the Brothers and myself is presented elsewhere in this book.

Organizing and summarizing the material proved to be a complex and challenging task. The organization and categorization of information was particularly exacting, given that the information was obtained from individuals belonging to subgroups with different levels of acculturation.

A particular type of health-seeking approach used by the Tarahumaras, particularly in Norogachi, was through handwritten letters asking for remedies for their ailments; this became one more source of information.

A very important objective of the data collection was to inquire about medicinal herbs and concepts of disease obtained from knowledgeable indigenous informants, such as Erasmo Palma Tuchéachi. Mr. Palma, who incidentally helped me learn the Tarahumara language, also collaborated in drafting *Chá okó*, a manual of healing terms in the Rarámuri language. He started by providing additional information about the thirty-four herbs and other remedies used by the Tarahumaras that I had previously described in my book *Cerocahui*. Then he shared information about many among more than one hundred identified by me during my stay in Norogachi. Fifty of these plants and herbs also had been described in Pennington's writings. Mr. Palma volunteered additional information, such as synonyms for the names of plants and their uses and even provided plant specimens. Other sources of information helped to confirm whether the data that had been collected regarding uses was widely accepted by the Tarahumaras as a group and not just by individual connoisseurs such as Palma. For this purpose, collaboration was obtained from Luis Moreno Bakasórare, another Tarahumara. Although not as knowledgeable as Palma, Moreno knew much about herbs and healing rituals. In addition, in Norogachi, I found another valuable informant, Mrs. Guadalupe Córdova de Gardea, a mestizo woman married to a Rarámuri. She provided information independently that corresponded with information obtained from the Tarahumara informants—for instance, a list of Mexican folk remedies also present in the Tarahumaras' list of remedies.

I left Norogachi recruited by the INI and assigned to its clinic at Samachique. INI is a Mexican government institution dedicated to the health and welfare of the Indian peoples of Mexico. In addition to providing medical care at the clinic—an eight-bed rustic hospital located at the convergence of the eastern outlet of the grand canyons and the Gran Visión Road—I supervised dispensaries located in Indian communities such as Aboréachi, Cusárare, Choguita, Wawachiki, and Sewérachi. I also had the responsibility of visiting the INI boarding schools, both in the locations just mentioned as well as those at Mesa de la Yerbabuena in the Batopilas Canyon and Basíware. Later, I was assigned to the Guachochi Coordinating Center as supervisor of clinics in Samachique, Bakiríachi, Guachochi, and Turuachi. The clinics and dispensaries mentioned provided conventional preventive medical care as well as treatment. Interestingly, ailing Indians sought help from these resources as well as from methods derived from their own culture or a mixture of both.[19] These responsibilities gave me the opportunity to travel extensively throughout the Tarahumara region and observe local differences.

In 1980, I became director of the Center of Regional Studies at the Universidad Autónoma de Chihuahua and started a program of localization and rescue of archival documents throughout the state, simultaneously overseeing a Chihuahua basic bibliography project[20] and a research project on the migration of Tarahumaras to the city,[21] as well as organizing and conducting a seminar on the languages spoken by the indigenous people in the state of Chihuahua.[22]

In 1983, I joined the University of New Mexico as a psychiatry residency trainee. The multicultural atmosphere of Albuquerque and local interests in American Indians encouraged the continuation of my work on Tarahumara issues. At the same time, I enrolled in a master's degree program focusing on cross-cultural and mental health issues. At the conclusion of the psychiatric training, I relocated to McAllen, Texas. In this city I continued organizing the data in a publishable form and reviewing recent publications on the Tarahumara culture and medicinal plants.

Ecological Concepts of Health and Disease

It may be useful to outline the conceptual framework guiding the organization of the information collected from the Tarahumaras. Health status may be considered as a state of equilibrium between variables such as the environment, physical, and biological agents, and the individual. The first set of variables, the environment, includes the geographic characteristics of the habitat of the Tarahumaras and the other human groups sharing their territory. After all, culture is in some respects a response to the challenges of the environment.

The second includes the physical anthropological traits and cultural features of the Tarahumaras; that is, those of individuals who consider themselves members of the group. The Tarahumaras are assumed to have characteristics different from those of the other population of the area, the mestizos. The mestizos, a mixture of Indian and Spanish blood, although living within the same environment of the Tarahumaras, exhibit different cultural characteristics.

The third set of variables includes physical and biological factors, as well as events considered by the Tarahumaras to be "causes" of disease. These causes are, in their case, mostly cultural constructs.

This ecological approach helps to place disease as an important event within the environment and the community, affecting the individual as well as the group. Through this process, the diseases of the Tarahumaras are presented as they conceive them. An attempt is made to explain processes from an emic point of view; that is, it focuses on the intrinsic cultural distinctions that are meaningful to the members of their society and to the methods that they use to restore the lost equilibrium. Correspondences and dissimilarities with the established medical nosology are also discussed. Later, a conventional biomedical approach is used to describe the general state of health of the Rarámuri.

Historical Antecedents and the Ethnographic Present

A historical overview is covered in a section of the book. This may help readers to understand the ethnographic information. After all, ethnographies are like historical documents, describing the recent past of a community. In most ethnographic studies, an attempt is not made to connect historical events and the ethnographic present, but the facts of interest to the historian are often not those that may provide useful background to the ethnographer. This may be due to the fact that the historian's approach is usually chronographic and macrohistorical.

In this book, I attempt to describe how the Tarahumaras did cope with their environment and external cultural and political influences, illustrating the community's strategies to adapt, cope, and survive.[23] Cultures are dynamic and change, but there are also forces tending to perpetuate operative mechanisms. There is a certain inertia opposing change and some cultural traits are preserved even when they are no longer useful.

Following, some of what is known linguistically about the Tarahumaras is summarized. This aspect is fundamental to understand their relationship with extant and extinct groups that interacted with them. Also, a Tarahumara version of the origin and development of the group according to their own tradition is presented. Their legend of the Two Brothers is central to this section.

Finally, regarding Tarahumara healing practices, these are used as survival strategies without scientific basis but within the context of what Kroeber called the whole cultural pattern.[24] As an element of social cohesion, traditional healing plays a paramount role.

The Environment

The boundaries of Tarahumara country are defined, explaining how the area is not only the habitat of the Tarahumaras but also of other human groups. A division in regions describing the ethnic populations as well as features of the physical geography and geology is presented. The two major regions, the uplands, or sierra, and *las barrancas*, or canyon country, are described, including aspects of their characteristics of climate, flora, and fauna.

Distinctive Features of the Tarahumaras

Not all Tarahumaras dress in their characteristic traditional garments or speak their own language. Furthermore, some individuals with physical characteristics of the mestizo claim to be Tarahumaras and vice versa. Within the population of the region, a continuum may be observed in the biological and the acculturation dimensions, ranging from full-blooded monolingual Indians and, at the other end, Caucasian individuals or mainstream Mexicans. This is particularly important when generalizations about members of groups are made. The risk of observational biases, false attribution ascription, and erroneous generalizations has not been uncommon. A way to determine if an individual belongs to the ethnic group would be whether he or she is recognized by the group as one of them at the same time the individual recognizes himself or herself as member. In practice, this categorization is often difficult to apply. Determination on who is a Tarahumara may depend on comparing the individual along several dimensions. Many anthropological accounts about the Rarámuri offer a picture of cultural homogeneity, seemingly ignoring the existence of outstanding individuals who influence in a recognizable way the culture. Nevertheless, a special effort is made in this book to describe the physical traits, attire, language, religion, type of dwellings, and other aspects that appear to characterize the Tarahumaras.

Agents of Disease According to the Tarahumaras

This section discusses the common ailments of the Tarahumaras as conceived by them. They do not make a sharp distinction between the causes and the symptoms of the disease. The name of a disease usually represents a symptomatic complex as well as the assumed cause.[25] It may consist of just the name of one symptom, which, in their minds, characterizes the entire disease. Furthermore, their thinking is guided by an animistic conception in which things in nature have souls or consciousness. For them, there are no sharp boundaries between what is real and dreams. As we will see, the Tarahumara shaman believes that disease is due to the loss of a soul; he then "discovers" the source of a disease by examining his dreams in search of the place where the "lost soul" of the sick person may be found.

Some normal physiologic phenomena, such as pregnancy and childbirth, are included. Childbirth, they believe, is closely related to the mystery of life and death, not only because of its symbolic meaning but also because childbirth is an important source of morbidity and mortality among them. This is an aspect that has not received attention in the anthropological literature.

Ceremonial and Herbal Healing

In this section, the diverse functions played by herbal potions and ritual healing within their healing practices are described in detail.

Maneuvers characteristic of the various categories of healers, such as the *sipáame, waníame, towita,* or *sukurúame*—this last an almost mythical sorcerer—differ considerably from the remedies of common men and woman as well as between those of the categories of healers mentioned.

An important step of my endeavor is to present an inventory of the healing herbs and potions used by the Tarahumaras. All of these are described, including the terms more commonly used or recognized by the Tarahumaras. Incidentally, a common problem is that different names are applied to the same plants, and several plants are known by the same name. Also included are the

Tarahumara or Spanish synonyms, documenting their source, definitions, and etymology. The scientific binomial identification for plants and animals is then presented following, as closely as possible, Council of Biology Editors' guidelines.[26] Topics covered include botanical considerations, pharmacological data, and nonmedical and medicinal uses.

When a ceremonial has been identified surrounding the use of a particular plant, the pertinent description is included. The peyote ritual, Jíkuri, is presented in a separate chapter since considerable attention has been given to description of this celebration in which the cactus plays such a prominent role as well as the beliefs associated with it. The reader must realize that in the literature many aspects of the ceremony actually were borrowed from observations from groups other than the Rarámuri, particularly the Huichols. Nevertheless, the existence of few but important works about the Jíkuri in the anthropological literature must be acknowledged.[27]

The book closes presenting a summary of medical and epidemiological information illustrating the current health needs of the Tarahumara communities.

Throughout, an attempt has been made to establish a link between the history of the Tarahumara culture and their present. As mentioned, a paramount event took place when the Rarámuri abandoned belligerence and adopted passive resistance in confronting the Spanish intrusion. This group decision was associated with the creation of activities that reinforced their identity and tribal cohesion and resulted in the development of their healing practices, among other strategies.

ONE

The Tarahumara Ecological Habitat

Mexico is a federal representative republic composed of thirty-two states of which Chihuahua, in the northwest, is the largest with a surface area of 247,087 square kilometers.[1] Twenty-four out of the sixty-seven *municipios*, or counties, within the state shelter a considerable native population.[2] Historically, the Rarámuri people occupied the central municipios, while the Pimas Bajos, Jovas, and Varijíos lived in the northern counties and Tepehuans in the southernmost ones. Other groups that have had a distinct historical presence in the state include the extinct Tubares, as well as the Guazaparis, Témoris, Chínipas, and other possible Tarahumara subgroups. The recent and more controversial — as some scholars debate their existence — Masculai were also located within the area.

Tarahumara Country today, located in the southwestern portion of Chihuahua, is bounded by the meridians 108°50′ west and 106° west and the parallels 30° north and 25°40′ north. This land, 74,951 square kilometers, represents 30.33 percent of the state's surface area. The total population of Tarahumara Country, including the Tarahumara peoples, other Indian groups, mestizos, and whites who live there, is about 320,022. The precise number of Tarahumaras is not known; estimates have ranged between 75,000 and 100,000.[3] The number of Tarahumara-speaking individuals was estimated at 54,431 in 1990 and 62,555 in 1995. In 1997 the estimated Indian population of Chihuahua was 129,250.[4] Municipios located peripherally in Tarahumara Country are the largest in population, but they are less densely inhabited by Tarahumaras.

Tarahumara Country may be divided into two regions: the uplands, locally

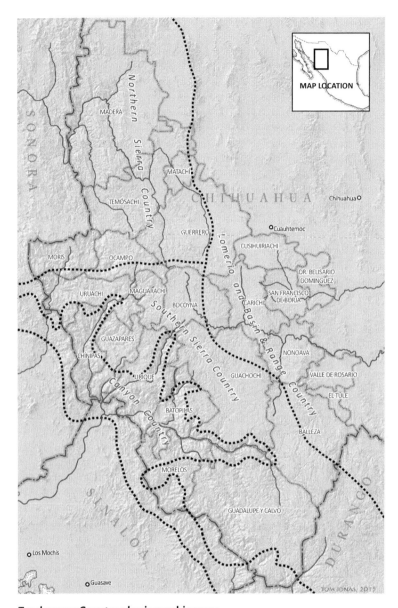

Tarahumara Country physiographic areas
The main physiographic areas—northern and southern sierra country; lomerío, basin, and range country; and canyon country—and municipios with Tarahumara population. (Map derived from La Peña [1948], Pennington [1963], Schmidt [1973] and Martínez et al. [1996]. Copyright © 2015, University of Oklahoma Press)

called the "sierra," in the north, northeast, and south; and canyon country, or the barrancas, in the southwest. The uplands consist of a succession of mountain ridges, hillocks, and peaks interspersed with valleys; they are part of the southern Chihuahuan segment of the Sierra Madre Occidental, the continuation of the Rocky Mountains in Mexico. Six hundred kilometers of the Sierra Madre in Chihuahua are lined up in a north-south direction, but in the southern county of Guadalupe y Calvo, in Tarahumara Country, the main chain turns slightly, acquiring a northwest-southeast direction. The principal summits in the state are Mohinora, 3,307 meters; Romúrachi, 2,985 meters; and Cerro de la Capellina, 2,571 meters. The average elevation of the Sierra is about 2,270 meters, or 7,447 feet, with the southern part being 300 to 600 meters higher than the northern part.[5] The uplands have been described as a rolling plateau, whose southern two-thirds show more rugged terrain, with mountains, gorges, and cliffs, than the relatively plain northern third.[6]

The geologic formations of the Sierra Tarahumaras evolved during the Paleozoic era, but the actual mountain building didn't take place until the Cenozoic era. Volcanic eruptions, which deposited millions of tons of ashes and lava, were the Sierra's main orogenic force.[7] The solidified volcanic deposits formed thick layers of volcanic tuff and rhyolites with deposits in some areas of crystallized materials such as free quartz, volcanic breccia, andesites, porphyries, and obsidian, as well as agate-like rocks and optic calcite. There were several metallogenic events, which resulted in deposits of gold, silver, and copper; these metals were a source of wealth during the Spanish colonial time.[8] Washing out gold and silver from the sands of the rivers, locally called *chiveo*, provides a marginal livelihood to some people today.

The characteristics of the rocks suggest a sequence of geological events, particularly on the banks of the upper Urique River. For instance, in Romichi, near Norogachi, sedimentary strata indicative of the primeval river course can be discovered under layers formed by a flow of lava. Embedded in these rocks are fossils of small animals trapped by the sediment. Study of this material may help to determine the sequence of the volcanic eruptions and the dates when the animals lived.[9]

Canyon Country, or the Barrancas

The canyons located in western Chihuahua exhibit grandiose beauty and awesome depth. An informal contest among writers has tried to establish which is the deepest canyon—the Barranca de Urique or the Grand Canyon in Arizona. It has been claimed that the Urique Canyon is more than 2,000 meters deep. Almada reported its depth as more than 1,524 meters, concluding that it was deeper than the Grand Canyon.[10] According to Gajdusek, the depth of the Barranca del Cobre upstream from Urique Canyon has been estimated to be even deeper, more than 2,438 meters deep. Wampler indicates that both canyons are more than 1,609 meters deep for considerable distances.[11] In the vicinity of Pamachi, the Barranca del Cobre has a depth exceeding 1,981 meters. Compare that to the depth of the Grand Canyon at its southern rim in Arizona, near the national park headquarters—it is 1,609 meters.[12]

The bottom of canyon country at many points is only a few hundred meters above sea level. The climate is tropical, as are the fauna and flora. The topography of canyon country, which includes a variety of formations exposed over time, is due to the erosion of the volcanic tuff by streams. Natural dikes are prominent features of the middle sections of the Urique Canyon slopes, and great benches of volcanic material characterize the middle and upper canyon slopes of the Verde River.[13] Sedimentary rocks that were dragged away by rushing water and deposited in the lowlands left exposed hypabyssal formations within the canyons. Erosion also exposed deposits of metals, allowing for their exploitation. Mining centers were developed during colonial times at some of the areas of contact between volcanic materials and adjacent sediments.[14]

Hydrographic Features of the Area

A variety of rivers flow through Chihuahua and help support life in Tarahumara Country. Small streams originating in the uplands, close to the Continental Divide, join on the Pacific side, forming the deep canyon system of the Fuerte River. On the opposite side of the Divide, streams converge to form the Conchos River, which drains into the Gulf of Mexico. The northeastern portion of Tarahumara Country drains into three small endorheic basins, which end in Laguna de los Mexicanos, Laguna Bustillos, and Laguna Encinillas and do not continue to the sea.[15] The rapid streams that drain into the Conchos River re-

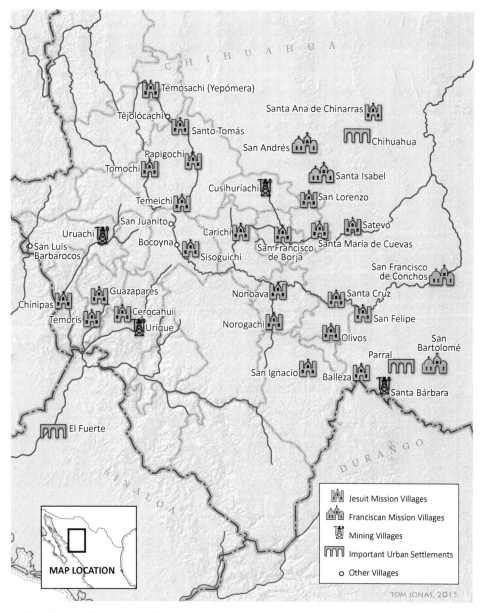

Tarahumara Country population centers
The main historical and present-day population centers, including locations of Jesuit and Franciscan missions and mining centers. (Copyright © 2015, University of Oklahoma Press)

main dry most of the year; only a few springs are full year-round. Unlike the Fuerte, the Conchos, the main affluent of the Rio Grande, did not cut deep canyons.[16] Its basin covers an area of 77,090 square kilometers, most of which is in the state of Chihuahua; a small portion is located in the state of Durango.[17] Several of the most representative and typical Tarahumara settlements—Sisoguichi, Creel, Bocoyna, Tewerichi, Narárachi, Bakéachi, and Wawachérare—are located by the headwaters of this river. The water of the Conchos River is used for irrigation in central Chihuahua and supports the flourishing agriculture of that zone.

Another important river in Tarahumara Country is Rio Sinaloa, or Petatlán, which flows out of the waters of the highest peaks of the Sierra. This river leaves the state of Chihuahua and enters the state of Sinaloa near the town of Tohayana. It then returns briefly to Chihuahua, then turns again to the southwest and meanders back through Sinaloa to reach the Gulf of California. The basin area of the Rio Sinaloa covers 13,300 square kilometers, including the southernmost Chihuahuan canyons of Barranca de Dolores.

The Verde River, known at various points along its course as San Miguel, San Ignacio, Río de la Sinforosa, and Wérachi, is the main affluent of the Fuerte River. It begins in Durango, then enters Chihuahua, gaining waters from the Arroyo de los Loera, Arroyo de Chinatú, and then the Batopilas River. Finally, it joins the Urique, Septentrión, and Chínipas Rivers, becoming the Fuerte River proper. The river then meanders through Sinaloa and finally empties into the Gulf of California.

The Fuerte River's basin occupies 36,275 square kilometers.[18] This river cuts impressive canyons, including the ones of Wérachi and Sinforosa. In this area, Tarahumara groups live in isolated hamlets, or rancherías. During winter these people migrate to the bottom of the canyons, which have a much warmer climate during this period. Located in the margins of the Batopilas River are the former mining towns of Batopilas and La Bufa. Among the most impressive canyons carved by this river are those of La Bufa canyon and Arroyo de Munérachi.

The most striking among these rivers is probably the Urique River. Its course runs parallel to the Batopilas River. The main canyon formed by the river is known successively as Copper Canyon, Tararecua, and Barranca de Urique.

The Septentrión River, also known as Arroyo de Cuiteco, is a stream with a relatively slow incline. The track of the railroad Chihuahua al Pacífico runs

The impressive Urique Canyon as seen from Puerto del Gallego. The central mountain is el Cerro de la Ventana. At the bottom center is the town of Urique and to the right, the hamlet of Guapalayna.

alongside. The Río Septentrión joins the stream of the Río Chínipas in the state of Sinaloa.

The Oteros, or Chínipas, River runs north to south along the old mission of the same name. It traverses Varijío country and carves the canyons of Concepción and Semuína. The impressive bridges of the railroad Ferrocarril Chihuahua al Pacífico cross this river. The Mayo River, known by the name of the Indian tribe living alongside its lower course, originates from the union of the arroyos of Moris and Candameña and has a basin surface of 13,750 square kilometers.[19] The Barrancas of Candameña and Mulatos, although not as deep as the Copper Canyon, are remarkable because of their vertical walls. Barranca de Candameña was carved by Arroyo del Santísimo, a stream that also forms the Basaséachi waterfall, often referred to as the highest single-fall waterfall in the world with about 400 meters in height.[20] On the Chihuahuan banks of the river and its affluents there are mining centers such as Uruachi, Nueva Unión, and Maguarichi. The Yaqui River takes its name from the Indian tribe that lives alongside the valley formed by the lower portion of its course. Yaqui's basin occupies an area of 72,630 square kilometers.[21] In Chihuahua, the Yaqui River is known as Papigochi or Sírupa. Along the banks of the Papigochi are located the only cities of the Sierra: Ciudad Guerrero, formerly Papigochi, and Ciudad Madera; and old towns such as Matachí and Temósachi.

Regional Climate

The Sierra Madre Occidental chain of mountains deflects winds carrying moisture, cooling the air and producing rain. At the Sierra's summits the annual precipitation averages 500 and 600 millimeters, while on its eastern side precipitation falls 300 to 500 millimeters. Farther to the east, one finds the arid and semiarid zones of the states of Chihuahua, Durango, Coahuila, and Zacatecas.[22] The Sierra acts as a barrier for the winds and humidity coming from the Pacific Ocean, including from summer and fall hurricanes and tropical storms. The rainy season usually begins in June and lasts until September. In winter, the rain comes in spells, from very slight sprinkling rain to true storms, locally known as *equipatas*. Equipatas bring in some instances more water in a single day than the whole rainy summer season. When the temperatures are low, an equipata may become a snowstorm.[23] Interregional precipitation variability is considerable. For example San Juanito, with only sixty-five rainy days per year, has a mean annual precipitation of 1,364 millimeters, but Ciudad Guerrero to the east with 107 rainy days has only 506 millimeters. As a mean, the land of the Tarahumaras receives 807 millimeters of precipitation per year.[24]

According to available data, the uplands have a mean annual temperature of 11.8° C, with highs of 21.4° C and lows of 2.52° C. On average, 181 days are free of frost per year. Temperatures ranging between 10 and 29 degrees below zero may be expected from mid-December to the end of February and some snow between November and May. The temperature and rain registry from Norogachi is representative of most of the Sierra with an annual mean temperature of 11.6° C. The warmest days of the year occur during the end of June and the coldest in January and February; the first frost falls around September 24, the last on March 30.

San Juanito, Guachochi and Madera are particularly cold zones. Interestingly, these are among the most densely inhabited areas in the Sierra.

The canyon country is an area with a tropical climate with an annual mean

Annual temperatures (in degrees Celsius) and rainfall (in millimeters) in Tarahumara Country

Station	Location/ Region	Altitude (m.)	Annual Mean	Highs Mean	Lows Mean	Extreme Max.	Extreme Min.	Rain
Balleza	26°57′N-106°21′W lomerío	1920	—	—	—	—	—	491
El Vergel	28°33′N-107°33′W uplands	2240	—	—	—	—	—	141
Batopilas	27°00′N-107°39′W canyon	501	24.0	30.1	18.0	44	−12	300
San Juanito	27°58′N-107°40′W uplands	2348	9.5	19.7	−1.3	33	−27	1364
Romúrachi	27°50′N-107°28′W uplands	2980	8.7	15.2	2.2	29	−26	485
Carichí	27°56′N-107°08′W lomerío	2038	13.7	22.9	4.5	3	−15	536
Chínipas	27°24′N-108°33′W canyon	515	22.2	28.9	16.0	42	−8	990
San Lorenzo	28°10′N-106°29′W lomerío	1600	17.9	27.7	7.9	40	−13	430
Norogachi	27°15′N-107°08′W uplands	2015	11.6	21.86	1.26	33	−15	509
Sikirichi	27°08′N-107°10′W uplands	2320	14.0	23.7	4.3	39.5	−15	536
Gpe y Calvo	26°06′N-106°58′W uplands	2316	13.1	22.5	3.7	37	−18	1094
Chinatú	26°12′N-106°59′W uplands	1982	13.4	24.0	2.3	37	−18	920

(*continued*)

Annual temperatures (in degrees Celsius) and rainfall (in millimeters) in Tarahumara Country (*continued*)

Station	Location/ Region	Altitude (m.)	Annual Mean	Highs Mean	Lows Mean	Extreme Max.	Extreme Min.	Rain
Guerrero	28°33′N-107°33′W uplands	2010	13.8	23.4	4.3	37	−18	906
Madera City	29°12′N-108°16′W uplands	2092	10.9	20.3	1.3	39	−22	606
M. Huracán	29°38′N-108°14′W uplands	2165	12.7	21.1	4.2	38	−18	734
San Juan	29°48′N-108°17′W uplands	2200	9.7	19.7	−0.3	37	−24	672
Tres Ojitos	29°58′N-108°31′W uplands	2600	8.8	14.8	2.3	36	−19	1016
El Poleo	29°24′N-108°28′W uplands	2300	10.3	18.2	2.1	33	−17	1064
Col. García	29°58′N-108°20′W uplands	2244	10.9	20.3	1.3	35	−20	570
Viv. Madera	29°10′N-108°08′W uplands	2200	10.8	19.4	2.1	34	−21	774
Nonoava	28°28′N-104°44′W uplands	1640	9.7	26.2	1.9	40	—	1490
Concheño	28°18′N 108°11′W uplands	2134	13.5	23.4	3.6	39.5	−13	1141
S. F. Borja	27°35′N-106°41′W uplands	2146	16.0	26.0	6.4	40	−10	601
Temósachi	28°58′N-107°50′W uplands	1858	13.0	23.3	2.5	38	−18	592

Annual temperatures (in degrees Celsius) and rainfall (in millimeters) in Tarahumara Country (*continued*)

Station	Location/ Region	Altitude (m.)	Annual Mean	Highs Mean	Lows Mean	Extreme Max.	Extreme Min.	Rain
Yepómera	29°01′N-107°49′W uplands	1900	12.0	22.3	1.3	—	−25	—
La Junta	29°45′N-107°58′W uplands	—	16.2	25.5	6.5	42	−11	475
Cerocahui	27°17′N-108°03′W uplands	1500	16.5	—	—	—	—	1165
Cuiteco	27°27′N-108°00′W uplands	1720	16.1	22.3	1.3	—	—	1074
Urique	27°13′N-107°55′W canyon	599	23.5	—	—	—	—	774

Source: Table prepared by computing data compiled by Alvarez (1970) except for Norogachi figures, which were calculated from data recorded by Salvador Herrera and Manuela Holguín de Herrera for Comisión del Río Fuerte from July 1975 to August 1976. Data for the three stations in Municipio de Urique were taken from Maderey (1985; 1989) whose sources were Servicio Metereológico Nacional and Secretaría de Programación y Presupuesto.

temperature of 24° C. Low-altitude places such as Batopilas, 501 meters above sea level; Morelos, 606; Urique, 599; Chínipas, 515; and Moris, 714, are located in between mountains with altitudes of over 2,000 meters above sea level. Temperature data from Batopilas may represent the norm for the whole canyon country. The weather is quite warm and with high humidity. In July the highs reach 44° C, and the average high temperature is 30.1° C. The average low is 18.0° C. There are approximately 300 days free of frost per year. The precipitation by year amounts to 632.2 millimeters.

Tarahumaras living close to the canyon rims reside on the high altitude areas in summer, and on the low altitude places in winter. The climate favors Mediterranean and tropical types of agriculture in the canyon floor, where coffee, oranges, lemons, limes, avocados, bananas, sugar cane, and tobacco are grown. In the uplands, the climate is mountain-cold temperate, with a summer rainy

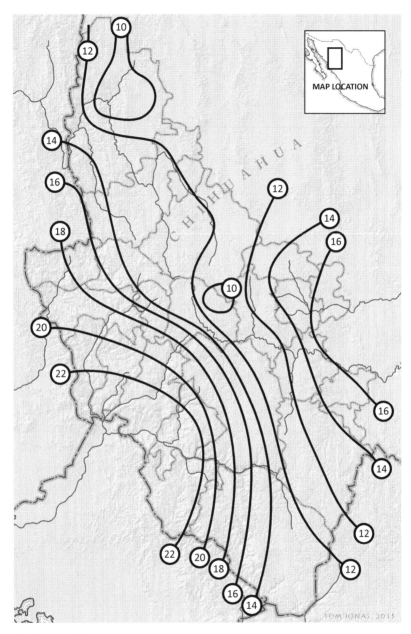

Tarahumara Country isotherms
Isotherms representing the annual mean temperatures. (Copyright © 2015, University of Oklahoma Press)

season and winter snow or equipata. Most technical reports seem to ignore the presence of winter rain, classifying the upland's climate as Köppen's BSkw (dry with a summer rainy season) or Cwb (temperate, close to subtropical with a dry winter). Just a narrow strip along Río Balleza is classified as BShw (subtropical, dry). The canyon country, between the altitudes of 600 to 1700 meters, has a subtropical climate, temperate with summer rains and dry in winter (Cwa).[25]

The Tarahumaras recognize only two seasons to the year in their country: *bará* or *las aguas* (the rainy season) and *romó* (winter).

The Soil

In the uplands, brown forest soils are common and in many places they have undergone the process known as "podzolisation" by which water drags humus and iron compounds into the ground, leaving an acidified surface composed of silica and silica-like compounds.[26] Soils throughout the Sierra have little depth; the superficial layer consists of partially decomposed pine leaves and other detritus. The underlying rock is mainly volcanic tuff or rhyolite and lacks natural fertilizing agents; the land therefore is poor, requiring fertilization for agricultural purposes.

The action of erosion is well apparent; many areas do not have soil covering and the rock (*laja*) is left bare. Deforestation has progressed at an alarming rate since the 1960s and does not seem to be abating.[27] Finally, slash-and-burn agriculture (*mawéchi*), inadequate drainage of farming lands, and overgrazing by goats accelerates erosion. However, black soils (*chernozem*) rich in humus, may be found atop the Sierra Madre surrounding the forest brown soils.[28] In the canyon country, the soil under *monte mojino* belongs to the lithosoles class.[29] The scarce soil among the rocks is worked out by hand, yielding a good crop the first year, followed by severely decreasing yields. At the canyon's bottom, alongside the river, the soil is usually sandy or muddy. The abundance of water and the fertilizing action of debris deposited by the river makes the soil more productive.

The Flora

Above 2,000 meters in altitude, pine forests are the characteristic vegetation and between 1,500 and 2,000 meters, mixed pine-oak is the norm.[30] In most

The *Pinus ponderosa* and oak trees (*Quercus* sp.) of this mixed pine-oak forest in Tarahumara Country.

areas, *Pinus ponderosa* is prevalent but at least ten other species of pine are also found. Incidentally, *okó* is the generic Rarámuri name for pine trees, but the Rarámuri know the different species and have names for each one of them. Near Cerocahui and in some middle elevation upland meadows, one may find stands of *Pseudotsuga mucronata* and *Abies durangensis* (Spanish *pinabete* or Tarahumara *mateó*). The less broken terrain of the northern Sierra made possible large forestry exploitation there (Madera City, Matachí, and Temósachi), but it is currently in decline. Evergreen oaks usually grow mixed with the pines except in the higher places where only pine trees grow. In the lowlands, instead of pine trees, climax oaks may be found. A third representative tree of the Sierra is the juniper, whose habitat extends beyond Tarahumara Country's boundaries. Interspersed in this mixed forest, the *madroños* with their bright red, peeling trunks and branches contrast with the green landscape. Other trees, such as ashes (*wisaró*) and willows are also common. In the eastern *lomeríos* country, characterized by hilly terrain, pine trees practically disappear, giving way to *huizache* (*wichá*), *mezquite* (*bejoké*), and *manzanilla* (*ku'wí*) and making the landscape similar to that of desert zones with occasional cacti such as *nopal*

(*Opuntia* sp.) and *Ferrocactus herrarae* and *Echinocactus otonis*. Ascending the Sierra from the northern deserts, the traveler crosses through dense clusters of *Opuntia cholla*.

Vegetation in the canyon country varies, depending on the altitude level of the land.[31] At the higher levels, it is similar to that of the uplands with specimens of mixed pines, such as *Pinaceae* and *Ericacea*. The middle slopes are characterized by thicket formations, locally known as *monte mojino*, composed mostly of thorny plants of the family *Leguminosae* (*Wichá*). This ecosystem is also known as known as *bosque de vinorama*.[32] Finally, in the bottom of canyons one finds remarkable trees such as *guamúchil*, *pochote*, or kapok tree (*Ceiba acuminata* or *Bombax* ceiba L.) *samo*, *zalate* (*Ficus carica silvestris*), *amapa* (*Tabebuia palmeri*), and the giant cacti characteristic of the region (*chawiró* or *chawé*).

The Fauna

Animals are for the Tarahumaras an essential part of their worldview. They are not only a source of nourishment and sometimes of medicinal remedies, but Tarahumara legends depict them interacting with humans in ways reminiscent of those seen in the higher Mesoamerican cultures.

Mammals

The great carnivores of the area have become nearly extinct. Nonetheless, stories are told about wolves, *narigochi* or *naríguri*, killing cattle in several places. People also still delight telling stories around the bonfire about sightings of *mawiyá*, the puma (*Felis concolor*), and *ojí*, bears (*Ursus horribilis*), which more than likely are now extinct. Minor canids are still common, such as *basachí* or *wasachí* (*Lyciscus* or *Canis latrans*, or coyotes). The fox, *keyóchi*, may be found in two varieties, brown and gray.[33] The Tarahumaras distinguish three kinds of skunks: *pasuchi*, *siyachi*, and *pasuchi kochi*. *Pasuchi* (*Mephitis bicolor*) is the common skunk; *siyachi* is the dwarf skunk, or *zorrillo enano*; and *pasuchi kochi*, or "pig-skunk," is the hog-nosed skunk.[34]

Domestic animals of European origin include livestock, such as cattle (cow, or *wasi*), bull (*toro*), ox (*bóisi*), goat (*chiba*), and sheep (*bo'wa*). Domestic pigs are known as *kochi*. Dogs (*o'koci*) are a mixed-breed cross of native American dogs and those brought by the Spaniards.

Besides mawiyá, worthy of mention are *rekamuchi*, a wild cat, and *ru'chí*, which is the bobcat (*Catus ferus*).[35] In bygone times, a leopard (*mawarí*) roamed along the canyon rims. *Musa* probably is a category represented by feral cats.

Deer are considered a serious threat to the crops. In a single night, a couple of hungry deer may destroy the whole cornfield (*milpa*), which otherwise would provide nutrition to a family for a year. The Tarahumara word for deer is *chumarí*, which probably comes from the verb *chumama*, to walk on tiptoe.

Rowí, the rabbit, is considered a pest, but is also appreciated as a delicious dish. This animal in folk stories is depicted as a messenger god. There are many varieties of rabbits—the dark, small, and nimble rabbit and the white rabbit, probably a feral rabbit. Hares and jackrabbits (*Lepus callotis* and *Lepus* sp.) are called *ruyé* or *ruwé*, abound in the vicinities of cornfields.

The *cholugo* (*Nasua narica*) resembles a raccoon (Spanish *coati*; Tarahumara *churé*) but differs in that it has a more slender body and a longer snout.[36] This brown-coated, gregarious animal has a ringed tail. Once, the cholugo was the most important source of animal fat for the Tarahumaras, but since the introduction of rifles it has largely disappeared.

Bats, or *sopichí*, are common during the rainy season, especially along the canyon slopes. Vampire bats are responsible for cases of rabies, human and bovine. Some informants claim that the word for bats, *sopichí*, comes from *sopetígame*, "the transformed ones," holding the belief that bats evolved from mice. Another version claims that sopichí means reincarnated, or "the appeared ones," an interpretation relating bats to the "ghosts" of dead people.

The Tarahumaras give mice (*Mus* sp.), the generic name *chikuri*, but they distinguish among several varieties—*chikuri pochíkuri*, a whitish mouse; *chikuri sitóchi*, a dwarf mouse; and *chikuri chu'arochi*, which lives among the rocks. Rats are called *rorí* and are classified according to their color—gray, red, brown or black. *Reposí* is the given name of the ordinary mole and *o'kíri* designates the *tuza arriera*, the "muleteer" mole.

The following species of squirrels have been reported: *Sciurus vulgaris*, *Sciurus aberti*, *Sciurus apache*, and *Sciurus* sp.[37] The Tarahumara generic term for squirrel is *chipawí*, but *chará* is the dwarf squirrel. *Chiká*, the squirrel of the fields, appears in the rainy season and is reputed among the Tarahumaras to be a tasty dish. *Chimorí* is the gray squirrel of the trees and a related species, *chimorí moro* or *siyóname* (blue squirrel), is distinguished by its gray-bluish color. *Lá'chimi*, or *lá'chimuri*, is a nimble yellow squirrel. *Chipawí* is considered a pest, a serious menace for the crops. *Chimó* (*chichimoco*) is the name of a ubiquitous terrestrial chipmunk.

Porcupines (*Erethizon dorsatum*), are known as *wechéame* or *kochi wechéame* ("thorny pigs"). A Tarahumara informant gave the name *cha bitiko kochi*, "the lying pigs," for the boar (*Sus acropha*), explaining that they are called so because they really are *tasi cho rewéame kochi*; that is, pigs that do not yet have a name.

Badgers (*batui*) and river otters, *churo* or *bajurí* (*Lutra annectens*), used to be common in mid-size streams but are now slowly becoming extinct. Otters are the object of peculiar beliefs, the fat from batui was used as an oily base for ointments and poultices.[38]

Apparently there is not a Tarahumara name for the armadillo (*Dasypus novemcinctus*) and in the area of Norogachi they are unknown. In Cerocahui and Guachochi, an old belief claims that the carapace of an armadillo can be used medicinally. A soup made with opossums was recommended for women who may have childbirth difficulties.[39]

Birds

The Tarahumara generic word for birds is *churugí*. Two commonly observed birds in this area are *chuyá*, a bluebird considered a threat to crops, and *rosáchi*, a whitish bluebird that lives among pine trees and lacks the characteristic crest of chuyá.[40] The *chuyépari* (*Catherpes* sp.) or *saltapared* is a little bird that is released from a cage or sack during the celebration of Holy Saturday to symbolize the Resurrection of Christ. *Chogéwari*, a bird known by mestizos as *pájaro viejo*, or *villista*, is a medium-size yellowish bird, used as the norm to compare the size of other birds; for example, *chiyowi*, an orange-yellow bird, is said to be bigger than chogéwari, while *chonchi*, a graceful spotted bird, is said to be as big as chogéwari. Ordinary sparrows, *Passer domesticus*, including a multicolored variety, are known as *wijútame*. A little blue bird, probably also a sparrow, is called *okiwi*. *Tochapi*, or in Spanish *chirulita*, is a small bird of white tail and lively singing. According to a myth, this bird is *ru'síwari*—the peyote stone where the spirit of Jíkuri lives, transformed into a bird. It is commonly believed that when a snowstorm approaches, the bird sings announcing its arrival.[41] Hummingbirds are known as *simuchí*.[42] The swallow, *Hirundo rustica*, is known as *sowé, soě*, or *sowépari*; however, some also call it *cotirini*, which is probably derived from the Spanish *golondrina*. *Osáruwi* is a little bird that builds a nest fashioned as a small raft. The *ripíruwi*, possibly a swift (*vencejo*), is a migratory bird that comes to the Sierra to lay and hatch its eggs. *Tucha* (killdeer, or *Charadrius vociferous*) may easily be spotted alongside the springs whistling a characteristic *tuiuí-tuiuí*; the kingfisher, in Spanish, *martín pescador*, is known as *bachárawi*. Several varieties of cardinals may be observed: *kwichi* or *mekwákari* is the cardinal with red belly and greenish-blue back; *tubisi* is all red; and *picho*, a variety with red chest, crest, and black and blue tail.[43]

The most striking species of woodpecker is the great *pitorreal* or *carpintero imperial* (*Campephilus imperialis*) first described by Lumholtz and now considered extinct.[44] The common woodpecker is known as *ayoka* or *wayóaka*; the redheaded spotted woodpecker (*Picus scalaris*) is known as *koracha*; and the spotted (*pinto*) woodpecker as *chikákari*. Finally, the yellow woodpecker is *kapichiri*.

Chiwí (*Meleagris* sp.) is the turkey, either domestic or wild. *Rechorí* is the common scaled quail. Quail (*Callipepla squamata*) is known as *pú*, but roadrunners (*Geococcyx californianus*) are usually also designated with the same word. Herons (*Ardea Herodias* and *Garceta candidisima*) are known as *wachó*, while cranes (*Grus* sp.) are known as *kurú or koró*, the name given because the crane croaks *"kúru, kurú."* Wild and domestic species of ducks and geese are known as *wasoná*; that is, *anas diazi*. A parrot, *guaca* or *cotorra serrana* (*Rynchopsitta pachyrhyncha*), is said to come to the Sierra in September and October from the canyons to eat *piñón*, the fruit of *Pinus cembroides*. They also show up in May to eat pine tree spikes and fruits of *aborí*. These parrots are known by the Indians as *kánari* or *ikánuri*.

Turtledoves (*Turtur* sp.) are known by the name *urí* of which the natives distinguish two varieties. Ordinary doves and wild pigeons are named *makawi*. The presence of band-tailed pigeons (*Columba fasciata*) and mourning doves (*Zanaldura macroura*) also has been reported.[45]

Hawks are collectively called *rawíwi*; these include *ayátorii*, which is a lead-bluish hawk, while a gray hawk with white chest (*Accipiter mexicanus*) is *rekágani*. A yellowish hawk is *wipiréchari*. A reputedly nonpredator hawk is *girichi*, while another nocturnal hawk, considered lazy because it does not build a nest but lays its eggs directly on the fields, is called *pimari*. Belonging to this list is *retósari*, a winter hawk.[46] The eagle (*Aquila chrysaetos*), an endangered species, is known as *ba'wé* or *wa'wé*, and the gray hawk, *aguililla* (*Buteo* sp.), as *kusá*.

Watósari is an eagle from canyon country distinguished by a white stripe, and many believe that its arrival forecasts frost.

The owl, *rutúkuri* (*Buho virginianus*), is supposed to announce an impending death, a belief shared by many other Amerindian groups throughout Mexico. Barn owls or *lechuzas* (*Glaucidium* or *Strix* sp.) are known as *okóturi* or *okóchuri*. The Tarahumaras consider these as birds surrounded by mystery as suggested by their enigmatic eyes and "language": *hú, hú*. The spotted owl (*Strix occidentalis*) is another endangered species. *Tápani* is the dwarf barn owl, characterized by its peculiar ears; a big, round-headed barn owl that likes to nest on the ground is called *gabósari*.

Vultures (*Cathartes aura*) are known as *wirú*; interestingly, Tarahumaras call airplanes by this name. Other predators of importance are crows, or *gorachi* (*Corvus* sp.) and the Mexican blackbirds (*Quiscalus macruous*) are called, as in the rest of Mexico, *chanate*. Finally, *wapíkuri* is a black bird with a white strip in the base of the tail.

Reptiles

A great diversity of snakes is common in the Sierra, but the number is much larger in canyon country. *Chachámuri* (*Crotalus atrox*), the "pig-headed" or royal rattlesnake, up to seventy centimeters in length, is the most feared of snakes in this land. *Sayawi* is the ordinary rattlesnake (*Crotalus* sp.). A poisonous snake, green with black spots, is known as *banagéame*. *Okósini* is another green or gray snake reputed to be very lethal. *Makúsuri*, known by the mestizos as *pichicuate* or *chipicuate*, is another snake thought by both groups to be very dangerous. *Bamagásuri* is a poisonous snake characterized by its red belly and white stripes on the head and *rawirísiri* is a dark yellowish one. The ordinary coral snake is *apósini* (*Elaps fulvius*).[47] *Wásini*, the blue coral snake, which may also be green or whitish yellow, is considered poisonous. A white collar is distinctive of these snakes. A beautiful snake is *wajumar*, a yellow and black or green and black serpent, known by the mestizos as *víbora ratonera*, or mice-hunter snake. *Rinórowa* is the Mexicans' *chirrionero* (*Masticophis flagellum*), which is reputed to suck from the cows' udders, in some instances killing them. This long snake, 1½ to 2½ meters long, when it is in heat, hangs from the tip of its tail, acting like a living whip.[48] A smaller chirrionero is *kujíriwi*, 1 meter long; this snake has stripes along its body. *No'pí* is a colorful snake that is said to live in the water springs. The ordinary garter snake, *kawiwá* (*Thamnophis sirtalis*), is black, gray, and white striped. The brown water snake is *bakúsuri*, while a yellowish-colored one is *bakarachi*.

Rochá is the common lizard, colored black or gray. *Rojoroki* is a beautiful brown and white spotted lizard. *Rurugípari* is a mud-colored, striped lizard. *Ruchagápari* and *yegóchuri* or *igóchuri* are lizards that live on fallen twigs and trunks of trees. The iguana (*Iguana rhinolpha*) prefers the canyon slopes and is known as *ropagókuri*, or in Spanish, *escupón*, because of its attributed defense—spitting at its enemies. The horned lizard (*Phrynosoma orbiculare*) is known as *láchimi*, while two species of turtles, earth- and water-borne, are known as *murí*.

Amphibians

Remó is a word applied to toads (*Bufo* sp.) and frogs (*Rana* sp.). *Remó siyóname* (*Rana halecina* or *Rana catebeiana*) is the green frog, which probably is not native from the region but is appreciated as a tasty dish. Tadpoles, Spanish *renacuajos*, are known under the names of *sibore* or *siyori*. A white-greenish frog is *baríki*, while *watákari* is a dark frog that lives among the rocks and is used to prepare a poison that reportedly may be lethal for animals or man. *Wikúwari* is the name given to its offspring.[49] The mud puppy (*Ambyostoma tigrinum*), in Nahuatl, *axolotl*, is an odd water animal known among Rarámuri as *re'así*.

Fish

The generic name for fish is *ro'chí*. *Awíki* and *matorecha* are much appreciated edible varieties. *Resé* is a little fish that can be found in the upper Río Conchos, while *sakachi* belongs in the upper Urique. In spite of unfavorable conditions in the upland streams, *ruchú*, a two-inch long dark fish, thrives well. *Maserowi* is probably the Tarahumara name for catfish.[50]

Invertebrates

Crayfish are known as *rinú* and *chachai*. *Sa'í* is the earthworm (*Lumbricus terrestris*), *guyene* (*Hirudo* sp.), the river leech, and *naráguri*, the ordinary snail. *Kuchíwari* primarily designates *Ascaris lumbricoides*, but other worms are often called by this name. *Rimíkuri* is a special ringed worm, while the green caterpillar is known as *rajewá*, from *rajema* to burn. The corn worm is *matéburi*. A favorite dish for the Tarahumaras, the agave worm, is *nowí*.

INSECTS   *Ruchumákari* is an insect, which appears to walk over the surface of water and is known by the mestizos as "mother of water," while *bakokómori* and *rinagápuri* are animals that crawl upside down under the surface of water. The centipede (*Scolopendra* sp.) is known as *ma'achíri*.

Common flies (*Musca domestica*) are called *se'orí*, but the name is also used to designate other insects. Bees (*Apis* sp.) are called *akáame se'orí*, that is, "sweet fly," while *se'orí rosákame*, or white fly, is the name applied to termites. Ants are called *sikói* or *sikuwi*. Bumblebees are *bi'kó*, and wasps, *weché*. *Wapará* is the name for hornets. Butterflies are known as *nakarówari*, "flying ears," and a flower-sucking butterfly is known as *ru'kábari*. Dragonflies are called *sikuro*.[51] Fireflies are known as *kupisi* and ladybugs are known as *machá*. Among the scarabs (*Cotinis mutabilis*) can be mentioned *iwéchi* or *mayate*, in Náhuatl, *máiatl*, and *toro* is a beetle that resembles a little bull. The stinky *Eleodes* sp., in Náhuatl, *pinacatl*, is known as *ujini*.

The winged grasshoppers are *masakari*, while the wingless are *ochípari*; *garabosi* is a big grasshopper, green and red, that often destroys bean crops. *Seeró* is the name of *Stagmomantis* (praying mantis).[52] *Kujúbari* is a particularly harmful grasshopper. Crickets are known as *rukúchuri* and cicadas, *nakáchari*.

SPIDERS   *Narúchare* (*Latrodectus mactans*) is the black widow, while *ro'ka* is a poisonous gray or brown spider characterized by its beautiful spider webs. There is a large, exceedingly poisonous spider called in Cerocahui *ramáchuri* or *ramáchari* and, by mestizos, *matavenado*, or deer killer. The awesome-looking tarantula is known as *sipurí* or *sipurá*. The scorpion is known as *ma'ikóani* (*Hadrurus mexicanus* or *Centruroides* sp.).

⤕ ⤕ ⤕

This is thus the ecological habitat of the Tarahumaras—an environment some have called hostile and harsh, but which contains plentiful natural resources that provided the basis of survival for them and their neighbors, the mestizos and other indigenous groups who have shared this land over the centuries.

TWO

A Historical Review of the Tarahumara People

The Tarahumaras, or Rarámuri, are remarkable people in many dimensions. Reviewing what is known of their origins may contribute to understanding why they are the way they are—or at least why they let observers see certain facets of their collective personality while zealously hiding others. The task is not easy: archaeological exploration of the area where they presently live or lived in the past is scanty; historical records made by religious and military explorers attend to the goals and ends of such chroniclers and are seen from their perspectives; the Rarámuri themselves lacked a written system to record their own version of their historical and legendary beginnings. Therefore, what follows is an attempt to put together what I have gathered, particularly those events that allow reading the Rarámuri story between the lines.

Early Settlements in the Clovis Complex

The territory historically considered the land of the Tarahumara or Rarámuri people has been inhabited since the Pleistocene by man, probably mammoth hunters from the Clovis complex. The group, distinguished by their fluted lanceolate projectile hunting points, appeared in North America 11,050 to 10,800 years before the present[1] and eventually entered the sierra, canyon, and adjacent basin and range of Tarahumara Country. Tools of the Clovis people have been found in sites throughout North and Central America but not in the Tarahumara region.[2] Mammoth remains found in the region, particularly in the

Papigochi valley and the Arroyo de Basuchil, suggest the presence of man, the main mammoth predator.[3]

Irwin-Williams proposed that from 7,000–6,000 B.C. until 3,500–1,000 B.C. peoples sharing the Cochise Desert cultural tradition populated the American Southwest and northern Mexico.[4] This tradition represents, for many authors, the matrix from which most cultures in the American Southwest and northern Mexico derived. Others consider the Cochise to be just a northern extension of a general northwestern Mexican hunter-gatherer culture.[5] Di Peso questions whether the Paleo-Indian hunter culture, the Clovis folks, actually antedated or followed the Cochise Desert gatherers in northern Mexico.[6] Next to consider is the establishment in the area of Chiricahua settlements. Chiricahua materials have been reported as far south as southern Chihuahua. Several authors state that the tool kits recovered from this culture support a mixed foraging economy; however, by 2,500 B.C., as Gummerman and Haury propose, agriculture was already spreading to the highland Mogollon area through the Sierra Madre corridor.[7] Maize cultivation diffused northward from the Tehuacán Valley and other ecosystems in Mesoamerica followed by crops such as squash[8] and beans. Maize agriculture not only made possible the developments of larger towns such as Casas Grandes on the northern boundary of Tarahumara Country, but also enriched the diet of the hinterland dwellers complementing the one derived from their hunter-gatherer exploits.[9]

Zingg, who conducted archeological investigations in the uplands near Norogachi and in the canyon country near Batopilas, postulated that the most ancient Tarahumara culture was marginally affiliated with the Basketmaker culture of the American Southwest.[10] Di Peso insists that the proto-Tarahumara culture was similar to that of the Anasazi Basketmakers.[11] Extrapolating from the findings of Ascher and Clune, he claims that they planted corn and made a crude brown ware (Alma Plain), some of which was decorated with red cross patterns.[12] Zingg labeled this culture as Río Fuerte Transitional Culture, which was followed by the Cave-Dweller phase, which lasted until the arrival of the Spaniards contacts and still survives in some areas.[13] Inspection of burial caves in the Norogachi area suggests that the burial pattern that Zingg considered typical of the Basketmaker period is not specific; for example, many burials that he would consider typical Transitional Rio Fuerte, with a flexed position in caverns and containing blankets of rabbit fur, antler awls, punches or needles, string apron skirts, and other representative articles, contain also plain ware brown pottery. In addition, modern ethnological and historical evidence suggests that interment in caves persisted until recent times and burial in granaries (as practiced during the Cave-Dwelling phase) was confined to certain areas. Little has been mentioned by the archaeologists about the clusters of roundhouses found in the Sierra.[14] This type of dwellings, called *kokoyome*, or *sonogori* in the region,[15] are different from the modern Tarahumara houses, or *ga'rí*, which are square or rectangular and apparently were in use during the Cave-Dweller phase. I had the opportunity to examine several sites with five to ten similar structures in the neighborhoods of Norogachi, Romichi, Kochérare, and Yegochi. The round structures appear inserted on the bare rock; they are two to five meters in diameter and have a central hole possibly used to set a fire. In the site of Romichi, one finds among these ruins arrowheads made of obsidian with harpoon features indicating their use for fishing.

At some point, the people from Chalchihuites in what is now the state of Zacatecas became the main agents of diffusion of Mesoamerican culture for northern Mexico and the American Southwest. However by A.D. 1,060 the diffusion center had moved to Casas Grandes.[16] The presence of shell (*vermetid*) necklace beads, intentional cranial deformation,[17] skull trepanning,[18] and other cultural traits in distinctly Tarahumara burial caves may reflect that the Tarahumara area was considerably influenced by the Casagrandians.

It is thus quite possible that the Tarahumara hinterland, located just at the southern boundary of the Casas Grandes culture, was used as a supplier of wood, corn, and other utilities to the northern urban settlers; conversely, the Rarámuri might have received cultural influences, some of which survive until the present.

The Linguistic Evidence

The present-day Northern Tepehuan and Pima Bajo cultures are very much like that of their Tarahumara neighbors in terms of material culture; however, their ethnic self-identification, ethos, and language define them as very different from each other. The available philological evidence supports the hypothesis that the Tarahumara language sprang from a southward Uto-Aztecan migration. This migration, as proposed by Hayden, might have derived first from the Amargosan complex and then migrated south and westward.[19]

The historic and modern Tarahumaras are bordered in the north and south by O'otam- or Piman-Tepiman-speaking groups. The Tepehuans of San Pablo (modern Balleza) recorded in historic times and the Northern Tepehuans inhabiting the southern Sierra beyond the canyons of Wérachi, San Ignacio, and Sinforosa occupy the southern boundary of the land of the Tarahumaras, while the Mountain Pima Bajos lived at its northern limits in or around the towns of Yécora, Maicoba, and Yepachi. The Tarahumaras and the closely related Tubar and Varijío languages thus cut as a wedge, running east to west, the continuity of the Piman-Tepiman family that extends along the Sierra Madre southward into Durango.

A plausible explanation for this distribution would be that the colonial and present location of the Tarahumaras resulted from a migratory movement which, coming either from central Chihuahua or, more possibly, from Sinaloa and Sonora, interrupted its advance while crossing Pima-Tepehuan country and settled there, keeping the northern and southern Pima-Tepiman speaking groups separated. If this hypothesis held true, Piman-speaking groups would represent a remnant from the occupants of the Sierra before the Tarahumaras' arrival.[20] Miller offers an alternative theory—that of a movement of the Pima-Tepiman-speaking through or around the Tarahumaras.[21] All in all, the Tarahumara language seems to be genetically closer to Opata and Cahitan languages than to the Piman family.[22] Continuity with such groups as the Opatas, Eudeves, and Jovas has to be considered too.

Unfortunately, little is known about Concho, a language that was spoken at the eastern boundary of the Tarahumaras. Linguists consider this language Uto-Aztecan and probably Taracahita, but, as Miller points out the evidence to support this is meager and limited to only a few extant words.[23] On the other hand, although Opata and Tarahumara were not mutually intelligible, Miller in his classification clusters Opata and Eudeve as a subgroup of Taracahita, and qualifies Jova as most probably being Taracahita too. As a matter of fact, Tarahumara word roots predict the meanings of many words from the Opata spoken in the colonial era.[24]

In the states of Sonora and Chihuahua, several place names are found ending with the particle -chi. Interestingly, these names, presumably Opata, Jova, or Eudeve, still can be interpreted etymologically in modern Tarahumara; for example, Bacadéhuachi, Tonichi, Aconchi, Bacóachi, Teuricachi, and many others. Drawing a line around the area containing such place names, a pear-shaped zone is obtained, which includes a great part of Sonora and extends itself almost to Naco on the Sonora-Arizona border pointing to the area proposed by some linguists as the origin site for the Sonoran Uto-Aztec languages. Tarahumara oral tradition appears to support this linguistic hypothesis. A Tarahumara legend claims that the Rarámuri came to occupy their present habitat from the northwest where they were close to *wa'rú ba'wechi*, the great water or the sea.[25]

Tarahumara lore suggests that there were close links between them and Mesoamerica. This is illustrated by their legend titled "God and the Devil" or "The Two Brothers" narrated by our Tarahumara informant, Erasmo Palma Tuchéachi. The legend originally recorded by Brambila in 1964 and published by me in 1974 has many Mesoamerican cultural elements and parallelisms common in other Mesoamerican ethnic groups.

"The Two Brothers" or "God and the Devil"

In ancient times, two brothers disputed over the control of the universe; the Younger Brother would later become God the Father and the Older Brother, the devil. They became hostile to each other and fought for the possession of heaven and earth. The Older Brother was about to win but the Younger Brother climbed on top of a poplar, escaping being caught and killed. Older Brother brought an axe and began to try to fell the tree to bring his brother down, but every time he did so the poplar would recover its original thickness. This made Older Brother very angry, so he stole Younger Brother's wife and ran away with her. When Younger Brother came down from the tree and found what had happened he told *wirú*, the vulture: Go atop the mountain and sniff so we'll find out where they went. The bird did so but could not detect the fugitives' smell. He then sent another vulture to the waterside. He did find them on the other side but was so exhausted that he could not fly back to report his finding. So he sent a little ant floating on a piece of bark. When the ant found Younger Brother, she crawled to his ear and communicated her finding. Then they called bees to help catching Older Brother, who by then had become a mountain lion. The lion tried to steal some honey from the honeycomb when Younger Brother gave the order to trap him. The bees let a plank fall on the mountain lion's paw and then all the bees covered the lion and stung him. When the lion was unconscious because of the bee stings, Younger Brother ordered the animals to gather firewood and asked another lion to "slice him open," and he did so, extracting Older Brother's heart. Before this happened Younger Brother had commanded wirú to position itself in the heights and tackle the heart. He sent *gorachi*, the crow, further up; *wa'wé*, the eagle, even further up; and finally he commanded *kusá*, the gray hawk, to stay almost at the entrance of heaven. Younger Brother then ordered the lion's heart to be thrown on the burning coals. As expected, the heart flew up and away. Wirú was easily overcome; so were gorachi and wa'wé. But the last relay, kusá, however, caught it right at the entrance of heaven and brought it back to be burned over the burning coals. Because all started like this in ancient times, so some still steal the wives of others.

The Colonial Era

The written history of the Tarahumaras began in 1588 with the first contacts of the Spanish conquistadores with inhabitants of the western Tarahumara region. Chronicles portrayed the Tarahumaras as an isolated group, but actually intertribal contacts were very significant before the arrival of the Europeans and continued throughout the colonial period.

Pennington outlined a territorial map of the land occupied by the Tarahumaras at the time of the contact with the Europeans.[26] The boundaries he proposed, however, might have extended beyond the limits he set. As an example, the Ranchería of Chuvíscar, recognized by the Spaniards since 1653 and located just 15 kilometers from Chihuahua City, represented the northeasternmost Tarahumara outpost.[27]

The ethnic groups bordering the area occupied by the Rarámuri constituted a rich mosaic of cultures; they included the Conchos to the east and north, Jovas and Pima Bajos to the northwest, Varijíos to the west, Tubares to the southwest, and Tepehuans to the south. Between the Varijíos and the Tubares, Pennington places the Guazaparis, Témoris, and Chínipas; even though these probably were Rarámuri subgroups, they were identified by the Jesuit chroniclers as distinct "nations."[28] The southwestern corner probably was a point of contact with the Huites, while in the northeast they were neighboring the Chinarras. Northwestern contact with the Opatas was probable, as was contact with the Cáhita-speaking groups to the west.

Surrounded by so many groups, it is not surprising that the Tarahumaras periodically engaged in intertribal conflicts. The intensity of their bellicose activities was reflected in some of their ceremonies in which the crania of beheaded enemies were used as symbols of victory and celebrated with ritual dances. Fonte, a missionary priest, related in 1608 how the valley of San Pablo (present-day Balleza) was inhabited by Tarahumaras and Tepehuans, who sometimes engaged in bellicose encounters and at other times entered into collaborative alliances.[29]

Mateo de Vesga in 1620 described an incursion of Tarahumaras and Tepehuans against Spanish farms in the region of Santa Bárbara.[30] Instances of belligerent encounters involving Tarahumaras and Opatas and Tarahumaras and Tepehuans were described as late as 1824 and 1840.[31]

The Contact

The Spaniards who first made contact with the Tarahumaras around 1588 were soldiers, miners, and missionaries. Upon arrival, the Spanish soldiers made clear their purposes, engaging in ostentatious displays of military power. An account of events occurring within this period describes how when Osorio, a Spanish commander, arrived at Chinipas with his soldiers, two thousand well-armed Indian warriors appeared on top of a hill threatening them. Osorio asked some of the more daring Indians to shoot at their leather shields with their bows and arrows. The arrows failed to pierce the shields. The Spaniards then fired at the shields with their gun-powered weapons, destroying the shields to the astonishment of the Tarahumaras.[32]

The missionaries were committed and courageous; they were not seeking riches but converts. According to one account, Father Juan Fonte's arrival in 1607 and 1611 to the valley of San Pablo, currently Balleza, was greeted with affection and encouragement to enter further into Indian land and was also conducive to the foundation of the first pueblo de misión in southeastern Tarahumara Country.[33] This strategy had one tactical advantage; it enabled the Tarahumaras to weigh and measure the strength of the enemy before deciding to take hostile action.[34]

Reducciones

The Spaniards tried to gain increasing control over the Indian population. Under instructions from Spain, they created *pueblos* (towns) and tried to resettle the Indians living scattered in the land. This tactic called *reducción*—resettlement—was detailed and specifically mandated by the Spanish "Laws of Indies," which were designed to govern their empire in the New World.[35] According to Neumann, a Jesuit missionary, the goals of the reducción were to prevent the Indians from living scattered "like beasts."[36] In actuality, this policy allowed better control of the population and facilitated barter activities, such as exchanging merchandise for grain. Later, recruitment of Indian labor to work in mines was an issue not overlooked. Another objective of the reducciones was to facilitate the evangelization and subjection of the natives to the Spanish crown. The resettlement to a great extent failed in Tarahumara Country. The Indians preferred to live in small hamlets. The coercive measures applied to implement the Spanish resettlement attempts later triggered indigenous belligerence.

Epidemic Diseases as Biological Weapon

A factor that enabled the conquest of the native peoples of America by the Europeans was the transmissible diseases brought to the New Continent by the newcomers. The natives, lacking immunity for these disorders, experienced very high rates of mortality.[37] Epidemics acted as an effective bacteriological weapon that decimated the Indian population, facilitating the Europeans' penetration. The lack of written records impedes accurate assessment of the initial damage caused by epidemics to the northern tribes of Mexico, including the Tarahumaras. However, the devastation most likely reached the magnitude of the 1521 smallpox epidemic in central Mexico that contributed to the defeat of the Aztec Empire. Years later, in 1576–1577, typhus and typhoid fever caused the death of an estimated three-quarters of all the Indian population in central Mexico.[38] These diseases took a large toll of lives as they spread into northern Mexico and the American Southwest.[39] Father Fonte's 1611 account noted that smallpox had arrived in Tarahumara Country even before the Spaniards had reached the area.[40] Interestingly, in facing these diseases, the Tarahumaras seemed to seek spiritual help rather than just physical healing. Shortly after the contact the role of the missionaries as physicians of souls and bodies was readily accepted by the Tarahumaras, reflecting the spiritual orientation of the group.[41] The resettlement policies of the Spaniards increased the contact with sick individuals, facilitating the transmission of diseases. Missionaries such as Father Julio Pascual played a heroic role assisting the sick. The Indians did not overlook this. In 1662, a "plague" decimated the Tarahumara population, particularly killing the youngsters.[42] More plagues followed in 1666 and 1668.[43] In 1695, Neumann reported a devastating epidemic interpreted as an omen predicting future uprisings.[44] In 1730, Glandorff, a missionary, described a plague that killed 800 children and 664 adults.[45] In 1833, cholera caused great mortality as the epidemic swept Chihuahua with further outbreaks in 1849 and 1851. Smallpox continued causing mortality among the Tarahumaras until contemporary times. Lumholtz, who traveled the area during the last decade of the nineteenth century, reported many smallpox casualties. In 1918, the Spanish influenza caused 4,057 deaths in Chihuahua, affecting indiscriminately Tarahumaras and mestizos.

The Jesuits

From their Tepehuan missions in the south and, later, from the mining centers of southern Chihuahua, the Jesuits penetrated the land of the Rarámuri; but they began their incursions from their missions in Sonora and Sinaloa and following the routes set by the Spaniard soldiers Osorio and Martínez de Hurdaide a few years before.[46] Around 1621, Father Juan Castini reached Chínipas from his base at Villa de Sinaloa.[47] By 1626, Fathers Julio Pasquale and Manuel Martínez had established several pueblos in that area. This enterprise had a tragic end. In 1632, rebellious Guazapari and Varijío Indians in a town named now Guadalupe Victoria killed the two priests. The Spaniards retaliated fiercely. Spanish soldiers and "loyal" Indians under the command of Captain Pedro de Perea slaughtered 800 Indians and resettled 400 more in the missions of Toro and Vaca, Sinaloa.[48] As a consequence of these events, missionary work was delayed until 1676, when a mission was reestablished in the valley of Chínipas.[49]

In eastern Tarahumara Country, missionary efforts were also interrupted in 1616 after a Tepehuan uprising and the deaths of Fathers Juan Fonte and Gerónimo de Moranta. It was not until twenty years later, in 1639, that Fathers Jerónimo Figueroa and José Pascual re-established the missions at San Pablo, which is present-day Balleza, and Huejotitán, from where they extended their missionary efforts to Santa Cruz, present Valle de Rosario; San Felipe de Conchos; San José; and then Satevó.

The territory of what was called Tarahumara Antigua or Misión de la Natividad lay in the Chihuahuan foothills and basin and range country and contained the best workable land and therefore the most suitable space for colonization. Penetration from the Tepehuan mission and Parral advanced as a wedge between the Concho Franciscan missions and the Tarahumara hinterland that would become later the mission of Alta Tarahumara.[50]

Mining Discoveries and Settlements

By 1564 the valley of San Bartolomé, the present-day Valle de Allende, had been explored by the Spaniards who found that the land had agricultural potential.[51] Shortly afterwards, the discovery of silver mines in Santa Bárbara, where a mission for the Tepehuan was located, created a need for working hands and food supplies for the mining center. As early as 1625, there were reports about Tarahumaras working in the Spanish farms of the valley of San Bartolomé.[52] Pennington notes that Conchos and Tarahumaras soon joined the original Tepehuan inhabitants of the valley. By the end of the century, San Bartolomé had become the granary that supplied the mining district of Parral. This development preceded the demographic revolution that would occur after the great bonanza of San José del Parral in 1630, an event that would deeply affect the destiny of the Indian peoples of northern Mexico. The discovery of the rich silver lodes north of Santa Bárbara in 1631 transformed the area into a booming center of mining, ranching, and agriculture.[53]

By 1638, Father Gaspar Contreras, a Jesuit, described a flourishing commerce between the Tarahumaras and the settlers of Parral involving exchanges of Indian agricultural produce for wool and other Spanish merchandise.[54] These activities contributed to the process of acculturation. Barely thirty years had passed since Fonte's *entrada* (entrance) and now Tarahumara men could be found working for a salary, trading with the Spaniards, and moving or commuting to work in Spanish mines, farms, and cattle ranches.

The settlement of Tarahumaras outside their traditional boundaries began at that point. Father Zapata's *relación* (report) of 1645 at San Miguel de las Bocas—present-day Villa Ocampo, Durango—commented that the Tarahumaras were gathered by the civil authorities to work in the farms of that region and were paid with food and clothes.[55] Tarahumara resettlements outside traditional Tarahumara territory, such as Guanaceví, also in the state of Durango, and others would follow.

Repartimientos

In the Tarahumara region, the colonial penetration was restricted to "points of contact"; that is, mines, missions, Spanish army posts, and haciendas.[56] The colonial policies included the reducción and the *repartimientos*.[57] Typically the Indians in a *pueblo de misión* would work for the priest or the community, but a predetermined number of Indians would be taken from these pueblos to work in the mines and other productive units. The great majority of the Tarahumara population, however, remained in the hinterland or in areas of only partial domination.

Although free, voluntary labor was the main source of labor in the mining centers of northern New Spain, some documents show that neither legal restraints, such as Spanish official pronouncements, nor the prohibitions of the Spanish *Audiencia* (colonial government body) fully eradicated forced labor or the modified repartimiento.[58] As a matter of fact, several reports indicate that some Tarahumaras were indeed subjected to forced labor.

A more regulated arrangement, the *encomienda*, specifically designed as a remedy to address the abuses of the repartimiento, was not established in Tarahumara Country. Infrequently, instances of subjection of Tarahumaras to slavery have appeared in the literature.[59] It was not until 1781 that Governor Teodoro de Croix successfully halted forced labor in Nueva Vizcaya, including the Tarahumara region.

A colonial strategy, which still survives today, resorted to the creation of needs that only outside goods and resources could satisfy: the need for tools, cloth, European animals, farming equipment, and even Spanish currency. To obtain these resources the Tarahumaras worked in the mines for a salary. Father Contreras wrote in 1638 that the Indians had been so pacified and domesticated that many worked for months at a time for the Spaniards. Father Zepeda related that the civil authorities used to gather Tarahumaras to work in the Spanish farms and that their work was paid with food or clothes.[60]

The Spanish military relied on *indios amigos* (friendly Indians) as soldiers, scouts, and messengers. For instance, the participation of Tarahumara and Sinaloa Indians in the army was indispensable in wars against other Indian groups.[61] Particularly valuable was the Indians' archery skills, a particularly useful resource for the Spaniards while fighting in rough terrain, where they could not maneuver their horses.

An institutionalized way of participation in the army was the military presidio, consisting of fortified settlements designed to control an area implementing the strategy called defensive colonization.[62] The presidios established at Huejoquilla, Cerro Gordo, and San Francisco de Conchos had crews of settlers, people from other races, and Tarahumara auxiliaries. The missionaries soon recognized the consequences of having together people from different ethnic backgrounds. Father Ignacio Javier Estrada commented that these centers contained Spaniards, mestizos, blacks, mulattos, and Indians from various regions.[63] Sexual exchanges of individuals of different ethnic backgrounds occurred involving Tarahumara men and women. He noted that the newborn Indians had become lighter and had lost their natural color. According to Father Pascual, missions progressed rapidly and steadily from 1630 until 1648 and some Tarahumaras then lived in the pueblos de misión and participated in the systems of repartimiento without much resistance.

Many Tarahumaras however did not accept the reducción-repartimiento system and abandoned the pueblos de misión looking for refuge in the canyons and remote areas—the hinterland. Jesuit writings reported these setbacks as *huída a los gentiles*, or flight to the gentiles. The Jesuits blamed it on the nostalgia of natives for their pagan customs and a drift to drunkenness and sexual excesses. They believed the flight to the gentiles was the work of Satan, called by them the Common Enemy.

The northeastern boundaries of Tarahumara Country were not static either. Griffen has hypothesized that in the Conchería, as Conchos disappeared, their abandoned settlements were occupied by Jumano-speaking Indians in the north and by the Tarahumaras in the west.[64]

By the end of the first half of the seventeen century, the Tarahumaras had already settled in missions established within Concho territory, such as Santiago Babonoyaba, and by the next century they had taken possession of the Concho mission at Namiquipa and were seen in large numbers in the area of Chihuahua City and Atotonilco in the eastern periphery of the Santa Bárbara-Parral mining district.[65]

The Great Uprisings

Having partially recovered from the human losses inflicted by epidemics and weighing their own strength, the Indians realized that the only alternative left to preserve their ways and freedom was rebellion. In fact, Neumann placed among the main causes of the rebellions the Spanish policy forcing the Tarahumaras to live in towns.[66] The Tarahumara uprisings that followed in many instances, however, coincided in timing with those of other Indian groups.[67] Most of the military and ecclesiastical documents of the time suggest that instigation and subversion played an important role. References to intertribal coordination abound in documents beginning with Father Fonte's relaciones of his *primera entrada* (first incursion). In 1651, Father Pascual described how the Tepehuans who had been allies of the Common Enemy (the devil) were quieted after the death of the *cacique* (tribal chief) of San Pablo (now Balleza), a Tepehuan, while the Tarahumaras continued their *tlatoles* (instigations).[68]

Based on their experience in the conquest of central Mexico, Spaniards had developed strategies to wage warfare against the Indians. One consisted of systematic destruction of individuals with the potential to become military leaders. This tactic was used not only by Guajardo Fajardo against Tepórame, the greatest Indian leader, but also later by Gutiérrez de Carrión, Gabriel del Castillo, and others.[69] In addition, forced resettlement and destruction of crops also were used.[70] The Indians apparently did not know how to handle this type of repression effectively. During the revolts of the 1650s and 1690s, the Tarahumara warfare scheme basically consisted of direct confrontations taking advantage of the height of cliffs or buttes where the terrain would impede the maneuvers of the Spanish horses and where they could neutralize the superiority of Spanish gunfire.[71] The Indians stormed missions or villages, rapidly withdrawing thereafter to their strongholds. In attacks led by Tepórame, the use of foot runners to coordinate attacks gave the Indians a considerable tactical advantage in the mountains. The Spaniards eventually learned to handle threats such as poisoned arrows, initially an effective Indian weapon.[72] A major military success of the Tarahumaras was the destruction on June 4, 1650, of the Mission of Papigochi, present-day Ciudad Guerrero, and two years later the Villa de Aguilar where the Tarahumaras slew Father Cornelio Beudin first and, later, Father Jácome Básile.

The reasons Governor Guajardo Fajardo had planned to create the Villa de Aguilar are difficult to explain.[73] The establishment of a village and not a mining operation, mission, or presidio, as well as the alleged intention of Guajardo Fajardo of creating a northern capital for Nueva Vizcaya, was a radical idea. Nonetheless, if this plan had succeeded, it would have created an important center of Spanish demographic development in the heart of Tarahumara Country. Bishop Evía y Valdés claimed that the foundation of Villa de Aguilar was the cause of more than 300 deaths.[74] One year later, he would increase this estimate to more than 600 dead.

As Decorme has noted, the revolt resulted in a twenty-year lapse in the development of missions in Tarahumara land.[75] After this lapse, missionary penetration became more extensive. Fathers José Tardá and Tomás de Guadalajara were prototypical missionaries of this period,[76] establishing a network of pueblos de misión that covered from southeastern Tarahumara Country to the Pimería Baja on the limits with Sonora. By 1678, they had founded eight pueblos de misión.[77] Almost simultaneously, the courageous missionaries Nicolás de Prado and Francisco Pécoro re-established the Mission of Chínipas that had remained abandoned since the killing of Fathers Martínez and Pascual.[78]

In 1683, the San Juan and Concepción mines were discovered at Cusihuiríachi and the next year the rich silver lodes of Coyachi. Massive discoveries of silver at Cusihuiríachi in 1687 would tilt the power balance toward the Spanish side. From these places, Spanish demographic pressure would be exerted from within. As it had happened during the development of the Parral mining district, not only labor but also necessities such as wood, coal, and food supplies would also increase the demand for Indian work.

The Indian Confederation

To understand the part that the Tarahumaras played in the generalized Indian revolts of the 1690s, it is necessary to examine what was happening with all the tribes of northern and northwestern Mexico. Caraveo has described the creation by several Indian groups from northern Chihuahua, Sonora, and southern Arizona what he calls the Indian Confederation of the 1690s.[79]

Although several scholars deny that the Tarahumaras participated, careful examination of Neumann's writings reveals that the uprising of the Tarahumaras was coordinated with other groups. Revolt was initiated by an attack by the Conchos on the mission at Yepómera in the Tarahumara region. The attack began on a day agreed upon with the other confederates in April of 1690. The next step of the uprising was the recruitment of Tarahumara pueblos in the area of Yepómera, Temósachi, Nawérachi, Sírupa, and Yepómera. In addition to giving the missionaries' cause another martyr, Father Diego Ortiz de Foronda, this attack ignited the Tarahumaras' revolt. Another group involved in this stage was the Jovas. Father Manuel Sánchez tried to summon military help but the Jovas killed him. The Indian revolt failed in the end and the existence of an intertribal coalition became apparent. Rebels who participated in the uprising and were apprehended declared that Conchos, Sumas, Janos, Jovas, Julimes, Chinarras, Acoclames, Tobosos, Chizos, and Apaches were accomplices of the Tarahumaras.[80]

Messianic Features of the Uprising

Although shamans, or *sukurúame*, had played an important role in the incitement to fight in previous uprisings, it was in the 1690s uprising that the messianic character of this participation reached an acme.[81] The shamans claimed having made a pact with malignant spirits and promised invulnerability against the Spanish bullets, preaching that combatants would resuscitate in the event of being killed.[82] With such a supernatural hold, they strengthened their authority over their people. Neumann's narrative also presents a messianic picture for the Spanish side. The uprising was forecast by a series of omens—the passing of a comet, a solar eclipse, and an earthquake.[83] These omens conferred a supernatural character to the events to come. An important trait of the Jesuit endeavors appeared to be a search for martyrdom by those who had overtly and consciously enrolled in this difficult missionary work with the hope of giving their lives for the cause of the Gospel. On the other hand, Neumann's chronicles depict General Fernández de Retana's actions with epic literary patterns like those of chivalric heroes in earlier centuries. Capitán Vasco, as Neumann calls the general, acquired in his narrative the epic stature of a Cortés or a Pizarro.[84]

The Tarahumaras continued resorting to the patterns of fighting and defensive strategies used in prior revolts but with a significant level or improvement and efficacy. Caraveo incidentally emphasized the important role that the mastery of horseback combat, a new skill for the Indians, represented for Indian warfare.[85]

Spicer sees in the rebellions of the 1690s a strong nativistic movement, which had been stimulated by the hundreds of deaths resulting from European diseases.[86] It was not only the Spaniards' technological superiority that would decide their eventual victory but also the experience gained by conquering other ethnic groups as well as suppressing the prior Tarahumara revolts. Of particular relevance was the decision, taken by General Fernández de Retana, about shifting the action from the Tobosos and other Indian groups to the Rarámuri. Retana was severely criticized by Spanish settlers arguing that the Tobosos and Conchos raids affected the transit of people and goods from Chihuahuan cities to the province of Nuevo México. Spicer contends that the repression exerted by Retana, killing sixty Tarahumara leaders, beheading thirty of them and impaling their heads on sticks along the road from Cocomórachi to Yepómera actually precipitated a more generalized revolt.[87]

The uprisings of the last decade of the seventeenth century assumed particular importance for the migration patterns of the Tarahumaras. The flight to join the gentiles became more than a symbol of noncompliance with religious conversion; it was a sign of open rebellion. Anyone abandoning his pueblo automatically was considered a rebel. Neumann relates that these escapes were mass movements involving not only individuals but also whole rancherías and even pueblos. This indicates that the displacements must have relatively well organized since such actions involved logistics such as the transportation of food supplies, families, and flocks.

Better organization on the Indian side brought about modifications in Spanish strategy. The Spaniards had to recruit larger numbers of Indian auxiliaries and, after controlling the centers of insurrection, resettlements and the extermination of rebel leaders were more drastically undertaken.[88]

Tarahumara Strategy Change

Other Indian groups that stubbornly continued warfare after their defeat in the 1690s were signing their own death sentence. This in spite of partial victories such as those enjoyed by the Apaches. The Tarahumaras, on the other hand, changed radically their strategy at the beginning of the eighteenth century: Tarahumara belligerence turned into passive resistance. They became tolerant of the penetration and development of missions, pretending submission. Nevertheless, the tribal spirit would adhere firmly to Tarahumara cultural values and effectively preserve the moral independence of the tribe. This passive resistance has continued, essentially unchanged, until the present, and it has secured the survival of the group.

From that point on, some cultural aspects would be zealously guarded. The Tarahumaras demonstrated a remarkable ability to hide their ceremonial practices from the missionaries and other authorities. It would not be until recently that Tarahumara customs, such as the bajíachi and tutuguri, came back to light without fear of repression, while fiestas del Jíkuri continued being hidden.

The Expansion of the Jesuits

During the first half of the eighteenth century, the Jesuits extended their missionary network covering most of the Tarahumara territory and became established in pueblos de misión, many of them with agricultural and cattle-raising potential. There was a slow decline in the population of communities, especially after 1725. The total Tarahumara population in 1759 was reported at over eighteen thousand, probably a conservative estimate.[89] As Chihuahua City developed, the Jesuits established there a *colegio* (school) in 1718 which, in addition to attending to the educational needs of the dependents of miners, also educated the children of Indian leaders. The institution was also a center to teach Indian languages to missionaries. Among their scholarly activities, Juan de Esteyneffer, a Jesuit brother, published the *Florilegio Medicinal*, a medical treatise that was distributed and used extensively in "these remote regions."[90]

Some authors have expressed their impression that, by the second part of the century the missionaries abandoned their high values and goals and became bourgeoisie; that is, akin to the members of the higher social classes.[91] Documents from 1730 and later describe conflicts relating to land ownership between the growing haciendas, small properties, and the pueblos de misión.

The Jesuits' Expulsion

The Jesuits were expelled from the missions in 1767 by an edict of Charles III of Spain. Franciscan friars arrived to replace them. The new missionaries, with few exceptions, were not gifted chroniclers as were the Jesuits, but they had missionary zeal and administrative talents.[92] According to Bancroft and Spicer, the Tarahumara missions deteriorated rapidly after the expulsion of the Jesuits.[93] Unfortunately, the impact of the Franciscans has not been carefully documented. It is likely that the Franciscan devotion to religious instruction and ritual may have left its mark.

An area in which Franciscans did more than the Jesuits is in the recording of medicinal herbs. For instance, the relaciones of 1777 of Father Falcón Mariano and Father Urbina compiled useful lists of medicinal plants. This might have been, as Vogel observed, because the Jesuits had avoided writing about Indian herbal remedies since they considered them linked to idolatry.[94]

During the last part of the eighteenth century, a massive displacement of Tarahumaras from what was the Baja Tarahumara mission, Satevó, Papigochi, Tomochi, Nonoava, and San Francisco de Borja to the Sierra occurred. According to Pennington, this was a gradual withdrawal of people to the west and the south into a more rugged, higher-level, pine-covered country, which is where the Tarahumaras live today.[95] From examination of available documents, the disappearance of Tarahumara culture from the areas first mentioned was also due to assimilation with other groups, miscegenation, and mortality.

The Nineteenth Century

To say that the Tarahumaras were little affected by the independence of Mexico from Spain in 1821 is probably an accurate statement. For them, the ceremonies of proclamation of the independence at villages and *reales de minas* (royal mines) must have appeared like any other of the colonizers' events, completely alien to them. Nevertheless, after independence, the innumerable revolts, coups d'état, and other political convulsions that characterized nineteenth-century Mexico had important repercussions on the region. Many Tarahumaras were recruited as army auxiliaries.

After independence, the missions were secularized in 1834 and with secularization attention to the Indians practically disappeared. During this period, many new mining centers were opened throughout the Tarahumara region, contributing to population redistribution in the area. Historians have given little consideration to the Indian participation in the continuous revolutions that afflicted Mexico in those years in spite of the many battles that occurred within Tarahumara territory.

From early colonial times, western-type medical care was dispensed to workers of mines. The missionaries also played a role as physicians dispensing this care. The mining towns of the last part of the nineteenth century brought to these remote areas medical practices from places as distant as Paris and London. Mestizos and even the Tarahumaras accepted and adopted some of these medical knowledge and procedures. A stratified society was evolving in the mining centers consisting of a dominant mine-owner group, *patrones* (bosses), and a mass of workers, tradesmen, and artisans. Initially, Spaniards and Creoles (*criollos*) were the patrones but by 1825 British nationals arrived at the mining centers and took the lead in many important mining ventures. The English, American, and Mexican bosses and their families had to be attended by qualified physicians who by this time also attended the workers and other residents of the mining center, including the Tarahumaras. It is remarkable that in this setting a highly sophisticated Austrian-waltz dancing society blossomed at the bottom of canyons, some as deep as the Grand Canyon.[96]

During this period, important medical advances occurred, such as a smallpox vaccine discovered by English physicians Jemison and Cheine that was dispensed at Guadalupe y Calvo in 1840. By the end of the century, important events—such as the strike of miners at Pinos Altos in 1883 and the revolt of Tomochi in 1895—precursors of the coming Mexican Revolution, took place within Tarahumara Country. Tarahumaras participated only marginally in these movements. In Tomochi, Cruz Chávez and his people rebelled against the dictatorship of Porfirio Díaz; in spite of an initial victory, the overwhelmingly superior federal army crushed the movement.

Lumholtz has summarized the socioeconomic situation of the Tarahumaras by the end of the nineteenth century as follows: "Most of the descendants of the former sovereigns of the realm have been reduced to earning a precarious living by working for the white and mixed-breed usurpers on their ranches or in their mines."[97]

The Twentieth Century

The twentieth century characterized itself as being a time of greater transformations in all spheres: political, social, economic, cultural, and religious. With the turn of the century, after a brief presence of the Josephine fathers, the Jesuits returned. Slowed by religious persecutions (1925–1934), the region then saw the reopening of churches, the reorganization of the old pueblos de mission, and the appearance of institutions not known until then in Tarahumara Country, such as boarding schools, hospitals, airports, and radio stations. Some Jesuits, such as Father Galván, introduced western medical procedures among the Tarahumaras.[98] But, most important, the Jesuits did not work alone anymore. They relied on missionary nuns from other religious orders, such as the Siervas del Sagrado Corazon de Jesús y de los Pobres, and later the Marist brothers, the Sisters of Mercy of Saint Charles Borromeaus, and others.

The Jesuits created the boarding schools, the *tehuecados* and *tohuisados*, girls' and boys' schools respectively. Adults were considered uneducable for all practical purposes. Tarahumara children were separated from their parents at an early age and kept under the care of the nuns. Graduates of the boarding schools were encouraged to marry among themselves and to create their own settlements called *colonias* where a new Christian society was expected to evolve. The Jesuits arrived sooner than the official Mexican school system at the concept of sheltered schools and bicultural-bilingual education. In addition, the Jesuits later implemented creative programs such as the radio schools. Broadcasting from Sisoguichi, they could teach Tarahumaras in very remote corners of their country. Father Díaz Infante began this ambitious program in 1955. Unfortunately, the program died in the 1980s from attrition.[99]

Mexican journalist Ricardo Jordán has called the land occupied by the Tarahumaras "warfare longitude" because the different Mexican political factions used this land to fight against each other during the armed conflicts of the twentieth century. In addition to the Tomochi uprising, several battles were fought in the area under Pancho Villa, Pascual Orozco, and other Mexican revolutionary leaders during the Mexican Revolution of 1910–1924. These battles took their toll of Tarahumara lives and maintained the land in chaos until the end of the religious persecutions in the late 1930s.

Chihuahua's state government opened a primary school in Tónachi in 1899 and another in Rochéachi in 1903.[100] The Mexican federal government arrived late to the scene. With antecedents in some ill-fated projects and in the experience accumulated when it confiscated the mission's schools during the persecution, it promoted a "Supreme Council of the Tarahumara Race." Like the Jesuits, the government chose the boarding school, coed-version, as its main educational institution. Curiously, the colonial dream of Guajardo Fajardo of a colonization center in the Tarahumara land was realized in modern Guachochi.

From 1950 on, the political process in Tarahumara Country has been characterized by the gaining of terrain by the state from the church and the demographic explosion of the mestizo component of the population throughout the land: practically all of the old pueblos de misión are now mestizo villages. The demise of mining and its replacement with another industry, lumbering, has characterized this stage.[101] Sawmill industries came with a scheme similar to that of the mining center; but the attached human settlements were of a more ephemeral nature, and any benefits, including medical services, have been very limited.[102]

Champion postulated that the acculturation forces acted with various degrees of intensity throughout the twentieth century in three distinct periods.[103] The first period (1890–1914) was characterized by increased pressure on the Tarahumaras, widening the opportunities to learn the Mexican cultural traits and the assimilation of many individuals into Mexican society. The second stage (1914–1934) was characterized by a decrease in pressure on the Tarahumaras to abandon the traditional Indian ways and by a relatively low rate of assimilation. Champion claims that the Tarahumaras of this period are the parents of the present-day Tarahumaras. The third period (1935–1955) was characterized by a new increase in the pressures to abandon Tarahumara customs. Possibly, the trend posited by Champion until 1955 persists until today.

Nonetheless, recently, cross-cultural marital unions appear to have increased, deriving especially from the modernization of the church and state operations, primarily in the schools hiring teachers or bilingual promoters. Their work is paid at much higher rates than any other excepting the large commercial activities and some of the illegal drug-related activities in the area. Currently, it is more common for an educated mestizo woman to become involved with bilingual Tarahumaras, who as teachers, educators, boarding school alumni, *ejido* authorities, or health *promotores* (nurses and public health workers) have better economic and social perspectives than suitors of her own culture.[104]

Another factor that has produced dramatic changes in Tarahumara territory during the last part of the twentieth century is the development of communications. In addition to a booming tourist industry, these factors have fostered an intense migration of Tarahumaras to the cities outside their traditional land.[105] Migration tendencies are reinforced by a highly acculturating education that teaches Tarahumara children the benefits of modern life. The boom in construction in the cities and the economic decay of the countryside are also important factors that have contributed to this cultural revolution.

Mestizo-Tarahumara Relations

The mestizo label is usually applied to persons having both Indian and European ancestry. In Tarahumara Country, mestizos have a wide range of racial and ethnic characteristics as well as acculturation levels. Describing in detail these characteristics is beyond the main scope of this work. The Tarahumaras usually consider mestizos to be wicked people; in spite of this, many mestizos treat the Tarahumaras kindly. The mestizos, with a few exceptions, consider themselves of higher social standing or more enlightened. The mestizos present themselves as *gente de razón*, meaning "rational or sound-thinking people." Mestizos and Tarahumaras tend to live apart. The former live along the major trails and villages; the latter tend to live in widely scattered hamlets or side valleys. Nevertheless, each culture has influenced the other. More and more Tarahumaras are adopting Mexican cultural patterns. Even language is reciprocally influenced.

A special interaction is apparent in the use of medicinal plants and in the beliefs concerning health and disease. The mestizo population has grown and continues to increase considerably faster than the Tarahumara population, probably due to differences in infant mortality and access to health resources. The interest in working for money and the tendency to congregate in larger towns have impacted many Tarahumaras, who have moved to towns such as Creel, San Juanito, and Guachochi. There are no signs that Tarahumara towns are or will be evolving.

In general, the Tarahumaras and in particular their natural resources are being exploited by mainstream Mexicans.[106] Cultural defensive strategies have succeeded in preserving Rarámuri cultural values but have not prevented the exploitation of their people.

THREE

Rarámuri, the People and Their Culture

In the first chapters of this book I have made reference to the Rarámuri, the mestizos, and other cultural groups in the area. I believe it may be pertinent to make explicit here some of the distinctive features attributed to the Rarámuri. The scientific validity of some of the distinctions has not been tested; however, my impressions may help to understand the perceptions of those alien to their culture.

For a native of the region trekking down a dirt path in the sierra, the individuals belonging to the Tarahumara people appear so distinctive that just observing the tracks left by an individual from their group may help to identify who went through. It is said that a Rarámuri leaves a footprint with the impression of the tips of his toes pointing outward from the axis of his body. On the other hand, mestizos' toes tend to point forward, parallel with the axis of their body. Tarahumara footprints also appear more uniformly impressed, suggesting a certain waddling while walking, with a preponderance of the action from hips and thighs over that of legs and feet.

Some local informants claim that the body scent of the Tarahumaras may help to identify them. Their scent may be affected by variables of cultural significance. Their smoky smell may be due to the fact that they cook and heat their dwelling with firewood inside. The smoke is handled by inefficient, low-draft chimneys. Their ga'rí, or dwellings, therefore are always filled with smoke. The combination of the aromas of *chopé*, a resinous pine wood, used to start the fire, and *kú*, their regular firewood, impregnate their bodies, clothes, and utensils. *Kobishi* (toasted corn) and batari (corn beer) also have characteristic smells.

It is therefore possible to identify Tarahumara individuals by their scent even if they are away from their own environment.

Tarahumara cooking and brewing differs from that of other groups in the area. Their food flavors are distinctive. Rarámuri brew *tesgüino*, their corn beer or batari, giving it a unique flavor. It differs in taste from the same beverage prepared by the mestizos. The Tarahumara *reméke* (tortillas) and kobishi (pinole, or ground, toasted corn) also differ in taste, depending on whether Tarahumaras or mestizos prepare them.

Their gestural behavior also has peculiarities. Tarahumaras, mestizos, and outsiders, in this order, greet others ranging from gently touching the tip of the other person's fingers, the Indian salute, to the finger-crushing handshake of individuals affiliated with the northern Mexico, Chihuahua, mainstream. In *Cerocahui: Una Comunidad en la Tarahumara*, I described the typical greeting of the canyon country Tarahumara people: their salutation begins by a repeated light touch on the shoulder of the other person with the extended hand.[1] This is followed by a light touch of the tips of the fingers of the right hand of the other person, quickly withdrawing the hand. Parents encourage their children to greet their elders in this fashion to demonstrate their respect. A variant of this greeting, perhaps the original Rarámuri salutation, is still practiced, especially by elderly Rarámuri, in the neighborhood of Norogachi. This consists of touching with the tip of the fingers the assumed site of the heart (*surá*) of the other person. This salute creates some incidents with individuals from other cultures. I have witnessed an old Tarahumara greeting a young nun, touching her chest with his finger as she flushed with embarrassment. Subtle differences also may be picked up from their speech by people familiar with their dialects.

The attire and physical appearance of the Tarahumaras are characteristic, and their skin is darker than that of mestizos. They typically wear colorful garments. A Tarahumara man often is observed with his hands behind his back, walking several feet behind his wife, whose waddling gait is amplified by the swinging of the several skirts she wears one on top of the other according to their custom. Anthropologists have relied on appearance and anthropometric indices to characterize the group. Vivó classified the Rarámuri among the sub-dolichocephalous (having a head longer than wide) of tall stature groups of northern Mexico.[2] In colonial days, Ratkay described the Tarahumaras of Carichí as follows: "As a rule they are dark in color, almost black. They have sparkling eyes, and almost all of them are strong and healthy of body. One

Typical demeanor of a Tarahumara man smiling for the camera.

sees only a few who are by nature lame, blind, deaf, or dumb. Those who are thus afflicted become so only through some accident."[3] Father Neumann, Ratkay's contemporary, noted, "This people are swarthy of color but not black."[4] Lumholtz observed that the Tarahumaras "are more muscular than their North American cousins and their skin color is light chocolate-brown."[5] The darkest complexions are seen in the highlands near Guachochi. In the higher altitudes, the people also develop higher statures and are more muscular than in the lower portions of the country.

More recently, some authors, such as González Rodríguez, have noted that the Tarahumara build is strong and slender.[6] Their average height is 1.65 meters (five feet, five inches) for the males and 1.58 meters (five feet, two inches) for the females, although there are taller and shorter individuals. Their complexion is bronze-like and they have prominent cheekbones, large foreheads, straight noses, small chins, and dark eyes. Their hair is thick, straight, and dark. The males are beardless and do not have much hair on their bodies. It is quite rare for their hair to turn gray or for them to go bald. Their physique is well proportioned and suited for hard working.

Many Tarahumaras dress in characteristic, distinctive garments. Men wear a shirt (*napatza*) with ample sleeves. White and red are the favored colors, but it is not uncommon to see green and purple shirts. A breechcloth (*tagora* or *sitagora*) is tied to the waist by the means of an exquisitely woven woolen girdle (*púra*). In some regions, the Rarámuri prefer the tagora ending at the back as a triangle; in others, the square tail is preferred. Occasionally, the hanging part of the cloth is nicely embroidered. A rectangular piece of cloth, the *cotense*, with a hole for the head, is worn over the napatza and tagora; this falls freely in front and back, usually down to the thighs. This piece may also be embroidered. The hair, shoulder length, is kept in place by an elaborately arranged cloth headband (*coyera* or *cowera*). Three-hole sandals (*aká*), with soles made from discarded rubber tires and tied with straps of leather, complete the masculine attire. Western and canyon country Tarahumara men do not wear breechcloths but simple western pants and shirts. The white *calzón*, pants made of coarse cotton, which used to be common, have disappeared. A tightly handwoven hat is sometimes worn.[7] Women wear a wider and shorter napatza; a ribbon is sewn

Tarahumara men listening respectfully to a *nawésari* being addressed to them by the *siríame*.

Blond Tarahumara children from Kwechi (one standing and one being carried). Compare them to Beto, the little boy looking up at them, who has the typical Tarahumara features and skin color.

Tarahumara girls playing.

Tarahumara women sunbathing (*rasúkima*).

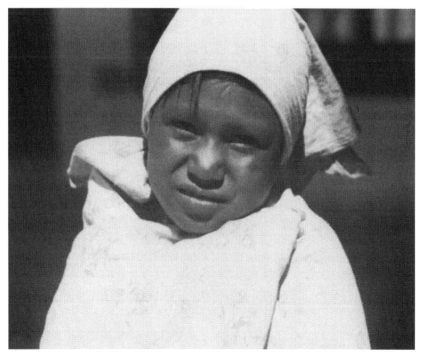

Face of a Tarahumara girl.

Young Tarahumara woman holding her child at the door of the INI clinic in Samachiki.

or the blouse itself is stitched under the breasts. When they are working in the fields, under a burning sun, the women may go with their chests bare. From two to twenty skirts (*sipucha*) are worn simultaneously tied to the waist by the means of a *pura* or *púraka*, or woven girdle. Tarahumaras claim that many skirts accent beauty. Traditionally, the skirts fall to the ankle but the young prefer them much shorter. In the canyon country, women's garments are also colorful but not with the profusion of colors of the skirts of their *weri* cousins from the uplands. Many like a black shawl reminiscent of those seen in southern Mexico. Although nowadays many women wear tire sandals or cheap, industrially manufactured plastic shoes, in the past they, unlike the men, usually went barefoot. Children dress, as in many other cultures, like the adults. Finally, a heavy wool blanket, the most remarkable of all Tarahumara handcrafts, completes their attire. To wear it, they hold a corner over their left shoulder, rolling the blanket over their bodies, leaving out the right arm, the result giving them a dignified appearance.

The Rarámuri Cave Shelters and Ga'rí

Some Tarahumaras live in caves (*resochi*) like their ancestors. Surprisingly, a cave in their particular environment has some advantages beyond protection from inclement weather. These shelters sometimes offer surprising conveniences. I have observed caves with small water springs and even one with a hot water spring in one part of the cave and one with cold water in another area of the same cave. In the past, large caves were used by people of high social standing as one seen in Narárachi. Occasionally, habitations are built at the cave's entrance to add additional space. Smaller caves are used as granaries, storage areas, or as a refuge for their flocks. However, most of their dwellings are improvised huts, some built with logs leaning against a rocky wall or against each other as a gable-on-the-ground. Wood or stone houses (ga'rí) are characteristic of the upland area. These consist of small single rooms erected on flat ground, made of stone or built like log cabins. The roof is made of *kanoa*, logs carved, squared, and grooved from stripped saplings placed alternately to form a gabled roof. Beneath the roof there may be a flattened mud ceiling.[8] The *kanoa* roof technique is also used in larger structures, such as the old colonial churches and *komerachi* (community building).

Stone houses have a peculiar chimney. As the house is erected, a pine trunk is placed through the walls, across the length of the building, at a height somewhat greater than one-half of the structure's height, Stones are placed one upon the other crosspiece to build the chimney draft, known as *kutéga* (throat), ending at the roof of the dwelling with the cracks between the stones filled with mud.[9] The arrangement becomes a low-draft chimney that allows the smoke to escape to the outside when there is no wind. This is an improvement over the open fires used in caves and in the more primitive houses. During the second half of the twentieth century, large oil drums, cut in half and connected to the outside with metal tubing, became increasingly popular and they may ease the exposure to smoke inside of their houses. However, the Tarahumaras spend most of the day outdoors ga'rí are used mostly for cooking, as shelter to spend cold or rainy days, and as storage. Most domestic activities take place in the front patio, a rustic porch, or outdoors around the house. Tarahumaras secure the doors of their houses, especially their *rikuá* or *sonogori* (grain houses), with ingenuously designed wooden locks.[10] Actually, the Tarahumaras are most parsimonious in the use of natural resources. The house-building techniques just described are also used to erect smaller structures such as chicken houses and storage facilities.

In the canyon country and western Tarahumara Country, wooden cabins with wood tile hip roofs are common in place of the traditional ga'rí. Some have a space between the roof and the walls as an arrangement that allows extra ventilation, necessary in the severely hot weather. In many areas, adobe houses are built that may incorporate some of the stone-house building techniques described. In general, the presence of windows in a house is a sign of acculturation.

Tarahumaras usually do not build fences around their properties. However, the presence of mestizo predatory behavior has pressed them to fence their property. For this purpose they use long pine poles joined by posts or simply widely angulated over the next pole.

The Rarámuri Temperament and Life Rhythm

The Tarahumaras are frequently described as seminomadic people. This term, however, may only be applicable to those Indians living on the rims of the canyons who move their homes to the bottomland during winter.[11] The increasing migration of Tarahumaras to the cities is permanent and may not be considered nomadic. However, physical mobility within their demanding mountainous environment does play an important role in the development of the physical fitness necessary in their strenuous daily activities.

The everyday life rhythm of the Rarámuri is slow with a monotonous continuity. Almost all of what they do follows this pattern. Their dances last several hours, giving the impression that they will never end. The kickball races last several days. Tarahumara music grasps one melodic theme, repeating it without variation for twenty minutes to an hour until a new theme begins. Civil and religious ceremonies are also characterized by this slow-motion rhythm. Every activity gives the impression that it will never begin but once begun, it seems as if it will never end. Much patience is needed to watch a kickball race, a peyote celebration, or a simple dance. Their slow initiation causes the impression that Tarahumaras are lazy and lacking in motivation, but it is not that they do not want to act; they just want to do it slowly.[12] Actually, this kind of temperament is necessary to cope with the harsh environment and the precariousness of life in the Chihuahuan Sierra. This has enabled them to survive and "tame" their environment. They appear to have found that even the least-significant-looking stone, herb, or living thing in their mountains and canyons has a specific function. This may be as a remedy, to catch fish, to feed them or their animals, induce hallucinations, build the resonance box of a violin, or as food. Their stamina and perseverance is proverbial. Many authors have described the Indian's legendary ability to run after a deer until the beast falls exhausted and may be killed with sticks and rocks. More impressive than the physical fitness required to hunt is their knowledge about the deer's habits, which they use to their advantage.

Psychological Traits and Normative Behavior

The Tarahumaras have characteristic psychological traits; they appear shy and inhibited. Some authors have remarked that it is only while inebriated with tesgüino that their usually shy, quiet, and retiring demeanor becomes loud and boisterous.[13] Other observers note that among themselves or with those who have won their confidence, they are more expressive, lively, and joyful.[14] My experience, after more than ten years of living among these people, supports this last observation. Their apparent reticence and timidity seems to function as a defense against violent confrontations with those identified as potential aggressors.

The subjective life of the Rarámuri is rich. For them dreaming (*rimúa*) has considerable significance.[15] They believe that through the experience of dreaming they receive messages from beyond.[16] Dreams are often for them premonitory messages.[17]

Since colonial days, it was noted that the Tarahumaras were naturally peace-loving and seemed never to quarrel among themselves. Their honesty has been long praised beginning with the first observers who made contact with them. They appear to be loyal and respectful of others' property. It has been said that they would rather die of hunger than to steal from others.[18] In modern times, several authors have remarked that homicide and violent crime are practically unknown among the Tarahumaras.[19] Kennedy suggests that the Tarahumara quest for individual independence and commitment to nonviolence are strongly internalized.[20] Only within the framework of a *tesgüinada*, or *bajía-*

chi, and therefore while under the effects of alcohol do they exhibit disruptive behavior. This particular instance, the bajíachi, is almost the only opportunity they have to release aggressive impulses. Fighting almost never occurs outside of the bajíachi. Aggressive behavior is labeled *chabochi*—belonging to the mestizo—or devilish. Displays of aggression are considered evil or corrupt.

The Tarahumara ethical principles reflect a profound respect for others. Even children's opinions and attitudes are taken seriously and their persons respected. Their tolerance of variance in sexual preferences should be noted. Male homosexuality, or *re'né*, meaning inverted, is seen as peculiar or out of the ordinary but accommodation is made for those displaying this behavior. Lesbian unions are tolerated if the participant individuals assume a clear dual role, with one of the partners showing dominance. Pedophiles and rapists are considered *ichápochi* or *oparúame*; that is, evil. Sexual transgressions such as adultery, bigamy, or seduction are severely punished, but the underlying factor that determines the severity of the punishment is the amount of anger or aggression involved.

The Transcendental Beliefs of the Rarámuri

The religious beliefs of the Tarahumaras are basically monotheistic, but some minor gods appear to be incorporated in their belief system.[21] Many syncretistic elements resulting from their exposure to Catholic missionaries are apparent. Christian Tarahumaras call themselves *pagótuame* (*pagoma*=to wash, *ru* or *tu*=activating particle, *ame*=gerund), "the baptized ones." Their main deity is *Onorúame*, "the one who is the Father." He made all things and gave them to be used to sustain their lives. In the more acculturated communities, Onorúame is identified with the Christian God the Father. Among the less acculturated, the Sun (Rayénari) is considered the main deity. Elements of sun worship have survived in the paintings with which the converted Rarámuri have decorated the walls of the colonial temples and in some handcrafts and their embroidery on pieces of clothing such as the napatza and tagora.

In the less acculturated communities, the cult to Eyerúame, "the one who is the Mother," still exists. She is identified with Mechá, the moon. In the more acculturated communities, this cult has become a Marian devotion. The syncretism of the moon and the Virgin Mary is expressed in phrases such as *keti eyé panina betéame*, "our Mother who lives above."

The counterpart of Onorúame is re'ré betéame, "the one who lives below." This mythical character is comparable to the Christian devil, or Ryáboro. According to an old tradition, Onorúame and *reré betéame*, "the one who lives below," were brothers, as shown in the legend quoted early in this book. It should be noted that there are many differences and ambiguities among different Tarahumara groups. Some informants acknowledge that the Catholic Virgin and Eyerúame are the same person. Others declare that Jesus is the son of the Sun. This makes it difficult to place some of their ceremonies and beliefs, such as Jíkuri and the spirit of peyote, within in the scheme of their system of beliefs. Jíkuri is clearly not identified as God the Father. Lumholtz recorded that Jíkuri was considered the brother of Tata Dios (Papa God), actually his twin, and therefore was called uncle. It has been observed that the *sipáame*, or shaman, who officiates in Jíkuri ceremonies frequently thanks God during the celebration for having given this deity to the Tarahumaras.

In earlier times, some natural phenomena apparently were considered deities. Neumann, writing more than two hundred years ago, observed that newborn children were ritualistically dedicated to the thunder and lightning.[22] Failure to dedicate the baby during the first thunderstorm after birth or failing to honor the sun and the moon with ordinary "inebriety sessions" was thought to make the child likely to be stricken by lightning. The devout Rarámuri considered themselves good friends of the thunderbolt, which then was perceived as a secondary god or mythical being.

The Tarahumaras believe in an afterlife; this is inherent in their rituals to symbolically nourish the dead. They consider themselves to be children of God destined to go to heaven after death. *Ichápochi* are the devil's children who will go to the underground after death. Ill-behaved Tarahumaras will go below as a punishment for their sins. Heaven is called Rewegáchi, or "the place of the ones who have a name." This expression may have been influenced by an early Catholic belief that baptism—that is, to have a name—was required to deserve entering and dwelling in heaven. Curiously, the gentile (*cimarrones*) or non-Christianized Tarahumaras, less exposed to the missionary indoctrination, believe that they are the columns or pillars of Rewegáchi and that if they are baptized, therefore, betraying their traditional beliefs, the heavens will fall down. Merrill has reported the Tarahumaras' belief in various levels of heaven and hell, but our informants rarely reported such compartmentalization of the afterworld.[23]

The Tarahumara Economic System

Rarámuri survival is dependent on a precarious economy based on agriculture and herding. Interestingly, their sheep and goats are used primarily as a source of fertilizer rather than a source of meat. The flocks provide some wool and a little milk, used to make cheese. Only occasionally, particularly during religious festivities, these animals become a source of meat. Most families depend primarily on small plots of cultivated land for subsistence. When adverse climatic conditions, such as drought, early freezes, or flash flooding, reduces or wipe out their harvest, starvation becomes an imminent threat.[24]

In the sixteenth century, when the Spaniards arrived in the region, they found the Tarahumaras dependent on a hunting-gathering economy. Like many other tribes in northern Mexico, they planted corn, beans, and squash, but they relied mostly on gathering activities. Certainly, the introduction of European animals—oxen as a traction force to pull the Egyptian plow and goats as a source of fertilizer—determined the definite agriculturalist character of the modern Tarahumaras. Otherwise, their means of subsistence have changed little through time. They still gather wild fruits, plants, and seeds and hunt small game, especially during drought years.

The Tarahumaras practice slash-and-burn clearing of the land, called *rozas* or *mawéchi*. This practice has been much criticized by those arguing that it contributes substantially to deforestation and the erosion of land. In spite of this practice, many scholars praise the Tarahumaras' respect for the natural resources available to them.[25] Most of Tarahumara agriculture takes place in floodplains or terraces, as the creation of cultivated lands on the forested mountain slopes or on top of *cordones* (flat-topped mountains and mesas) is quite limited. The lack of appropriate drainage of cornfields is less likely to contribute to erosion than what has been previously reported.[26]

One of the most characteristic features in the Tarahumara handling of personal resources is *kórima*. Kórima, simply defined, is sharing food with a visitor.[27] The concept implies that the poor have a right to receive a portion of what the more fortunate have.[28] The concept of kórima connotes cooperation rather than charity.[29] The obligation of sharing edible goods contrasts with the sharing of material goods, which is not considered something they are compelled to do. Acceptance of kórima implies eventual repayment or reciprocity as soon as the receiver's circumstances allow.[30] Nevertheless, Tarahumaras asking for kórima in the streets of Chihuahua City are actually begging and therefore no expected reciprocity is implied. The Tarahumaras from Cerocahui are very critical of those who beg for alms in the cities.[31] Similar criticisms are expressed by the Rarámuri of Norogachi about beggars from Narárachi, Wawatzérare, and Bakéachi.

The Tarahumara Language

The Tarahumara or Rarámuri language is considered a member of the Uto-Aztecan language family or, more precisely, the Sonoran or Opata-Pima-Tarahumara branch.[32] The language is spoken by approximately 62,550 individuals and is considered one of the most thoroughly studied Indian languages of North America.[33] Some authors group the Tarahumara and Varijío and the two extant Cahita languages, Mayo and Yaqui, into the Taracahita branch of the Sonoran group of languages. Nevertheless, the classification remains speculative since there are no other living members of this last group. It is interesting that the zone of maximum diversity for Sonoran languages at the time of the first contact with the Spaniards was located in Tarahumara territory.[34] Recently, five Tarahumara dialects—west, north, summit, center, and south—have been postulated.[35]

Tarahumara Social Organization

The basic unit of the Tarahumaras' social organization is the nuclear family—the husband, wife, and their children. Often, children from prior marriages, adoptive children, and cousins live as one family under the same roof. Sexual role differentiation is clear regarding work and family activities. To outside observers, the father is the head of the family. However, the wife commands considerable authority in family affairs. This probably is a consequence of the independent control of the property that women bring to the marriage. In their system, the wife retains ownership of any property that she brings to the marriage. While the father engages in public activities, he rarely makes decisions on economic matters without the concurrence of his spouse.[36]

Their system of kinship terminology defines the degree of closeness of a relationship within the family structure.[37] Nuclear families often evolve from the wife's family according to a matrilocal system.[38] The parents give their daughters in marriage on condition that the husband will live with the wife's parents and to a certain degree become subject or subordinate to them. Most married couples have lived with the wife's parents at the beginning of their union. A minority leave their parents to settle independently. Remnants of a matrilineal family system also have been described, in which the mother's older brother is considered the closest relative after the parents. Several authors have noted some family instability but this probably is due to an elevated mortality. I believe that all in all Tarahumara marriage is relatively stable.

It should be noted that the Rarámuri hamlets, or rancherías, consist mostly of members of one extended family.

Tarahumara Political Organization

The political or social power structure of the Tarahumaras includes two types of governing officials: one is the tribal or pueblo officials and the other consists of individuals who belong to external institutions. Among the latter we include Catholic priests, the ejido officers, and the officers of native organizations. These external officers also include municipal, state, and federal officials. This last group makes decisions that have an impact on the Indian communities and enjoys a high status, often yielding more effective power than the tribal officials. Some individuals do not hold a particular office but command influence and power well above that of official authorities. However, often it is the position, rather than the man, who is treated with special consideration.[39] Each official plays a diversity of roles, an *owirúame*, for example, is primarily a healer but he may share values and apply ministrations closer to those services belonging to those of a minister or priest. Similarly, many functions pertaining to civilian authority also operate in social contexts of a religious, magical, or healing nature.

The principal tribal officers, in order of importance, are the *siríame* (*gobernador* or *gobernadorcillo*), the *generari* (general), the *kapitano* (captain), *mayora* (mayor), *alwasili* (*alguacil* or constable) and *sontarosi* (*soldados* or soldiers). The officials listed usually designate the *tenanche* and *chapeyó*, who will be responsible for tribal celebrations. Although the names of most of these positions have a terminology that derives from the colonial Spanish hierarchical structure, their actual Rarámuri roles and functions differ considerably from the Spanish or Mexican political or administrative systems.

The Siríame Grande

The *siríame grande* (greatest governor) is the highest authority within Tarahumara society, but his power is limited by the simultaneous existence of Mexican state and federal authorities and ejido officials. The jurisdiction or area of influence of the siríame is the pueblo with geographical boundaries that have remained practically unchanged since colonial times. The Tarahumara pueblo therefore is not a town; it is an area defined by the siríame's jurisdiction.[40] It includes several rancherías, or hamlets. Norogachi with its five pueblos (*mariana póbora*) is unique. In this case, Norogachi, Chogita, Pawichiki, Tetawichi, and Papajichi form a group of autonomous pueblos. They have no jurisdictional authority over each other, but they participate together in many ceremonies, such as in the major religious feasts and Sunday meetings.

The siríames usually are respectable members of the community, good speakers, known for their personal integrity. Mature or old men are selected for this position because they are considered wise. This is a necessary quality in a siríame because he will not only play the role of judge in the trials for personal misdeeds or offensive behavior but also will have to exert his judgment in settling inheritance disputes.[41] As an informant noted, the siríame has to speak well, make long and appealing *nawésari* (sermons or discourses). He has to pursue the well-being of his people, have skills to govern, and, very importantly, has to recognize when is the right time to issue an order or command.[42] The siríame grande (*wa'rura siríame*) has several deputies, *gobernadores chicos*, who represent the Tarahumaras from large rancherías.[43]

The siríame has the duty of organizing his subjects and the responsibility of dispensing justice, including settling marital disputes. The siríame appoints his collaborators and officers, placing them under his authority. They are made accountable considering the importance of the tasks assigned and the degree of authority he has delegated to each person.[44]

Through his involvement in the various ceremonies—nawésari, batari, yúmari, and so forth—the siríame assures that participants observe good behavior and help each other. He is responsible for informing his people about matters and events that affect them individually or as a group. The siríame is also responsible for keeping and maintaining social peace. He plays oversight functions through the major religious celebrations and fiestas, supervising ceremonies such as the Jikuri, *bakánowa*, and other healing practices. He is responsible for the maintenance of public order, which incidentally is easily perturbed on such occasions. He prevents deviations in the conduct of the events, assuring that they do not become perverted and used for witchcraft.

Court Trials

Outside observers have been impressed by the dignified way in which decisions are made and justice is imparted among the Rarámuri. Public trials are conducted under the authority of the siríame. These are awe-inspiring public events conducted by a group of modestly attired pueblo elders, grave men holding their canes (*tesora*), sometimes sitting in front of crumbling adobe walls and buildings. "In this inauspicious setting highly respected and implicitly obeyed decisions are made with a solemnity, which under these circumstances appear sublime."[45]

The range of disputes, misdemeanors, and crimes judged by the officers of these courts include theft, mistreatment of children, spouse desertion, and irregular sexual relationships, including adultery. Gossip, brawls, and inflicting injuries are also tried. Unauthorized use of land or disposition of the animal flock of others are offenses subject to trial. Also considered are family disputes, divorce, and inheritance matters. Theft is a crime aggravated by the fact that it affects the prestige of the Rarámuri community where it is committed.[46] In modern times, serious crimes, particularly homicide, are turned over to the Mexican authorities. The court transactions have certain particularities. For instance, inebriation is considered a mitigating factor for crimes committed. Offenses such as theft, direct defiance of the Mexican authorities or that of their own officials, and homicide are viewed sternly.[47]

The trials take place after or during Sunday gatherings or within the time framework of festivities or celebrations. Occasionally, there are serial trials for a single case. The place where these sessions are held is usually the komerachi, an official place of reunion for the authorities.[48] Individuals accused of committing a crime or named in a trial complaint are summoned or arrested, a task entrusted to a kapitano, a sort of sergeant at arms. Many times the accused is brought to the place with a display of what appears to be unnecessary force. Those who wish to initiate the court action confer beforehand with the siríame to press charges, but also to suggest possible solutions and recommend the type and severity of the punishment that the accused should receive.

The siríame summons as many witnesses as possible to the trial since he knows that part of the process in the trial of those accused will involve supporters of both the accuser and the accused.[49] Once the assembly gathers, the accused stands facing the officials and the people behind the accuser usually forming two groups, one against, the other in favor. The trial begins with the siríame delivering a short discourse—nawésari—and then asking the defendant: Why do they have you here? He almost routinely answers that he does not know. Then a person will recite the accusations. For outsiders, the trial is difficult to follow since everybody seems to be talking at the same time, but the Rarámuri democratic process requires participation by all members of the community. The content of what each participant presents is often confusing. They may talk about what they consider proof of the action for which the culprit has been accused. Information is offered about alleged crimes and irregularities that the defendant or his accusers or the judges have committed in the past. Occasionally, individuals with oratorical ability participate in a role similar to that of lawyers for one or the other party. Commonly, witnesses as well as other participants offer arguments trying to incriminate one or the other party.

After the public exercise just mentioned, the evidence is carefully examined. Emotional or circumstantial evidence often is considered more important than factual testimony. For instance, culpability may be decided on circumstantial evidence, such as having seen the accused in a place that he rarely visits. Or why the accused used an alternative route to go home on that day. Previous accusations and convictions also may be considered incriminating.[50]

The role of "expert" testimony is particularly important. It involves, for instance, the exceptional ability of many Tarahumaras to read tracks left by men and animals on the ground with a precision that could only be matched by actually having visually witnessed the events. The evidence is usually passionately debated, attempting to accumulate arguments against the accusers more than actually challenging directly the evidence offered by the other party. It seems at times as if the substance of the allegation against the defendant is ignored or only occasionally addressed.

Once it appears that a decision is close, a process of bargaining begins: for example, it was not five goats that the defendant took from the owner but only one. As bargaining proceeds, he may acknowledge he took two goats and will go to three or four only if he is cornered in such a way that he cannot conceal his culpability. This bargaining becomes critical once the point is reached where punishment is considered.

In a bigamy trial, several criteria are used to determine who is the legitimate spouse. The first is how many children each woman has, which union is older, and last to consider is the means each woman has to support herself and her children.[51] Since the plaintiff is usually an abandoned wife, if her complaint prevails, the siríame may command the man to return to her. Many times, the other woman will be assigned another husband right there during the trial proceedings. Some siríames are remembered for their Solomonic or very ingenious decisions.

A feature of the Rarámuri trials is their expediency. Verdict, sentence, punishment, and remediation follow one after the other once the arguments are concluded. The concept of retribution is essential in these trials. Punishment, jail, or flogging were seen as a form of retribution. Incidentally, in earlier times, flogging was a standard punishment. More recently, the sentence may be determined in terms of a number of days that the defendant should work for his accuser. It is here where the most protracted bargaining takes place. Usually, the plaintiff requests a high and unrealistic number of days, knowing that he will have to moderate his demands at the end of the bargaining period. The defendant will argue that working so long for the plaintiff will deprive his own family of support causing them, who after all are innocent, an undue hardship. Also it is routinely argued that the crime does not deserve such harsh punishment.

Placing someone in a yoke is a form of punishment that has been abandoned but still is remembered and talked about. Only the head was placed in the yoke to punish a mild offense. A more severe offense would require a hand also to be clamped and both hands were clamped for the most severe cases. According to some informants, some siríames state that they could still order someone to be whipped if in a case they thought it was necessary. The death penalty has been definitely abandoned. In general, no matter the nature of the punishment, including turning the offender over to the Mexican authorities, shaming is the most effective means for social control in this culture.[52] A recent punishment modality is the imposition of fines to accompany restitution in cases of thievery. Adultery trials usually end with the repudiation of the woman involved and a fine imposed on the man. Jail is used primarily to keep dangerous felons who will be surrendered to the Mexican authorities or awaiting trial and not as actual punishment. Apparently, jail is avoided as a punishment because it requires punishing an innocent person at the same time, the jailer who has to stay guarding the prisoner. Jail is, however, a preferred punishment for those who cause disruptions during fiestas and large gatherings.

At the end of each trial, the siríame asks the contending parties to shake hands as a gesture indicating the restoration of peace and both parts are exhorted to live in a brotherly manner—*mapuregá najirémaga jaro*—and to avoid further disputes in the future.

Other Tribal Officials

The *generari* is an official appointed by the siríame. Prior to selecting the individual for this position the siríame canvasses the people searching for a person suitable for this position. Popular support is considered indispensable to carry out the responsibilities of this job. Also taken into consideration is a prior good performance as a kapitano. Incidentally, being a kapitano increases the likelihood of being later selected as a siríame. The generari wields considerable authority; he presides over the assembly and delivers the nawésari to the people when the siríame is absent or indisposed. He may even stand in for the siríame to invest new authorities. Ordinarily, he will perform only duties or orders commissioned by the siríame. The generari usually exerts his authority through the *ibagapitano* (captains, plural of kapitano). In some pueblos, for example, in Narárachi, the position generari is not considered necessary and his functions are carried about by the siríame or are assigned to the ibagapitano. In some circumstances, such as the mariana póbora, the position is considered indispensable.

The mayora is an official who has children and single youth as his primary responsibility. He is expected to counsel, reprimand, and punish those who display unacceptable behavior. He assures that children go to school and learn the catechism, which historically was his primary function. The more diligent mayora monitor children's school performance. Mayora are responsible for overseeing family interactions, encouraging harmony and the maintenance of moral principles in family affairs. An interesting role of his is to be the tribe's matchmaker and counselor. He helps to reunite separated couples, finds mates for single persons, and asks for the hand of a prospective wife. He also looks for a suitable bride or bridegroom for the widowed.[53] In a similar fashion to the siríame, the mayora delivers homilies (*nawésari*), particularly during wedding celebrations, at the beginning of a tesgüinada, or when asked to give advice to children or adults on morality. He advises newlyweds to be faithful to each other and to continue respecting the authority of their elders, including that of their mother-in-laws. Tarahumara couples may live together for a time before getting formally married; then the mayora will encourage them to be married by a Catholic priest. The mayora coordinates his activities with the church, encouraging parents and godparents to take children to the priest to be baptized. The mayora oversees the activities of the women assigned as tenanches who are responsible for offering incense during religious observances and as housekeepers who maintain the interior of the church. Merrill has observed the *maromas* or tenanches performing these same tasks in Basíware and Rejogochi.[54]

The kapitano had been thought to be just a messenger of the gobernador.[55] In actuality, they do have considerably more responsibilities. Their function is to maintain oversight assuring the maintenance of domestic peace of families. The number of ibagápitano in a pueblo varies widely, depending on the area and current needs. In Norogachi, there are twelve "grand" ibagápitano and one additional minor kapitano for each ranchería, totaling about fifty. In another pueblo, there was one kapitano for each siríame, *teniente* (lieutenant), and *alikante*.[56] Many of their duties consist of the enforcement of tribal norms, but they are also responsible for announcing forthcoming events, festivities, or meetings, and so forth, in every hamlet. Their activities include maintaining order in circumstances in which people congregate, such as the Jíkuri fiestas, tesgüinadas, and other festivities. The ibagápitano localize and bring to the *presidencia* (office of municipal or regional president) or komerachi those persons who have been suspected of having committed crimes or were witnesses of these actions. On Sundays, during the nawésari, the ibagápitano are made responsible for gathering the people and maintaining order. In Norogachi, their distinctive weapon and sign of authority is a spear with a bayonet attached to it.

Other officials, the arguasiri or alguaciles, are assigned the functions of policemen or jailers. In Norogachi, there are usually four arguasiri. They are in charge of watching over the persons jailed. On occasion, the task is assigned to a kapitano or the generari if the jailed individual is prominent or is thought to be dangerous. The arguasiri are responsible for the safekeeping of the tesora (staffs or batons) and bring them to any place where the siríame officiates. The arguasiri, sometimes called *ropiri*—whips—use this tool as an indispensable part of their equipment, which incidentally has other functions, such as keeping dogs out of the temple. In earlier times, they acted as *verdugos*, or executioners, and continued responsibility for administering corporal punishment and clamping people in yokes after capital punishment disappeared.[57]

In some places, as in Narárachi, there are other officers, the *sontarosi*, who use old-fashioned rifles as a symbol of their authority. In Norogachi, the ibagápitano perform all the functions assigned to sontarosi; in other places, such as Basíware, there are from five to ten regular sontarosi.[58]

Temporary functionaries are usually appointed to assist in prominent religious events. Their tenure is limited to the duration of the celebration. These individuals, responsible for organizing festivities, are known as *fiesteros* or tenanches. *Chapeyó* are those in charge of the dances. Other important individuals are the *monarko*, or main dancers, and their counterparts during the Lent celebrations, called *wa'rura pariseo*.

The Nawésari Ceremony

Nawésari usually takes place following the Sunday mass or during any large meeting or period of festivities. The ceremony begins with the siríame standing on the outside steps in front of the door of a church. Usually, the men exit the temple and form a circle in the esplanade in front of the building. The women occupy the center of the courtyard, showing their respect to the siríame by remaining with their back toward the church's door and turning around very slowly until they stand facing him. The arguasiri initiate the proceedings by giving a tesora to the siríame. As mentioned earlier, when not in use, these staffs are symbols of authority and are kept in a safe place, closely guarded by the Tarahumaras. The staff, a symbol of authority, is used each time the siríame speaks to the community. The tesora is also held when the siríame is acting as a judge or inducting someone into a public job.

The siríame delivers the nawésari to those present; this homily usually lasts twenty-five or more minutes. He commonly displays oratory skills with a fluent verbal delivery, rather formal in contrast with the everyday communication of the people. His speech is characterized by the use of highly dignified, complex expressions and elegant rhetoric. Cryptic messages are inserted during his delivery.[59] The ordinary person may not be able to fully grasp the rhetorical flavor

Nawésari ceremony. Agustín Palma, the siríame of Norogachi and a great orator, is addressing the sermon to his people. Pictured immediately to the left are the siríame from Choguita and Pawichiki. Father Mauricio Rivera, S.J., is seen to the right and Father Carlos Díaz Infante at the far right.

of the speech, but they listen respectfully and can relate the message, not missing a single word. School-age children appear to grasp the main message of the homily with ease. Children are exposed almost from birth and weekly to these sessions, a source of traditions and normative behavior advice.

The nawésari usually includes an introduction, discussion of religious beliefs, communication of moral precepts, announcements, information about recent events or significant past happenings, and closure. Throughout these, the audience is addressed as "the ones who have nothing" or "the little and poor." It is highlighted that God has given them all they possess. Throughout, tradition is emphasized, implying that that the way things should be is not a free option but that thins should be done the same way they have been done since ancient times.

In practically all nawésari, Rarámuri-Catholic syncretisms are apparent. God, they believe, put us on the earth and gave people the breath of life. God is trustworthy and cares for all. He rejoices when the Rarámuri perform good deeds and will punish those who do evil. He lives in the "place above" and created us from nothingness. The siríame repeatedly makes normative suggestions, such as "one must act righteously because that is God's will" or "one must believe all that the 'holy father' says," referring to what the Catholic priest says while preaching in churches. Spreading gossip or to bad-mouth others is proscribed[60] as well as to argue or fight. Disorderly conduct and scandalizing behavior during events such as the tesgüinada deserve divine punishment. Adultery and sexual behavior outside marriage are condemned. Irreverent attitudes of youngsters to the tribal traditions and customs and absences from weekly gatherings and being inattentive are also negatively sanctioned.

Warnings and announcements are among the most characteristic elements of the nawésari. The siríame proclaims that a messenger, generally a white woman or a little child, descended from heaven or came from a distant place like Narárachi to command that the Tarahumaras sacrifice an animal—most often a white one—to God. These sacrifices are to prevent catastrophic happenings, such as hail, snowstorms, and floods.

General news ranges from announcing religious festivities to political events, such as asking people to assemble for the visit of a *wa'rura* (important political figure, such as the president), and ejido-affairs notices, such as that corn or money will be distributed to the people.

Prior to the closing of a nawésari, statements are made again about Tarahumara poverty and declarations about the need to preserve traditions. The siríame concludes his discourse with the statement, "This way I have spoken." Generally, the people give their approval, shouting, "*ga'rá juku*" ("It's all right").

The Catholic Missions

The magnitude of the religious, educational, and relief activities of the Catholic missionaries has been considerable; so has also been the weight of the priests' opinions, suggestions, and wishes in communities or pueblos and the high social standing attached to priesthood. In contrast with this, until recently no Tarahumara could aspire to reach this position. Father Jesús Hielo Vega, a Tarahumara from Rituchi, was ordained as a priest on April 13, 1975.[61] Unfortunately, he perished tragically at Cerocahui in 1986. However, a native *rezandero* (person who says prayers) takes over some of the chores of the priest in pueblos and rancherías where a priest is not available. The functions actually performed by the Tarahumaras are never a complete substitution for those of a priest. The Tarahumaras who aspire to prestige and power with their people do participate in liturgical chores.

Catholic beliefs do reach the core of the Tarahumara mind except among the gentile or *cimarrón* communities—that is, those who have not adopted the Catholic faith—but even then negation of subordination to the church defines their identity.

The degree of importance given by Tarahumaras to the role played by individual priests varies from community to community. An extraordinary example is the case of Father Ernesto Uranga, a Jesuit missionary who mastered their language and deeply integrated himself into the Tarahumara community.[62] (Uranga died in 1966 piloting an airplane in the Sierra. In communities such as Narárachi, many Tarahumaras still expect him to return. The people often ask visitors about him. In Pamachi, his name, given as Santo Urango, has been incorporated in their prayers along with other Catholic saints.) However, what the Tarahumaras expect from their priests is limited. Baptism is the only Catholic sacrament considered by them to be strictly necessary. Catholic baptism may be administered by individuals other than an ordained priest, although they prefer to have the priest. A priest may also be required to pray for the dead and to preach on Sundays. The need for a priest is striking in important celebrations, such as the end-of-the-year gatherings, or *noríruachi*. Priests are expected to administer sacraments and celebrate the mass.[63] They are also expected to lead processions, hear confessions, and lead the rosary prayers. In some regions, the blessing of water, lands, and animals are also functions expected from the priest.

Ejido Authorities as a Social Power Organization

Parallel social power structures in the land of the Rarámuri, such as the ejido, must be mentioned. Ejido is a system of land grants and resources given in trust to the people. The plan originated following the Mexican Revolution as part of massive land reform. The program was initiated by President Lázaro Cárdenas and was continued by the administrations that followed him. The original idea was to make land available for farming to those dispossessed. Parcels of land called ejidos were assigned to the people individually or collectively to farm. The government retained ownership of the land. Almost all the ejidos created in Tarahumara Country were *ejidos forestales*; that is, areas designated for lumber harvesting. The land owned for agricultural and cattle-raising purposes was informally exempted and not made part of the ejido system. Small parcel ownership was allowed, respecting traditional local rules of ownership and inheritance. This situation often created conflict with the ejido boundaries. The ejido authorities, rules and regulations, and decisions often came in conflict with traditional Tarahumara land ownership principles.[64] Furthermore, lumber harvesting became the most important source of wealth within the economy of the area. The ejido authorities usually had a high economic and political standing. Important to consider is that the ejido membership did not discriminate between mestizos or Tarahumaras. Individuals from both groups could become ejido members, called *ejidatarios*.

A decision from the ejido authority could be and often was in conflict with the Tarahumara tribal rules. According to Mexican law, the highest ejido authority is an assembly of ejido members. The elected officials of this body constitute an administrative council. An ejido that owned a sawmill, such as Norogachi and many other pueblos, customarily elected a Tarahumara as president of the administrative council. The positions of secretary and treasurer were occupied usually by mestizos. In areas such as Pawichiki and Wawachiki all the ejido authorities were Tarahumaras, but even there the secretary or treasurer was a mestizo. This person therefore becomes a de facto highest authority of the ejido given that he controls the finances. Furthermore, usually a Mexican administrator was responsible to hire and fire workers in lumber fields, road construction, timber transportation, and the sawmill operation. This individual therefore becomes the most influential person in the ejido. The result is that the Tarahumaras continued to be excluded from power and authority in their own land.

Consejo Supremo de la Raza Tarahumara: The Highest Council of the Tarahumaras

A number of attempts have been made by the Mexican state and federal government, the missions, and individuals to give effective representation to the Tarahumaras. Leadership from one of their own people was thought to be advantageous given that the Rarámuri population is not large and lives widely scattered throughout the land. Such leadership could foster effective use of material and health resources and cultural homogeneity and prevent cultural decline, especially in the periphery of Tarahumara Country and in areas with large mestizo settlements. Such authority could also keep in check abuses from local caciques. With this in mind, the Mexican government sponsored in 1938 the creation of a council with received the high-sounding name of Consejo Supremo de la Raza Tarahumara. This effort had some favorable effects. Tarahumara people from many pueblos frequently approach the council requesting investigation and arbitration of internal disputes and address of problems. The presence of council representatives at pueblo gatherings, especially on occasions when authorities are designated and installed, has increased their acceptance and moderating influence. Events sponsored by this body have encouraged the development of a strong feeling of affiliation among Tarahumaras who live at considerable distances from each other.

Father Ernesto Uranga's Council, the Mariana Póbora

It is worthwhile to mention the attempt to develop a project of social organization independently from the Council of the Tarahumaras. Father Ernesto Uranga, the Jesuit missionary, tried to develop a confederation of pueblos named Mariana Póbora meaning "the five towns" (Norogachi, Papajichi, Choguita, Tetawichi, and Pawichiki). This organizational activity however was restricted to the Norogachi area, using as a base traditional mechanisms of authority from the Indians. This attempt took advantage of the closeness and cultural homogeneity of these pueblos, promoting a clear definition of positions of authority and empowering functions that had been deteriorating over time, such as the role of the generari. This effort attracted considerable interest from the Tarahumaras but it was interrupted by Father Uranga's untimely death.

The Role of Federal, State, and Municipal Authorities

Mexico as a country is a federation with a representative, democratic, and republican government. The Mexican constitution establishes three levels of government—federal, state, and municipal. Voters elect all officials at the three levels. Representatives from these three levels are present in the region. Historically, the level of presence and activity of the Mexican government was determined by political and military needs arising outside of the region. These included the handling of occasional uprisings, the defense against drifting groups and controlling uprisings after the Mexican Revolution.

The country under Spanish rule and later during the development of mining operations required administrative structures to regulate the exploitation of these resources and the imposition and collection of taxes. The federal government has been relatively slow to respond to the needs of the Indian and mestizo populations. At various periods in history, attention to the Indian communities was almost entirely left in the hands of the Catholic missionaries. The present structure of government in the region, with few exceptions, dates back to the administrations of Presidents Calles and Cárdenas, a time of conflict in state-church relationships that did not allow coordination or cooperation with the long-established missions. Describing the complexity of Mexican government participation in the area would go far beyond the scope of this work, but some salient features of these agencies need to be mentioned.

The National Indigenous Institute, or INI, is a governmental organization created to cover a wide spectrum of the needs of the indigenous population of Mexico. During the second half of the twentieth century, many of its original responsibilities were assumed by other agencies from the Mexican federal government: regulation of land-ownership by the Agrarian Reform Secretariat and forest exploitation by Profortarah (Productos Forestales de la Tarahumara), another federal agency. The area of health was specifically addressed by the INI's health system, although its activities were often superseded by coverage provided by the Federal Secretariat of Health and Welfare that operates in the state of Chihuahua through a state-federal compact. The expansion of such dependencies within these remote areas led to efforts of coordination implemented during President Echeverria's administration and later by President López Portillo's COPLAMAR (Coordinación General del Plan Nacional de Zonas Deprimidas y Grupos Marginados). The intention of these high-profile plans by these administrations was not to release INI from its role as the main provider for the Indians' needs but to make available to the Indians and other marginal groups a full array of resources from the federal government, such as the Mexican Institute of Social Security (IMSS), with all its resources for the provision of universal health care. Many small hospitals called COPLAMAR-IMSS were founded under the auspices of these institutions.[65] In general, attention to the needs of the Indian population has continued to be a federal responsibility. President Salinas's government promoted in 1988 another umbrella program, SOLIDARIDAD, and in 1995, President Zedillo's administration transferred the responsibility of health care to the states.

The Mexican Army is also an important resource representing the federal government in Tarahumara Country. The army was practically withdrawn from the region at the conclusion of the Mexican Revolution except for sporadic interventions. The fast proliferation of opium poppy cultivation by outsiders in the 1970s increasingly required control measures well beyond the capacities of local police. A federal campaign led by the army and the Policía Judicial Federal called "Operación Cóndor" led to a significant escalation of the military presence in the zone. Their activities had a definite impact in many areas of human activity in the area.

Some projects implemented by the state of Chihuahua, begun in the 1970s, such as the expansion of a road network called Plan Gran Visión, had multiple positive as well as negative consequences in the region. The State Coordination of the Tarahumara created by Chihuahua Governor Baeza and the state legislature in 1987 and later expanded under governor Patricio Martínez in 2000–2004, gave the state a presence in the region to a degree never seen before.

The importance of the role of municipal authorities has fluctuated in Tarahumara Country. As a result of historical and economic factors, municipal townships were usually established in mining towns, which included offices for mine registry, tax collection, and the awarding of licenses and permits. After the decay of the mines, the political structures created to meet their needs remained but their resources decreased substantially.[66] The shift of the regional economy from mining to timber exploitation made many municipal offices located in old mining towns obsolete, contrasting with the needs of the growing settlements devoted to timber harvest processing or shipping. The municipality of Guachochi is an exception; from the outset, it was created with the specific purpose of creating an indigenous municipality.[67] It should be noted that the Tarahumaras, although they have their own tribal authorities, are also subject to the authority of municipal, state, and national governments.[68]

Informal Leadership

Informal leaders play a very important role in Tarahumara social structure, in some instances, greater than that of the formally recognized officers of the tribe, ejido, or national government. The "leader" role might resemble that of a kind of wise elder whose social participation may be hardly apparent but whose actions carried about by his designates are recognized by the people and their prominent influence in the town's life is widely acknowledged.

FOUR

Affiliative Social Activities of the Tarahumara People

In a country split by deep canyons and mountains, the Tarahumara display remarkable cultural and linguistic homogeneity and cohesiveness. The activities and events described in this section have as a common denominator their taking place at the ranchería, family, or interfamily level, as opposed to the grand celebrations that include a whole pueblo or group of pueblos.

The Bajíachi: Institutionalized Drinking

The *bajíachi*, or in Spanish *tesgüinada*, is one of the most important forms of social interaction within the Tarahumara culture. The term bajíachi means "place or time to drink." During these occasions, batari or tesgüino, a corn beer, is consumed by a group of men and women. Bajiachi events celebrated widely throughout the land create an overlapping social network that facilitates communication, sharing of traditions, and group norms.[1] These events contribute to strengthening cultural norms and homogeneity. Elders have the opportunity to share their wisdom by telling their stories and legends and giving advice to the young.[2] The bajiachi plays other important functions: these collective drinking events are organized as a form of compensation for the individuals who participate in collective work. Building homes, fences, and farming involve group activity. Collective labor is an important element in the economic structure of this culture. It also plays a time-out function that sometimes gives occasion to dysfunctional behavior, such as acts of violence and displays of unacceptable sexual behavior.[3]

The bajíachi requires logistical planning. Preparation of the drink consumed in these events requires several days. The ingredients—corn, water, and various herbs—are placed in large clay containers to ferment. The fermentation takes several days and the drink has to be consumed when it is considered ripe. This exact point in time has to be estimated with some accuracy because once the beverage is ready, it cannot be stored for later consumption or individual use. The men and women who will be attending the event have to estimate the day when the drink will be ready and the group will meet. Last-minute notification is not feasible given the dispersion of their dwellings.

Kickball Races, or Rarajípari

The Tarahumara kickball races, or *rarajípari*, have attracted considerable attention because of the remarkable display of physical stamina exhibited in such contests. In these races, contestants run continuously 80 kilometers, and not infrequently 160 kilometers, along dirt trails in the mountains. A race of over 700 kilometers, representing more than a week of running, has been reported.[4] Participants may cover up to 160 kilometers in twenty-four hours. Races lasting forty-eight hours are not uncommon. Data from field research indicate that experienced kickball racers are capable of energy expenditures of more than 10,000 kilocalories/24h, a figure that represents the upper limits of human voluntary work effort.[5] Physiologic studies on Tarahumara runners have noted that that a runner's blood pressure drops as his pulse rate increases, a change that enables the heart to increase cardiac output. Like all mountain people, their oxygen transport capacity is high and they have a high concentration of hemoglobin and a high hematocrit, or percentage of red cells in the blood. Reportedly, their cholesterol levels were low.[6]

According to local tradition, the Rarámuri have had their kickball races from the very beginning, "just after God created them." However, historical and archaeological data suggest that the races may be of relatively recent origin. The oldest reference to rarajípari is from a mission priest, Father Steffel, in 1791 who believed that the races as a cultural event had begun in the late seventeenth century. Earlier writings by Neumann and Ratkay on the Tarahumaras do not mention the races.[7] Other scholars believe that these races are an old cultural feature.[8] The very term Rarámuri, which means "fleet feet," implies that they saw themselves as a tribe of runners. The term however may not refer to kickball racing but to their pride in their running ability and proficiency as military messengers. It has been suggested that rarajípari replaced Indian games played by the Rarámuri with rubber balls and that the change was promoted by the missionaries to eliminate the heavy betting characterizing the latter.[9] Finally, work on the Casas Grandes Indians indicates that games with rubber balls and kickball racing with wooden or stone balls were practiced by the members of this culture.[10]

Rarajípari is a contest between two teams. Each team consists of one runner with an assistant or assistants. Most of the races take place during the rainy season, *ba'rá*, when the corn crops demand relatively little attention. The people attending these races support the team from their own village or community, led by a *chokéame*, or race organizer. A central feature of the race is the activities of the *sakéame*, or bet makers.

The runners must propel a wooden ball only with their feet as they run in a preselected course. In the women's version of the race, the running course is shorter and instead of kicking a ball the runners toss a ring made from fiber (*se*ré*) wrapped with cloth. They use a stick with a curved end to toss the hoop as they run.

Two basic types of race courses are used: a simple out-and-back trail or a circular course that combines a number of laps to form a figure eight or cloverleaf. The starting point is usually at the center where spectators gather and where the *tari* (a "mound of bets") is positioned. The trails are relatively clear from brush to make finding the ball easier. Sometimes, these contests involve more than one race, resembling track meets, including contests for men and women of various ages and even for people with handicaps. The racers are carefully matched for age, ability, and previous running record. This makes fair and interesting contests. Frequently, races matching runners who share the same special condition or handicap have been observed. For instance, I have observed races for women with advanced pregnancies or for those who have recently given birth. Sometimes, the races take place at night. The spectators run alongside the racers, carrying torches to provide light. The most important races—for example, one involving runners reputed to be the fastest, such as those from Cabórachi, against the ones known for the greatest endurance, such as those from the Sisoguichi-Panalachi area—may require months or even years to organize.

In general, the racers have a characteristically lean body and small muscle mass not unlike other endurance runners. On the other hand, exceptions are common. I observed a runner with one leg shorter than the other by at least four inches, the consequence of an old, poorly healed fracture, running against a healthy rival. The tolerance for the demands of running such long distances has raised questions as to whether these capacities are inborn, acquired through training, or both. The Tarahumaras are mountain people. Tending their flocks provides conditioning for racing because it requires running and climbing through very rough and difficult terrains. Physical conditioning begins early. Rarámuri children begin herding pigs when they are four years of age, and it is not unusual to see six-year-olds skilled at tending sheep and goats. Seven-year-olds practice launching the ball night and day and organize competitions among themselves. As they get older, they travel from hamlet to hamlet, running instead of walking and kicking a ball as they go.

It appears that runners are more likely to drop out of contests because of superstitions and fear than from physical limitations. Runners usually take steps to avoid possible injury. Many use traditional remedies or magic amulets, such as a button of peyote tied to the runner's back, infusions of various herbs, or rubbing their legs with myrtle oil. Complex ministrations are often employed to supposedly prevent witchcraft or poisoning. Accidents, cramps, trauma, and emotional displays may be observed in this sport, as in any other comparable contest. Common reasons to drop from a race are leg cramps or abdominal pain, which may be simulated by a "bought" runner, given that bets are involved in these contests. Nevertheless, it is more common to observe stoic runners who continue in the contest with their feet bleeding, leaving bloody footprints on the trail.

Tutuguri and Yúmari Dances

Scholars have tried to differentiate between tutuguri and *yúmari* dances, given that their characteristics and aims of both overlap considerably. It appears that in earlier times differences existed that were important.[11] The tutuguri or *rutúburi* dance is a central element in ceremonies of offerings to the transcendental power. According to scholars, dance is the main form of prayer for the Tarahumaras. During the offering ceremonies, an animal considered clean is sacrificed. Goats, rabbits, jackrabbits, deer, or squirrels are immolated but never snakes, pigs, or dogs. Another theme is an acknowledgment that God gave men and women all that they possess; therefore, the people should correspond by returning some of it to him. In some localities, the ceremony is still addressed to the sun god, Rayénari, and the moon goddess, Mechá. In many areas the concept of *konéma*, or the act of giving nourishment to God, is incorporated. The tutuguri dance is also observed within the context of other ceremonies that intend to prevent sickness, heal, feed the dead, or ask for bountiful crops. It also may be performed expecting to influence the sometimes ill effects attributed to the sun, moon, or stars.

During the tutuguri, a singer, the *wikaráame*, plays a rattle (*sáuraka*) while chanting a monotonous melody and dancing back and forth across a square of ground that was cleared and swept for the occasion. On the east side of this area, a small altar is placed in front of a cross symbolizing the Holy Trinity or, in places not under the church's influence, the sun, the moon, and the morning star or the four winds. The cross is covered with a piece of cloth decorated with necklaces such as those that women wear. The dance is initiated as the wikaráame goes in circles around the cross moving his sáuraka and tracing in the air symbolic crosses around the four sides of the cross. Usually, two other crosses are placed at both sides of the main one. At this point, the offerings are placed on the altar.

Erasmo Palma, a Tarahumara from Tuchéachi, described and interpreted the ceremony incorporating some of his mythical beliefs and cultural perceptions. His narrative presented in free translation goes as follows: "This is the way the Rarámuri were taught by their *anayáwari* (ancestors). White men were made by the one who lives below. Rarámuri, on the other hand, were created by the one who lives above. Serpents were created by the one who lives below; pigs also. On the contrary, deer and rabbits come from above; horses and donkeys, from below; the cow belongs above. Whites, it is said, are from below, since they are abusive with the Rarámuri; they make the Rarámuri suffer. For this reason, it is said that white men are sons of the one who lives below. But the one who lives below was not capable of giving them their breath; it was given to them by the Father. The Rarámuri dance the *rutuburi* (tutuguri) when certain animals are killed. They never dance it when a pig is killed; likewise, they never dance rutuburi when a jackrabbit, dog, snake, or donkey, is killed. They do dance the rutuburi when they kill rabbits to offer them to the Father (Onorúame). Other animals, like goats, chicken, cows, and bulls are offered to God too. This way, whichever animal is good, it is also said of it that it was made by the Father. This is the Rarámuri faith; the rutuburi is danced to honor God and is danced to speak to God. White men, on the other hand, it is said never offer anything to God when they are about to eat. Because of this—because white men eat without making offerings to God—it is said that they proceed from below. When the Rarámuri drink, they do so offering first the drink to God."[12]

The tutuguri ceremony proceeds as blood, broth, and batari or any liquid food or herbal concoction is sprinkled on a cross. Solid food portions also are thrown or moved in circles in front of the cross, symbolizing *konéma*, or feeding God. Then, food is presented and covered with a piece of cloth.[13] At a given time, most of those present join the wikaráame, with men standing in line at his right side, women on the left. For hours, they dance with small steps and a slight limp going east-west, west-east through the whole dancing area.

Close to the end of the ceremony, the step changes and the line of women run with little steps counterclockwise around the cross while the men do it clockwise. The wikaráame may then approach the cross and the offerings in a similar way to that at the beginning of the dance and he rattles his sáuraka toward the four cardinal points.

FIVE

Great Life Occasions and Ceremonies
Birth and Death among the Tarahumaras

Although at first glance, the Tarahumaras may appear detached from the transcendental events of life and death, a closer look reveals the significance these events have on their worldview and how the rituals attached to them have relevance and provide continuity in their pursuit of daily life.

Tarahumara Pregnancy and Childbirth

The Rarámuri approach pregnancy and childbirth in ways that would appear rather casual to the outsider. The women's daily activities, even those demanding significant physical effort, change little during pregnancy. It is not uncommon to see women in an advanced state of pregnancy competing in a stick-and-hoop race. Even more striking, the women competing in these races are carefully matched in regard to the stage of pregnancy just as they match the contestants with other types of handicap to make the races even. It is commonly mentioned by those familiar with the Tarahumaras that a woman may be in the middle of a trip to the town when labor begins and she stops just to deliver the baby and almost immediately continues walking as if nothing had happened. The other is that the woman, feeling the moment of delivery is close, may climb to the top of a hill or run to the bottom of a gorge to have the baby in relative solitude. There is a story often mentioned in Norogachi about a female runner who interrupted her race to deliver a baby and completed participation in the race a few days later. Regardless of these stories, delivery usually occurs when the woman is in the middle of her usual chores. Sexual activity is not interrupted until the eve of the delivery.

A pregnant woman is called in Tarahumara *sapéame* or *ropéame*. The first term is also used to describe obese persons (*sapá*=flesh); while the second alludes to a large abdomen (*ropá*=belly). Within the social context of a bajíachi, when the behavior of the group is uninhibited by the influence of alcohol, mention of the words *sapéame* or *ropéame*, whether applied to men or women, elicits smiles and even laughter among the audience.

The multiple sipucha, or skirts, worn by Tarahumara women and their relatively casual attitude make pregnancy inconspicuous until it is very advanced. Moreover, no preparations are usually taken at home to be ready for the delivery.

During pregnancy, symptoms frequent among urban females, such as morning sickness, cravings for certain foods, and malaise, are not mentioned by Tarahumara women. A missing menstrual period is often considered a sign of illness and treated with herbal remedies rather than suspected as a sign of pregnancy. Infertility is certainly not a happy event among them. In contrast with what seems to be the case among mestizos, pregnancy is considered natural rather than a disease. Pregnancy is welcomed even if it is the result of rape or may create other inconveniences.

There is a belief that God permits every woman to have twelve children, after which no more may be born.[1] This contrasts with the fact that a family with two or three children is the norm, a pattern that has not changed through the years. A survey carried out in Norogachi in 1977 showed that the mean number of deliveries within the life span of Rarámuri women was from eight to ten.[2] However, as a general rule, the Tarahumara family is small due to a high infant mortality rate. Women usually remarry shortly after becoming widowed, seemingly assuring that fertility continues.

Tarahumara women know that pregnancies last nine lunar phases. The approximate date of the expected birth is related to the moon phases. For instance, if the conception is thought to have occurred on a full moon, the baby is expected by the new moon of the ninth lunar month; if the baby is not born at that time, then it will be born at the time of the following crescent.[3] The starting point for the count is either the date of a remembered sexual intercourse or the first missing menstrual period. There are other beliefs associated with the moon phases. The crescent cycle—that is from the new moon to the full moon—is defined as the time when the moon "is gaining strength." During this period, the moon is supposed to share its strength with people and other living beings; for instance, if one is going to cut a tree, its wood is expected to be harder than usual; if a baby is born, it will be stronger. The opposite is also believed to be true; from the full moon to new moon, the moon progressively "loses strength," so while it would be the ideal time to chop wood or catch a deer, a baby born during this cycle will be weak. Eclipses are especially feared. The belief is shared that a cleft palate and cleft lip are the result of the moon having eaten up the lip and/or palate of the baby born during an eclipse.[4] Interestingly, few persons know their date of birth unless they were born during a holiday. After a couple of years have passed, parents seem to have forgotten the birth dates of their offspring.

Preparation for Childbirth

Some ceremonies performed before the actual birth of the baby have been described. One of these curing ceremonies—rituals performed to heal or prevent illnesses—consists of passing a flaming torch over the head of each member of the family, burning a bit of hair. This is done to cut the invisible string that connects the baby with heaven. Until it is cut, the child cannot be born. This ceremony also serves to keep lightning and windstorms away.[5] The family may be cured with a sucking tube or with crosses marked with tesgüino. After the ceremony child and family may receive a *wikubema*, or cleansing ceremony.

Behavior during Labor

Women in labor tend to seek solitude and shelter in the depth of a gorge or in high places in the mountains; they may just stop on their walk to the village and have the baby by the side of the road. During labor, they adopt an upright position and this seems to be the reason why labor is called *rikínama*, which means "to go downward, to descend."[6] Delivery also is designated *nawama*, or "to come, to arrive." To give birth is expressed with the verb *nawárima*, derived from the word just quoted. Many women deliver the child alone without any help and some actively avoid all company. However, nowadays many Rarámuri women no longer run to the deepest part of a gorge or to the top of a mountain as they used to do, and home deliveries are becoming common. The behavior of the Rarámuri woman contrasts with that of mestizos, who usually seek the companionship of old women, particularly those considered knowledgeable and wise. In some areas, as labor pains begin, enemas of *manzanilla* (chamomile) or *malva* (mallow) are administered.[7] Herbal potions are given to facilitate labor and ease pain. Certain Tarahumara women who from prior experience anticipate a difficult childbirth may take potions intended to prevent long and painful deliveries or excessive bleeding.

Once the time of birth is close, a pole (two meters or longer) is placed leaning on a wall if the birth is taking place inside the house or against a tree or a rock if outdoors. The woman places one of her arms over the inclined pole and rests the weight of her body on it. The incline of the pole (about 45 degrees with respect to the floor) appears to be useful during labor since the woman may lean on the pole from sitting-up or upright positions. The pole is a good support when contractions are taking place in which case the pole is strongly grasped to help the contractions or to rest while leaning on it.[8]

During delivery, the woman remains fully dressed, removing only her sandals. The girdle that secures her skirts may be stretched or loosened during labor. If the woman is assisted by others, the edge of the skirt (sipucha) or the blouse (napatza) is pulled down, protecting her modesty throughout the process. Rubbing and massages, when used, are gently delivered, after assuring that her private parts are well covered. From time to time, water and kobisi (parched corn) are given to the woman. Kobisi is usually the only food given to women during labor, but regular food is provided if she requests it. As the baby appears, he is gently deposited in the dirt floor. The length of childbirth seems to be shorter than in mestizo women. Perineum tears seldom occur, even among women giving birth for the first time, probably due to the vertical position adopted during labor. Maternal mortality on the other hand is high. Bringing a child into the world is a serious hazard for these women due to the frequency of puerperal sepsis and other complications.

Intervention of the Owirúame

The assistance of a shaman is sought only when serious problems are expected. Intervention of an *owirúame* may include use of herbal remedies, but the physical maneuvers he applies to assist the delivery are what is mainly expected. The owirúame may apply a variety of massages, including one applying strong thrusts on the bottom of the uterus, similar to the Kristeller's maneuver, and he may even attempt to perform the obstetric manipulation known as an "external cephalic version"—rotating a breech or side-lying fetus so that it can go down the birth canal—which they call "to accommodate the baby." The shaman places himself behind the woman and manipulates her abdomen from the back, sometimes compressing or pushing down, reinforcing the womb's contraction. Ointments derived from several types of animal fat, such as that of rattlesnakes or cholugos, or oil of *arrayanes* (myrtle) and olive oil are used, including commercial ointments. Several other poultices may also be utilized with the intention of facilitating massage and helping the child descend.

Due to the position of the woman, kneeling or crouching, the owirúame cannot receive the baby himself; an assistant or a woman therefore performs this function as the baby lands directly on the ground or in a nest of clothes prepared beforehand. Although some of our informants claim to know the maneuver consisting of pulling back the woman's perineum to ease the expulsion, the owirúame does not touch either the perineum or the crowning baby. The parameter used to decide that a laboring woman needs further help is usually the fact that she has already had a prolonged labor; however, the more acculturated women do seek with increasing frequency the help of mestizo midwives or even come to Clínica San Carlos in Norogachi. When the birth is taking too much time, the family and the surrounding community deliberates to decide what further steps to take; usually, the idea is not to take the patient to another place but to bring in help. The alternatives are usually three: bring in a doctor or nurse from the closest hospital, a mestizo midwife, or a Tarahumara shaman. A decision is taken only after having exhausted local knowledge; that is, after having tried suggestions offered by the more experienced people in the neighborhood. Old women are usually respected for their knowledge about childbirth matters; sometimes they will recommend a beverage or even attempt some form of massage to help the laboring woman, but usually they just stand by waiting for something to happen.

Husband's Participation in the Delivery Process

The husband (*kuná*) plays a relatively secondary role throughout the childbirth process. The husband may be dispatched to go and look for all the goods that his wife may need—water, clothes for the baby—or to deliver messages. He may go and bring the shaman, a physician, or a midwife if they feel one is needed. This sometimes includes arranging transportation for them or taking the woman to a clinic. At times, a distraught husband will show up at a clinic with a saddled horse for the doctor or nurse to ride. They may also contact, for the purpose of availing transportation, a local merchant who owns a pickup truck. Less acculturated young Tarahumara men may go through the delivery process sunk in deep panic, anxiously repeating, "*Ke tashi rikínare ta* kúchi*," meaning "The little one does not want to come down."

Whether the woman is inside the house or at the bottom of a gorge, the man keeps himself away but close, seemingly aware that he may be suddenly summoned to help with some task. If there is no owirúame in the region, another man, usually a relative or a friend, may be asked to help the woman to deliver the baby by pressing her abdomen. The husband, even though not formally banned, never plays this role. After the delivery, the husband will keep a period of sexual and work abstinence during which he will not work for three days, believing that his axe would break, or the horns of his ox would fall off, or he would break a leg if he breaks this rule. After the third day, he will take a bath and return to his chores.

Afterbirth

Usually another woman, the shaman helper, or the mother herself picks up the baby who has landed in a nest of rags or on the plain dirt floor. The umbilical cord is then cut with scissors, usually the same ones used to shear their sheep.[9] The umbilical cord is cut long enough to be tied on itself and on the side of the baby; the placental side may be tied or just left alone, allowing the placental blood to flow until the placental vascular bed is empty. A few Tarahumaras follow the mestizo custom of tying the umbilical cord with a shoelace. Although the placenta is rarely pulled to accelerate its detachment, retention or excessive bleeding after delivery is common. When placental retention occurs, usually they wait for the spontaneous expulsion, often waiting too long, before going for help. A piece of the umbilical cord, about six inches long, is usually buried "so that the child will not be stupid."[10] I have observed Tarahumaras burying the cord and the rest of the membranes close to their house. The reason given to me is to avoid the *kimara* (placenta) from being eaten by the dogs. The placenta sometimes is left hanging on a nearby pine branch for some ill-defined ritual purpose.

The newborn is cleansed with warm water, but usually the baby is not fully bathed unless he appears dirty or covered with blood. The baby is then dressed and wrapped in a blanket.

The Fire Baptism

Fire baptism, a designation of unclear origin is a native naming ceremony. Among the contemporary Tarahumaras, this event is controversial. Some informants claim that everyone has received a "water name"—that is, a Catholic baptism—as well as a "fire name," an ancient ritual. Some assert that the fire baptism is a custom that survives only among the gentile Tarahumaras, who reject the Catholic baptism. When this topic is mentioned in Tarahumara gatherings, it usually elicits an extensive discussion of anecdotal and often contradictory claims. Some speak cryptically about it as if they were apprehensive about disclosing a secret ritual, comparable to the Jíkuri ceremony. Others would irreverently joke about fire naming. In the Norogachi area, most Tarahumaras bluntly deny having a fire name. All of this may reflect an effort to conceal something that the missionary clergy would not accept. In instances where a fire baptism is celebrated, an owirúame or *owétzaka* is called to officiate. The ceremony is said to have the power to prevent misfortune or bad luck; whether one thinks of this ceremony as a "curing ritual," meant to heal or prevent illness, or as a true native sacrament, its simplicity is captivating. First, the owirúame passes a burning stick of *chopé* (resinous wood) or smoldering corncobs over the head of the newborn, drawing crosses and performing circular motions over the baby's head. This is done to destroy the thread or string that supposedly connects the soul of the child to a pine tree.

Funeral Ceremonies: The Nutéa, Nutékeri, or Nutékima

The *nutéa* is a rite of passage intended to assure an uneventful journey to the dead. The words *nutéa, nutékeri,* or *nutékima* literally mean "saving food for the dead to eat." It is believed that a dead person has to walk a long way before entering into *rewegáchi*, "the place of those who have a name," and nourishment is required for the journey. These food-offering ceremonies proceed as follows: Under the cover of nocturnal darkness, somewhat attenuated by the fires lit in several places, dancing of the yúmari is initiated. Drinking of batari is done, increasing in pace with time. After considerable drinking and dancing, a group takes a blanket or piece of cloth that was owned by the dead person. Several individuals grab each corner of the blanket and it is well secured; they strongly shake it as if trying to throw something upwards. Having shaken it, they appear pleased, believing that they have sent one soul to heaven.[11] The owirúame and all the dancers in the patio give a farewell to the deceased. Then, a man and a woman covered with blankets and carrying house utensils dance around a cross. After a while, all those present dance and leave the ground. The closest relatives are expected to participate in the ceremony. At the end, they take off their *coyeras* or *coweras* (headbands) and some of their dress and throw them into the patio. Meanwhile, the owirúame breaks a small wooden cross and throws the fragments high and away from him. Drinking batari then continues.[12]

In these ceremonies, food is placed before one or several small wooden crosses representing the soul of the dead person. The owirúame, the "one who heals," that is, a shaman, ministers the ceremony, addresses the dead man's soul, and offers the food. Meanwhile in a square area of ground outside the house set aside for this purpose, the tutuguri is danced to "lift up" the soul of the dead person.

The first nutéa is performed a few days after the demise of a person and it is repeated at intervals of one year—three times if the deceased is a man and four times if it is a woman. Women, it is assumed, walk slower than men so it takes longer to take them to heaven.

An interesting feature of the nutéa is that some of the male dancers (*matachines*) appear wearing female clothes.

SIX

Major Festivities of the Tarahumaras

The Rarámuri celebrate with enthusiasm the major Catholic religious festivities, particularly those taking place during Lent, Holy Week, and at the end of one year and beginning of the next—Our Lady of the Pillar, Our Lady of Guadalupe, Christmas, New Year, Day of the Magi, and Candlemas—and in some areas, the patron saint of a town. These events—the center of the Tarahumara social life—are celebrated even in places relatively inaccessible or seldom visited by missionaries. Major festivities, unlike those taking place at a ranchería or group of rancherías, involve the entire pueblo or even a group of pueblos. In recent times, these celebrations have become important tourist attractions.

Interestingly, sometimes people do not seem to know or remember what a particular fiesta is commemorating. Nevertheless, these social functions provide a time reference for the initiation of important activities such as plowing, seeding, and harvesting. During these celebrations, people gather in large numbers, putting aside all other activities to participate and celebrate. These gatherings facilitate exchanges of information and provide a framework for some of the events already described, such as the indigenous court trials, nawesari, and elections of tribal officers. The gatherings also provide a setting for kickball races and other recreational activities; for instance, in Guapalayna, races are staged during Holy Week before the beginning of the religious events. The celebrations also give men and women opportunities to meet and select mating partners.

Noríruachi

Noríruachi—the turns—is the name of the religious festivities that take place during Lent and continue through Holy Week. The timing of these events follows closely the Christian liturgical calendar. The festivities incorporate several religious themes and make some references to historical events from Spain. In spite of the content of these celebrations, many pre-Hispanic antecedents and indigenous influences are also well apparent.

The beginning of Lent is announced throughout Tarahumara Country by the pervasive rhythms of the *rampora* (drums) and their resounding echoing in the mountain ranges. Not only do musicians accompanying groups of dancers beat the drums, but every Tarahumara who owns a drum beats it, even as they walk from one hamlet to the next. During this time, festivities such as the tutuguri, bajíachi, and jíkuri also take place, all accompanied by the beating of the rampora and the flute (*kusera*). The steps of the dances, the body painting of many participants, and the profound appeal to the whole population suggest elements of ceremonies that existed before the arrival of the Europeans, touching deeply the mind of the Indians. Palm Sunday is designated Rakirúkuachi by the Indians. Maundy Thursday and Good Friday are also celebrated according to the Catholic liturgical calendar. Ceremonies including processions, the Stations of the Cross prayers, and even reenactments of the journey of Jesus to be crucified are carried out, ending in a hill appropriately designated Calvary Mount. In several communities, central elements of these religious themes have been modified or have disappeared. For instance, the figure of Jesus may not be included. Instead, the story of Judas is reenacted. Interestingly, references to Indian myths, fertility, death, and resurrection are incorporated. The *pariseo* and *bascorero* dances that are elements of these celebrations accompany the reenactment of battles between the Moors and Christians, which were introduced by the missionaries. A tradition defended in Samachique claims that the rampora reenacts or represents the sound of the beating of Jesus while the sound of the kusera reenacts the crying of the Virgin Mary. Arches made of pine branches clearly represent the Stations of the Cross. The shouting of the dancers is said to be the cry of the Arabs, or Moors, remembered at this time.

Early during Lent, teams or groups of dancers are assembled. In Norogachi, for instance, teams representing the friends of Jesus are designated *pintos*,

A group of *pintos* from Santa Cruz-Colalechi. They are known as friends of Jesus.

while others, the pariseos, represent the enemies of Jesus. In Samachique, the friends of Jesus are the Moors, an interesting incongruence. Both groups, but particularly the pariseos, wander through the hamlets dancing.

In Norogachi, pariseos arrive on Fridays and dance near the church or go on a pilgrimage from Papajichi to Pawichiki. The pariseos show up in towns, dance a few times, and run away, acting as if they were being chased and were trying to avoid being caught. The pariseos dress in ordinary clothes—napatza, or blouse; tagora, or breeches; and cowera, or headband—but they also wear a headpiece made of turkey feathers. They carry a spade or spear (*sakérowa*) carved from wood, and their bodies are painted with lime. Pintos wear only the tagora and cowera, leaving their torsos bare. They wear a red cloth tied on their waists with a flap hanging from their back. Their faces, chest, back, and limbs have large white spots painted with lime. A *wa'rubera pariseo* (chieftain) or *bandériame* (banner carrier) leads the dance movements of the group, waving a red flag. *Moros* (Moors) in Samachique use a different pattern of body and face paint than pariseos. An important feature of their facial painting is

A group of *pariseo* (pharisee) dancers from Norogachi. They are enemies of Jesus and friends of Judas.

the cross on the forehead, which marks them as friends of Jesus.[1] Finally, the *sontarsi* (soldiers) have no body painting but wear instead a headpiece made of turkey feathers.

Several processions are part of the celebrations of Rakirúkuachi. The local officers assign men ahead of time to gather palm branches from their home areas in the canyons. The branches are then taken to the mission priest, who blesses them. In his absence, this is done by the rezandero or sometimes the rezandera. The palm branches are distributed among the people by the siríame and the ibagápitano. The palm branches are handled with great care and reverence by the people during the processions reenacting the triumphal entry of Jesus into Jerusalem. In places where the missionaries' presence has been lasting, pictorial images commemorating the resurrection of Jesus are displayed during the processions. The dancers portraying the friends of Jesus and in Norogachi the pintos take part in the processions accompanied by the beat of their drums.

In the canyon country pueblos, men and women gather on Monday during

Holy Week and spend the day competing in kickball races and other games such as *najarápuri*, a native wrestling contest. The presence of some inebriated Rarámuri in these gatherings indicates that some seem to have anticipated the bajíachi imbibing that usually takes place on Thursday and Friday. Prior to all the events mentioned, the streets are swept and arches are erected from pine branches with their pine needles. In Samachique and the canyon towns, these branches are decorated with cones or rosettes of *soko* (yucca), which are also hung at the door of the temple.

On Holy Thursday in Norogachi, groups of pariseos arrive in the early morning and once the arches are erected, the processions begin. Arches are used as reference points where the procession participants stop to pray the Stations of the Cross while dancers perform around them. In villages poorly urbanized but having a church, such as Chogita, Pawichiki, Papajichi, Narárachi, Wawachérare, and Guapalayna, the procession participants just walk around the temple. In larger communities, full-fledged reenactments of the route taken by Jesus to Calvary to be crucified take place. In Norogachi, Cerocahui, and Narárachi, the processions begin marching on Thursday, while in Samachique they wait until midnight to begin.

Each procession is led by a marshal (*mariscal*) chosen for the occasion. In some places, he carries an old Spanish colonial sword as a symbol of his authority or social standing. If a Catholic priest is present, he usually heads the processions, alongside the marshal, preceded by carriers of a large cross and candles. If there is not a missionary present, the rezandero takes his place. The rezandero or the priest is followed by several respected Tarahumara women known as tenanches or maromas, who carry the incense and the images of the saints of the pueblo. The participants carry noisemakers, matracas (noisemakers), and whips that had been kept in custody at the sacristy of the temple.[2]

At each stop of the processions, the drummers stop beating their ramporas and the rezandero recites, from his memory of their sound, a few barely intelligible prayers that from a distance might sound like the rosary but whose words and meaning have deteriorated. After the prayers, the matracas and the drums resume their noise. In some pueblos, rather than having processions, the rezandero prays outside on each of the corners of the temple while people listen, standing as they do for the nawésari, women in the center with men beside them.

Tenanches with the saints of the pueblo (*Samachiki*).

All the active participants in the processions have been selected the first day of Lent and each has been designated a specific function. In some places, they are known collectively as the *morocapa*. In the middle are women; on the sides, men; and escorting them on both sides of the procession are the pariseos. This select group is followed by the crowd. The number of processions occurring on Good Thursday and Good Friday varies, but all of them walk around the temple counterclockwise, except one that walks clockwise.

In some places, for example in Aboréachi, the "enemies" of Jesus, the pariseos, suddenly appear, go inside the temple, and display gestures of contempt before the altar and the sacred symbols, leaving then as hastily as they arrived. In earlier times, the pariseos stole articles found unattended in houses and fields, including pots, clothing, and even cattle, which they shepherded and cared for until the time of reading of "the Last Will of Judas" when the goods and animals were returned to their owners. This curious tradition is still maintained in Cerocahui but it is on the wane.

At dusk the wakes, or *velaciones*, take place. For these ceremonies, an altar-like device is set up. Two men, armed with spears, will stand guard at the right and left sides of the altar. They represent Roman soldiers and take turns through the night. In some areas, the people know that the ceremony represents Jesus as a prisoner on Thursday and his demise on Friday. In other places, for instance Aboréachi, the ceremony is styled as a funeral on both days, with violins and flutes playing very sad tunes seemingly ignoring the sequence of events.

Very early on Good Friday, the pariseos flee to the bushes where they make a straw figure that represents Judas—Judo—with a rather large phallus that seems to be the remnant of a fertility rite what is supposed to increase the fecundity of their people. Also a Judasa—she-Judas—and, in some places, even Juditos—child-Judases—and Judas's dog are made. For the Tarahumara people it would be inconceivable that Judas would wander around all alone in the world without a family and a dog. The same day, the Judas is indicted, accused of drinking and copulating with men and women, and his wife is accused of engaging in adulterous actions. It is important to note that according to their beliefs, Judas and the pariseos are supposed to be evil. Everything they do is supposed to be sinful and scandalous. The Judas figure is taken by the pariseos to the temple several times during the day, exhibiting its large penis before the sacred symbols. In Samachique, the pariseos enter the town under the seventh arch, dancing from right to left, trying to scandalize all those present.

At noon, the Crucifixion procession begins. An image of Jesus is covered with a red cloth. In Cerocahui, a remembrance of the moment when Jesus, on his way to be crucified, encounters his mother, has special relevance: an image of Jesus, carried by the men on the way to be crucified, and one of Mary, carried by the women, join at one point where the images "kiss each other" and continue on the road to Mount Calvary. In Norogachi, they also have male and female processions but in this case they join each other behind the house of a prominent person and continue together. In Samachique, the pariseos stop to rest at noon. In each house where they stop, food is collected, which is placed in a container to be eaten in the evening. In Samachique, two evening processions take place, one of which proceeds in a counterclockwise direction.

At night, the groups of friends and foes of Jesus alternately dance at the temple's atrium while inside of the church the wake of Jesus takes place. In Norogachi, the wake ceremony is impressive because a human-size figure wrapped in a blanket, looking like a real corpse prepared for burial, is displayed. In Aboréachi, they keep guard over an old statuette of St. John the Baptist missing both arms. The image despite its missing arms, is the town's most revered symbol as the saint of the town.

The celebrations of Sábado de Gloria (Holy Saturday) begin very early in the morning. The *bascoreros* (dancers who perform an Easter dance) are painted with characteristic body designs and the musicians assemble. In some places, such as Samachique, the celebrants stay in the temple and begin dancing at sunrise; some children are painted as bascoreros and dance throughout the morning; in other locations they do not appear until Glory is sung. In all towns, the mood of the people becomes increasingly expectant and bursts into joy when the resurrection of Christ is announced. Also early, the friends of Jesus, the sontarosi, and in some areas, the pintos or moros, who have tried in jest to take Judas away from the pariseos with the ostensive intention of apprehending him and bringing him for punishment, go into action. A group of men goes into the hills to catch small birds, which will be released later when Glory is sung. In Samachique, the people gather by noon in the temple, where pariseos and morocapas arrive and dance around the bascorero. The pariseos continue to display a daring attitude toward the moros. They eventually enter the temple, knocking their wooden swords on the floor as they march in. The moros arrive with the bascoreros, and the pariseos run away. The pariseos continue dancing around the temple, making Judas dance with them while still challenging the other groups of moros and sontorosi. The birds and ashes are brought in, Glory is sung, and the cloth covering the image of Jesus is removed.

The guards lie down in rows on the floor of the church in Guapalayna and at one point they jump from their recumbent positions and run, making much noise as if in panic. They do it with such vigor that no one dares to step in front of them as they leave the temple, running wildly. At this point, the "friends" of Jesus free the birds and throw ashes into the air; the pariseos knock down and remove the arches that had been built, and the bells toll with the matracas adding to the noise.

Easter Sunday begins with a joyous dance, called a *pascora* or *pascola*. In Norogachi, bascorero dancers have their bodies painted with a rather complex pattern of concentric crosses, dots, and lines in black and red colors. The patterns drawn on the body of one dancer are like negatives of those drawn on another; what is black on one is red on the other and vice versa. A rapid pace with small jumps characterizes the dance with the dancers following each other with steps that are compared with those of the deer during courtship as the male follows its mate. The dancers perform following a path in a figure eight. A small drum and a violin are played as the men dance with their *chanírusi*, or hawk bells, tied to their feet.

At this point, the end of Judas is near. The friends of Jesus finally get a hold of him and take him away from his guards.[3] The hatred for the betrayer Judas reaches a climax, and the figure of Judas is "killed." In Norogachi, he is stoned; in Narárachi, he is repeatedly shot with an old rifle; in Aboréachi and Basíware, he is killed by shooting arrows at him; in Guapalayna, spears are used; and in Papajichi, bayonets are the weapon. The spears go through the straw figure, with fragments flying all over. Finally in Samachique, Judas's arms and legs are pulled one by one until the figure comes apart. The Judasa, Judas's children, and their dog suffer the same fate. All that is left is then burned.

In Samachique, a formal competition of *najarápuri* (wrestling) is held as a closing event. Individuals of the different teams try to take down their opponents holding only the *púraka* (woven belt) of their adversaries. The bascoreros wrestle first; then any member of each band of pariseos or moros wrestles anyone from the opposite party. Then the winners are acknowledged by the audience.

In places where a priest is available at the end of the pascora, a Mass is celebrated and following it the siríame pronounces a special farewell nawésari. In Samachique, the naming of new tribal authorities—siríame, mayoras, ibagápitano, sergeants, and sontarosi—marks the end of the feast.

Once the celebrations end, the people go to their hamlets where they drink batari. In Samachique, the moros and pariseos continue their contests, this time trying to see who can drink more. When all the batari is finished, the men and women go to their homes and shortly after all the events end, they begin planting corn. A new cycle then begins.

End-of-the-Year Cycle

In Norogachi, the end-of-the-year cycle starts on October 12 with the feast of Our Lady of the Pillar, a solemnity known as Pirárochi. This is the first fiesta de matachines of the year. Then comes Warúpachi (Our Lady of Guadalupe's Day) on December 12; in Sisoguichi, December 8, the Immaculate Conception, is also celebrated with a gathering and matachines. Next comes Christmas (Nabirachi) on December 25. Some pueblos celebrate the end of the Christmas season (Pascua de Navidad) on January 6, the Day of the Magi or Reyesi, while others prefer to celebrate Candlemas, Día de la Candelaria or Telachi, on February 2, a feast commemorating the presentation of the infant Jesus Christ in the Temple, to end the cycle.

The matachín dances are an event originated from the missionaries' attempts to Christianize what they considered pagan Indian dances. These dances are performed mainly in church celebrations and, like the other Indian dances, are considered a form of prayer. The purposes of these celebrations are to honor God and the saints and *wekáwari tanepo*, "to beg forgiveness." The steps are relatively fixed and the music is monotonous, slow-paced and played with violin and rattle accompaniment. Drums and flutes, characteristically indigenous instruments, are not played. The matachín dancers wear elaborate costumes, which some observers have compared to North African designs. The main dancers (*monarkos*) wear an elaborate headpiece that in Norogachi consists of five rectangular mirrors glued or soldered together to form a box, with the mirror on top protruding at the back with a tail of ribbons attached to it. Large *paliacates* (handkerchiefs) are attached to the crown and cover the head, leaving only space for the face to show. Large pieces of colorful fabrics are tied to the waist with a púra and hang in front and behind the dancers like aprons, with pointed ends on back and front. The dancers wear pants with long socks to which hawk bells are attached.

A dominant figure of this season is the *chapeyó* (dance director) and several auxiliaries, whose role is to summon, one by one, all the dancers. In important occasions the number can be in the hundreds. The chapeyó is expected to provide batari after the celebration. During the dance, the chapeyó's sharp shouts will direct the dance sequences, changes, and periods of rest. The main chapeyó sometimes crown their heads with deer antlers and carry a whip as a symbol

Matachín dancers in Norogachi. The *monarko* (dance leader) is Erasmo Palma Tuchéchi, a superb dancer.

of authority. As opposed to the tenanche or *fiestero* (celebration sponsor), the chapeyó's responsibility is restricted to the dancers.

Dance Contests, or Encuentros

During the Christmas season, groups of matachín dancers from Norogachi meet groups from other pueblos on the outskirts of the village to engage in a jovial contest. Groups of dancers perform in front of each other, displaying their best steps and challenging each other. At a given moment, the groups run into each other with swinging elbows, pushing and pulling and replacing their rivals in the line. Repeatedly, groups of dancers who have been pushed out assemble in the place that the other group had occupied. When the groups have reassembled, one is declared the winner. More than just a contest, this appears to be a ceremonial reenactment of bellicose encounters in which the pueblos engaged in the past.

SEVEN

Loss-of-Health Conceptual Schemes of the Tarahumaras

The Tarahumaras have elaborated their own explanatory schemes trying to understand the presence and progression of disease. These elaborations, described in the following pages, have provided them with a background to create their healing practices. Assumed reciprocal relationships between disease and natural and supernatural forces are considered. Common animals from their environment, such as snakes, birds of prey, as well as mythical vipers and other beasts, are seen playing a role. Hallucination-inducing agents are sometimes part of the picture. Celestial phenomena, whirlwinds, falling stars, and water streams also impact from their point of view the morbid processes that western society calls illness or disease.

In this chapter, I will illustrate how disease is sometimes viewed by the Tarahumaras as a punishment for neglect of moral mandates or for ignoring cultural precepts. Also, it may occur as the result of lack of respect for objects, plants, or authorities considered sacred. Their symbols, amulets, and some of the ceremonies to ward off illness are described to complement a comprehensive view of the cultural strategies they have evolved to deal with the pain and anxiety of disease or with the reality of the closeness of death.

The elements of this drama are accompanied with Tarahumara language designations and their translation to make the discussion as close as possible to Tarahumara thinking. This device may help to illustrate the conceptual structure of their understanding of disease. It will be noted that sometimes their terms are borrowed from the Spanish language, indicating in some way an acculturation process that has been operating since the arrival of the Europeans to their land, a process that has not destroyed the cultural identity of the Tarahumaras.

The Cold and Heat Complex

A belief common among diverse ethnic groups is that heat and cold act as causes of disease and that a balance of "coldness" and "hotness" within the body needs to be maintained. In examining the concept of the indigenous groups of the American continent, questions are raised as to whether their ideas are pre-Columbian or reflect the ideas of the European newcomers.[1] As is well known, the Greeks, mainly those of the Aulic school of Pythagoras of Samos, did influence the Hippocratic School and later Galen and his followers.[2] They postulated a humoral theory of disease causation that related humors to their relative "intrinsic temperature." All of this is part of the pan-Euro-Asiatic cultural heritage, which incorporated three great medical traditions: the Ayurvedic of India, the Greek humoral pathology, and Chinese traditional medicine.[3] The Amerindian concepts of cold and heat may have been influenced by Europeans who held the humoral tradition. Therefore, an American version of Galenic ideas nurtured by Spanish physicians like Hernández and Steineffer cannot be ruled out.[4]

Other scholars have argued for the pre-Hispanic existence of a cold/heat complex, basing this on studies of Náhuatl (Aztec) values and the Náhuatl vision of the world.[5]

Missionary and mining-center physicians may have played a role in the diffusion of the European concepts of cold and heat. While many Tarahumaras indeed regard some diseases and their putative causes to be cold or hot and consider the herbs used to treat them as either cold or hot, seemingly trying to reach some balance, mestizos seem to give more importance to the "thermal" quality of plants and diseases and the correspondence between the properties of the plants and the diseases to be treated; that is, the properties of the plants are matched with those of the disease. A cold remedy will be given for a hot condition and vice versa.[6] Consequently, the degree to which a Tarahumara applies the concepts of cold or hot to plants or diseases may be considered as an indicator of acculturation.

Within the Tarahumara culture, cold is in general thought to be an agent or contributing factor of disease. The Tarahumaras commonly assume a cause-effect relationship between cold weather and illness, particularly illnesses affecting the respiratory tract, but the role of cold in pathogenesis for the Rarámuri also involves their conception of how cold affects their functioning or that of some of their organs or body parts. Popular beliefs about the Tarahumaras suggest that they have considerable tolerance to a cold environment, given their capacity to live in the climate of their mountainous land, which is cold except in the depths of canyon country. In spite of this inclement weather, most Tarahumara activities, including their family and social life, take place outdoors. The importance given to this climatic feature is reflected in the opening of their verbal exchanges characterized by the greeting phrases *wé ruruwá jipe*—"it is very cold today"—or *wé ruruwá nejé*—"I am very cold." The concept of cold as an important element of morbid processes has peculiarities. For example, for them cold may affect just one organ or part of the body. They for instance may complain about cold vision when this function is impaired or coldness of the bones, the stomach, or other organs when any of these are afflicted.

Iwigara Loss as a Morbid Condition

The concept of loss of the soul (*iwigara*) needs to be described in detail because it is central to the Tarahumara understanding of disease. Basically, the loss of the iwigara implies the separation of a nontangible part or parts of the individual from his or her material body. Loss of the iwigara therefore represents a rupture of the harmony between the *sapara* (the flesh, or the body) and the *iwigara*.[7] Loosening the connection between these elements may be followed by the spontaneous wandering of the iwigara, which according to their beliefs may be carried away by water streams, devoured by a mythical beast, or perhaps driven away by a malignant force. People often report seeing the iwigara leaving the person in the form of a butterfly or a star just before the onset of a malady.

Iwigara literally means "breath."[8] In attempting to translate the Christian concept of the soul into the Indians' language, Spanish missionaries chose words such as the Náhuatl *tonalli* (soul or breath) or Tarahumara *iwigara* instead of neologisms coined in their native languages. They apparently failed to realize that the meanings of these words differed from those intended for evangelization purposes. Nevertheless, syncretisms evolved that have survived until the present among indigenous groups and the Mexican population. As the Spaniards tried to understand Indian concepts such as the "loss of the soul," they applied soul/body or matter/spirit concepts alien to Indian thinking. The Indians, particularly the Mesoamerican groups, assumed the existence of a soul integrated by diverse entities corresponding to different functions interacting between themselves and the material body. This complex assumption leaves room for the possibility that the "soul"—or a part of it—could become detached and that the body, even though somewhat "devitalized," could survive at least temporarily without the lost part. American Indians have been exposed for several centuries to the axiomatic construct that the life principle and the soul are the same entity. They therefore have developed explanations for how the soul may wander away while the person remains alive, adapting their explanations to the Christian beliefs. The Tarahumaras thought of several ways to resolve the conflict; the first one postulates that iwigara is, like tonalli of their Mesoamerican brethren, a complex entity that may disintegrate into one or several parts, which wander away while a vital principle remains attached to the body, keeping it alive. The lost fragments either later return or the person finally will die.

In the second one, informants in Rejogochi postulate the existence of several souls. When the main soul wanders away, some of the other souls remain attached to the body of the diseased person keeping him or her alive.[9]

The third one, representing my own observations in Norogachi, Samachique, and Guachochi, posits that humans have only one soul, but this soul is not life itself but an entity that confers life on the body. Therefore, the constant presence of the soul is not necessary to keep the body alive; only if the wandering soul does not return to the body, the person will die.

An understanding of their concept of illness may be gained through an analysis of the ways in which Tarahumaras describe the phenomenon. Their language allows the speaker to use expressions that would not make sense in other languages; for example, their expression *pe mukúreke binoi*, or "He is dead just a little." To translate such a construct one needs to appeal to concepts as "devitalization," which permit gradation.

Once a person has suffered the loss of the iwigara by any or several of the mechanisms that will be discussed later, he or she will exhibit a lack of interest in the surrounding environment, which may have a sudden or gradual onset and reach different levels; for example, a nursing mother may become detached from her baby to such an extent that she will neglect to feed it. The person afflicted by the loss of the iwigara not only refuses to eat, but if food is forced, will throw up. Then, the patient may become lethargic or unable to sleep. The condition may be compared with a state of profound depression, melancholia, or withdrawal. The individual often gives the appearance of being "thinking all the time." In some cases, a limited form of communication may persist and the individual may express feelings about his or her impending death. In other instances, the person affected by this condition actually will want to die, although suicide is uncommon among the Tarahumaras.[10] On occasion, panic is present, but the most common pattern is a display of total indifference.

Often the relatives of the person affected by this condition suspect and express a strong conviction that the individual has been bewitched (*sipabúa*)— by a sukurúame or sorcerer. Sometimes, envy of the patient's wealth or sexual attributes is blamed for the condition. In other cases, the reason remains unexplained.

The following is an illustrative case. It affected an eight-year-old girl, a student in the boarding school at Norogachi. The girl had joined other children in a picnic by the riverside when she fell into the water. She was able to reach the shore but once out of the water she developed fever and her state of consciousness became progressively obtunded, reaching a comatose state. A team of American physicians, who happened to be visiting the mission clinic, examined the girl and diagnosed the problem as pneumonia. However, in spite of aggressive treatment, including antibiotics, the state of the girl continued to deteriorate. A Tarahumara aide in the clinic suggested contacting an owirúame, or shaman, to "bring the girl's soul back." Reluctantly, but seeing the girl's quickly approaching death, the nurses went along with the idea. Shortly after being taken to the shaman, the girl began to improve and soon she recovered fully.

The anhedonia, asthenia, and adynamia caused by iwigara usually progress to malnutrition and later, death. Parenteral nutrition may be useful to support the patient, but in most cases the remedy is insufficient; only an owirúame may stop the process through the procedures that will be analyzed later.[11]

Dreams and Sipabuma, or Casting Spells

Among the illness-inducing factors thought to cause iwigara loss is the belief that someone acting from a distance may intentionally or unintentionally cause a disorder, evil, or calamity to befall another person. Three terms are used by Tarahumara to name this phenomenon, *sipabuma, sipanema*, and *sukumea*, which mean "to bewitch or cast a spell."[12] The condition thus induced is contrasted with illness from natural causes. For instance, poisoning others with herbs is known as *pátewama*, or *pátetima*; otherwise the sometimes serious intoxication with *Datura* sp. (jimsonweed) could be identified as iwigara loss. Similarly, individuals smoking the herb *riwérame* and blowing the smoke into the face of a runner, supposedly to make him somnolent, are not considered by some as witchcraft or casting spells, while others would believe it was.

Occasionally observed are certain maneuvers that may be considered sipabuma; for instance, during a kickball race it has been described how a deceitful individual may spread flour on the racetrack to make a runner believe that pulverized human bones have been placed in his path.[13] This occurrence may be considered an instance of sipabuma or for that matter may relate to *chu'i* (the dead or their bones), which is discussed later.

An important belief among the Tarahumaras is the conviction that sipabuma may be induced through dreams (*rimuma*=to dream). Usually, the condition is attributed to a *sukurúame* (sorcerer), who, it is assumed, has dreamed about the afflicted person with the intention of causing him or her evil. In general, dreams are considered as forming part of a continuum with ordinary conscious life. It is thought that the shaman is able to abduct the soul of the subject while dreaming. It is also believed that the same effect may be obtained during a dreamlike state induced by jíkuri (peyote), which incidentally has hallucinogenic effects. In some instances of dream bewitching, the sukurúame may dream about the type of poison to be given to the person. In a similar way, the owirúame, or healer, may choose through dreams the remedies to be used for an ill person.

Snakes and Mythical Vipers

Living in an environment where several poisonous species of snakes are common, it is not surprising that the Tarahumaras see these reptiles as playing an important role in the causation of illness.[14] The relatively high incidence of snakebites certainly makes them in fact a prominent source of suffering. Among terrestrial snakes, *chachámuri* (royal or pigheaded rattlesnakes) are considered the most dangerous serpents, followed by *banagéame*, and *okósini*, which paradoxically are reputed to be more poisonous than *sayawi* (ordinary rattlesnakes).

Mythical snakes represent a powerful illness-inducing source. I will discuss below the role of fear or fright (*majawá*) as an assumed cause of illness: snakes usually are frightening. However the process is complex. For instance, finding a serpent while walking on a trail is frightening, but associated with this experience there may be also an assumption that the viper was placed there through evil machinations from a sukurúame.

Other significant beliefs are that mythological water serpents dwell in ponds and springs. These imaginary snakes are large and live in places where the river is deep. Only shamans can see them.

In bygone days, Tarahumaras ritually used to place food offerings for the master of the river, a large serpent named Walúla, a beast that made ugly noises. Every stream, water hole, and spring was believed to have a serpent that "causes the water to come out from the earth." These serpents are easily provoked; homes are built at some distance from the water to avoid them. People, when traveling, do not sleep near the streams due to this apprehension.[15]

One snake that the Rarámuri particularly fear is called *no'pi*, which inhabits smaller water springs, or *pawitibo*. This viper is described as small about one foot long with bright red, yellow, green, and blue rings. Like its larger relatives, it is blamed for eating the souls of little children.[16]

A story is told of a child throwing rocks into the pond making the water snake angry. The snake devoured the child's spirit and he fell ill but he was saved by throwing sacrificial offerings into the water and by taking a potion made from *Buddleia* sp. roots prepared by the shaman.[17]

If a child is not obedient when told *má ku ba*, "Let's go home," he will get sick, his soul will stay in the water, and the water snakes will steal and breastfeed it.[18]

Bad Omen Birds: Koremá, Rutúkuri, Sopichí, and Ginorá

Koremá, also called *goremá* or *oremá*, is a mythological nocturnal bird depicted as small and brightly colored with a flaming tail; it has a harsh cry, flies at night, and lives behind waterfalls. People think of it with considerable apprehension.[19] It is a common belief that the koremá is a wicked animal at the service of witches with the purpose of bringing harm to the Tarahumaras. This bird in some way is associated with falling stars or meteorites. The term *koremá* means literally "a burst of light, a falling star."[20] The koremá supposedly peers through crevices of doors and windows searching for children, looking for an opportunity to steal and eat their souls. The bird also assails those who are outdoors and lie down on their backs, striking their bodies and leaving bruises on their skin. Its victims experience aching, nosebleeds, drowsiness, weakness, fever, and delirium.[21]

Many Tarahumaras, when asked, deny the existence of koremá; however, most acknowledge believing in it when less inhibited during the communal drinking of the bajíachi.

Like many other Indian groups the Tarahumaras see nocturnal birds of prey as carriers of bad omens and harm. Nevertheless, sometimes healing powers are attributed to these birds. The one that seems to stand out is the owl (*rutúkuri*), or in Náhuatl, *tekólotl*. The rutúkuri is known as a bird that sings announcing death. It is not clear whether the singing is believed to cause iwigara loss or if it is perceived only as announcing bad outcomes. The importance of the owl in Tarahumara mythology is considerable; for instance, one of their most important dances is that called rutúkuri or tutuguri, discussed earlier.

Another nocturnal flying mammal with symbolic and illness-causing attributions is the *sopichí* (bat). According to tradition this animal is a transformed mouse. Old mice supposedly become bats. They foretell or presage the arrival of death and eventually will predict the end of the world.

Ginorá, the rainbow, although a celestial phenomenon, is attributed some of the powers of the night birds. Ginorá abducts children, or their souls, as koremá does. A respondent claimed that ginorá might impregnate women, causing them later to become infertile.[22]

Jíkuri, Bakánowa, and other Malefic Agents

The cultural significance and features of the jíkuri as a ceremony will be discussed in a later chapter. In this section, the illness-inducing attributes of jíkuri, the cactus also known as *peyote*, are discussed in relation to iwigara loss. There are several beliefs regarding the consequences of touching or mishandling the cactus or the instruments used to prepare jíkuri potions, such as the grinding stone (*matá*) or other instruments coming in contact with the cactus. There are also consequences attributed to touching or having sexual contact with a person who has just participated in the jíkuri ceremony. All of these consequences are not understood as a physical phenomenon such as poisoning. The illnesses are believed to be the result of offending the spirit of jíkuri or to the use of peyote by a sukurúame to cause harm. Zingg described as an example a case of a teniente, or tribal official, of Samachique.[23] He had been a remarkable light-footed runner in his youth. His prowess was attributed to carrying during the races a fragment of the cactus in his girdle.[24] His behavior was irreverent and he offended the plant, following which he lost his racing prowess. "The spirits of the plant" supposedly made him ill. He became jaundiced and died, overcome by fear.

A more severe disorder often mentioned by respondents is *desgracia del jíkuri*, consisting of respiratory arrest, sometimes seizures, and death. I witnessed how an individual who transgressed the ritual prescriptions of jíkuri became ill after being stared down by the sipáame, or shaman. The person fell to the ground convulsing and losing consciousness. The man recovered after he received cardiopulmonary resuscitation. A similar case was observed in Rejogochi. In this instance, the episode was attributed to *bakánowa*.[25] It should be mentioned that the effects attributed to bakánowa are similar to those attributed to jíkuri.

The illness-inducing power of jíkuri is also attributed to a rock known as *jíkuri re'tera, ru'síwari,* or *sukí*. Ru'síwari is a black or dark volcanic rock with spherical coalescent concretions. These are said by some Rarámuri to be the heads and organs of the rock. The concretions, when separated from the rock, have the size of small marbles. Finding ru'síwari in one's path is a bad omen and interpreted as a sign that a sukurúame surreptitiously has exerted his malefic powers. Like koremá, ru'síwari can also be found transformed into a nocturnal bird.[26] Ru'síwari is described as a rock that, when transformed into a bird, flies at a witch's command to make ill and kill those it is ordered to harm. As a malefic force, it penetrates the victim's body and slowly makes him ill. Only a *waníame*, or healer, may have the force expelled from the body.[27]

Death and the Departed: The Chu'í Taboo

Before I present a complex set of beliefs surrounding death and the dead, I should briefly mention some of the Rarámuri ideas about stability and poise.

For them, these qualities are very important. This is apparent even in their graceful stride as they walk or run. It has been said that there are few things that make a Tarahumara laugh more than seeing someone fall down.[28] In spite of this comment, a fall for them has broader consequences than those of the trauma. It challenges their pride in their physical prowess and stamina. Furthermore, they believe that the ties that fasten together the iwigara and sa'para may become loosened. This is a particular concern if the fall is close to a water stream where the iwigara may wander away and be carried away to the sea by the stream.

Contact with or the mere sight of individuals who are next to death, as well as looking at children with deformities, are supposed to cause ill effects. The house of a deceased person is considered no longer habitable and is often destroyed. Chu'í—that is, human bones and things connected with death or dying—is thought to be a powerful illness-causing agent affecting people and animals. Sheep and goats are considered particularly at risk if they come in contact with bones found in caves containing human remains. It is claimed that bones have a salty flavor that animals like to lick. As a result of this contact, they experience *engalgamiento*, an illness causing progressive apathy, loss of appetite, weight loss, and wasting. For this reason, shepherds (*neséroame*) clean the caves where their flocks are to spend the night, using long sticks to pick bones if found and dumping them in a creek. They claim that even just handling the bones with the sticks might cause faintness.[29]

Breathing dust coming from the caves that contained, or were thought to have contained, bones may cause a severe illness. This taboo is illustrated in a curious old story describing how the leader of a kickball race took a right human tibia that he had unearthed. He set it on the ground of the cave, placing in front of it a container with tesguino and crockery containing food. On both sides, he laid one of the kickballs used in the race and in front of it planted a cross. The food and corn beer appeared to be offerings to the dead, asking them to help win the race by weakening the adversaries.[30] Contact with human bones, it is believed, induces fatigue; to this end, some bones may be brought to the racing track to be hidden in places where the competing runners will have to pass.

The actions of chu'í are not limited to the direct contact with bones. Chu'í, in the guise of a ghost, is said to wander around frightening people, especially relatives of the deceased who have failed to complete the *nutea* or other ritual obligations. Furthermore, chu'í may cause harm to the Rarámuri without any reason other than mischievousness.

Closely related to chu'í are also the beliefs about *ardimientos*, or will-o'-the wisp (*ignis fatuus*). These fires, according to the Tarahumaras, point to places where treasures are buried or hidden. Digging for such treasures sometimes lead to human bones, which as mentioned are greatly feared. Those who find an ardimiento have to share the secret of its location with a mestizo who supposedly is not vulnerable to the curse and may be able to reach the hidden treasure site.

Whirlwinds (*ripibiri, ripiwiri*, or *ripiwari*) deserve special mention; as a source of illness and death, these phenomena are greatly feared by the Tarahumaras.[31] Those caught in a whirlwind may become severely ill or even die. The ripiwiri causes bruises, wounds, and pain.[32] It is said that those trapped by the wind may experience a sensation of coldness throughout the body and intense aching severe enough to make the person cry. The dead bodies of the victims of this wind give the appearance of having been physically overwhelmed by a great force.[33]

There are cultural prescriptions and rituals which, failure to obey them, they believe, will result in illness. For instance, when they move to a new place and take with them their stones to grind corn (matá), they follow a careful washing routine. Failing to do so may be followed by ominous consequences. The person could "die" from the waist down. Unable to move his or her feet, the lower extremities will rot. The individual may be able to breathe, talk, and move his arms but his eliminatory functions will not work and will perish. Those affected may survive a few weeks but no caretaker will dare to help them because of the smell of the rotting limbs.[34]

Making amends for wrong moral actions and violations or omissions of ritual prescriptions may prevent the iwigara loss. These forbidden actions may consist of expressions of envy, jealousness, or hatred. A particularly serious violation is ostentation of wealth or sexual potency. Amends to all these wrongs are particularly important if the offended person is a sukurúame or a wealthy and influential individual capable of hiring one. An ordinary person may be the source of iwigara loss if he dreams ill of another individual or, even worse, if the person appears in the dream as being dead. It is considered to offend a deceased person if the celebration or completion of the required series of nutea and Jíkuri are not done.

Adaptive Function of Traditional Beliefs, Amulets, and Taboos

It should be noted that traditional beliefs and taboos might have an adaptive value within the framework of the challenging environment in which the Tarahumaras live. The attribution of evil power to water streams, nocturnal birds, or distant places leads to practical safety rules. Increased precautions will then be taken while traveling along water streams that may in fact become mortal traps in flash floods. The avoidance of wandering at night and watchfulness of the predatory behavior of strangers are proactive. Most of the activities of this people are carried on outside; recommending not sleeping outdoors helps to avoid bites from rabies-carrying bats and malaria-infested mosquitoes, particularly in the barrancas.[35] Anxiety is the signal of danger and fear may become a protective device.[36] From an early age, children are taught through story-telling to avoid water springs and streams, making them aware of possible risks. The watchful behavior incorporated in their stories has been criticized as contributing to support a system of beliefs based on fear, but its sometimes adaptive usefulness cannot be ignored. The Tarahumaras often seek shelter for themselves and their flocks in caves; the chu'í belief encourages removal of detritus that may carry disease. If ru'siwari appears in their path, it will be avoided and others will be warned about the presence of such a hazard.

Objects serving the function of amulets are considered protective. A crucifix (*sukristo*) is the distinctive symbol of the owirúame and his power over the supernatural.[37] Necklaces made of *batagá*, pendants of *sitagapi* and *wasárowa* also fulfill this purpose, as does a wooden cross worn as a pendant. A handful of ro'sábari or a chunk of wasárowa under the pillow is thought to protect a person from chu'í and other evils. Tobacco (*wipá* or *wipáka*) is considered a powerful repellent of snakes and also a remedy for snakebite.

A curious protective device against the evil night birds mentioned above is to place a mirror on the bed's headpiece or tie it to one's neck when resting outdoors. The ominous bird supposedly will see its image reflected in the mirror and will fly away.[38] As a maneuver of appeasement, Tarahumaras consider making flattering comments about a bird when witnessing a falling star, believed to be the tail of the bird in the nocturnal sky.

Once a feared illness develops, the knowledge and skills of the owirúame are considered necessary to cure it. He is supposed to track the lost iwigara and find out if it has wandered away or was carried away by a magical bird, snake, whirlwind, or water stream. The owirúame is supposed to find the lost soul in a dream and return it to its owner. He usually acknowledges that he has located the iwigara in a distant place, exhausted from its long journey. He prepares herbal potions and/or ritual treatments destined to reinforce the body-soul ties. His advice on where the healing herbs have to be collected, by whom, and at what time appear to be more important than being specific about the plant that will be administered as a remedy.

Culture-Bound Illness Recognition and Progression

Within the Rarámuri system of beliefs, attention is given to the progression of morbid conditions. They have noted how these are not static; in their own terminology, they walk—*enama, eyénama, enároma* (to walk)—from one place in the body to another or from person to person. This concept reflects their concern with features such as contagiousness and tendency to relapse, including the endemic or epidemic characteristics of the illness. Diseases such as colds, skin sores, or boils are therefore said to "walk" within one person or to others, or in Spanish *andancia*. Sometimes this concept implies severity. For instance, a walking disease is considered to be short-lived while a disease that does not walk may be considered fixed, aggressive, and serious. Diseases also may "fall" or "hit." Hitting implies that the disease is severely damaging or repetitive; thus, disorders presenting themselves in the form of bouts or attacks are considered "hitters." Diseases "fall" when they appear on top of a pre-existent disorder, such as episodes of *taranta* and *alferesía*, which we will discuss later. A common cold may hit the individual; *pasmo* falls onto a wound. A malignancy may either hit or fall on the affected person if the disorder strikes on top of a pre-existing condition. Native healers therefore conceive diseases as having their own character.

The Mystery of Susto or Majawá

The terms *majawá* or *susto* are used referring to one or several conditions related to experiencing fright; however, this does not imply that the condition resulted just from the exposure to a fearful situation. Majawá is seen as a complex, somewhat mysterious condition. For example, when it is stated that "the child is sick because a cow scared him," the statement implies that beyond the concrete declared facts the idea that "something" supernatural was associated with the experience. This understanding has some similarity with the effects attributed to the fright induced by the presence of a deformed child or a sick person about to die. It is not just that fear has induced the disease, but simultaneously some unseen and mysterious force is believed to be operating. This imponderable element is sometimes recognized and described by the ill person or family and reported to the healer, or owirúame. At other times, the effects are just suspected or even consciously denied or mentioned as circumstantial or coincidental information. The following example illustrates this: A lady who was suffering a severe exacerbation of her joint pains, probably arthritic, mentioned that the acute pain began as she was crossing a stream along a stepping-stone bridge when she saw a water snake crawling in front of her. In reporting this event, she was not admitting that the snake had frightened her. She was just highlighting the association between this experience and the onset of the symptoms of illness.

Majawá may be associated with *surabuma*, a sensation experienced and described as if the heart has been pulled away, or *surawima*, as if the heart has left the person.[39] These terms apparently try to describe subjective sensations experienced in the chest area, probably elicited by a fast heartbeat. Majawá affects women and children frequently; in children, the manifestations of majawá, or *susto* as called by the mestizos, may consist of frequent loose bowel movements, continuous crying, irritability, sleeplessness—*tasi gochire rokó*—and pain. In other instances, there is apathy, general loss of strength, and listlessness. An important sign associated with this condition is depression of the fontanel, or *caída de la mollera*. Apparently, the term *majawá*, fear, was first used to define an emotional response and later it was considered a more complex set of phenomena that included subjective and physical symptoms, as well as a mystical or supernatural interpretation.

As an illness, majawá is a complex set of symptoms that may involve gastrointestinal and respiratory symptoms, as well as psychological symptoms such as apathy and withdrawal. While conventional medical treatment may alleviate some of the physical and psychological symptoms, majawá will progress to a state of severe depression if not addressed by culturally oriented healing approaches.

When the Tarahumaras travel, especially when going to a distant place, they fear being overcome by the symptoms of majawá, also called *majariki*. Those affected remain conscious but appear to be in deep thought, somnolent, and weak. The person often reports hearing imaginary sounds like cowbells and whistles. The general demeanor is apathetic and detached. Commonly mentioned is having dreams of being in unknown places with unfamiliar persons. To prevent this condition, travelers carry a fragment of the bark of *wasárowa* or drink a remedy prepared by scraping its bark in a glass of water. A cross may also be kept as a protective amulet, especially if possible exposure to chu'í (human remains) is also feared. If majawá is already present, other herbal remedies are used. In some cases, they resort to a body-painting ritual. The owirúame draws crosses or solar designs, mainly on the ritual points of healing: the forehead, ears, nape, shoulders, chest, back, abdomen, testicles, thighs, limbs, and heels, in this order. The painting commonly used is *sitákame*, a powdered ocher-colored rock. They may also resort to rituals, such as the Jíkuri ceremony, or bakánowa, an herb with mystical and hallucinogenic properties.

The outcome of a person affected by majawá will depend upon the type of syndrome the afflicted person is presenting and its severity. A child with gastroenteritis or pneumonic syndrome may respond well to conventional medical treatment, although it is important to remember that gastroenteritis and bronchopneumonia are respectively the top two causes of mortality among the Tarahumaras.

Towita or Caída de la Mollera

A cluster of symptoms in children, such as weight loss, sunken eyes, dry eyes and mouth, irritability with constant crying, rejection of food—spitting it or throwing it out—depression, and insomnia are often identified by native healers as *towita* or, in Spanish, *caída de la mollera*. A depressed fontanel is identified as a sign and the cause of the illness. If the fontanel does not appear depressed, the same condition is still considered, arguing that the fontanel is sunken from the inside, or *patzana*. Towita may however be considered by the native healer at times as a symptom or an indicator of other conditions, such as majawá or *romima* (*empacho*). Interestingly, in this situation, a mestizo healer (a *curandera*) may be called for help rather than a Rarámuri owirúame who also claims the skill to treat this disorder. The treatment consists of trying to lift the depressed fontanel by pushing the palate of the child with the fingers or suspending and shaking the child upside down. A patch may be placed on the fontanel site after the "lifting" has been done to keep the lifted fontanel in place.[40] Herbal remedies, such as *basigó* and *boldo,* are also frequently used. The healer seems to know that it is important to give fluids to the baby. Needless to say, many children do not survive these approaches since the underlying process is not addressed.

Cosa, Bola, or Latido

A condition experienced as if an object or mass is fixed or moving inside the body, in Spanish is called a *cosa* (a thing) or *bola* (a ball).[41] If difficulty breathing is present, this is attributed to the compression of the mass on the windpipe. A heartbeat-like pulsating sensation, *latido,* is sometimes reported.[42] These problems are considered to be very serious, particularly if the mass seems to be occupying a large area in the abdomen. If it feels localized in a small area, it is said that the condition *napawire*; that is, "got together." The person affected may report "weakness," described as a feeling of emptiness in the abdomen different from hunger, or a sensation of "butterflies" if a person is fearful or anxious. If the mass is mobile, it may be interpreted as a worm moving inside the body. Similarly, the consequences of a cosa, if it is identified as a "stone" will also be like that of ru'síwari. Finally, when there are multiple masses and they are of a small size, the condition is defined as "inside boils" or *wawana.*

This syndrome appears to be sometimes a subjective experience, while in other instances it may be a manifestation of serious underlying disorders such as tumors or even aneurisms. The native explanations may relieve some of the anxiety associated with being ill but obviously do not address the cause of the disorder. The explanatory ruminations of the owirúame may even delay effective treatment offered by conventional medical care. The handling of these conditions gives opportunities to see how those affected and their families sometimes try to reach conventional medical attention.

Bawana or Chá

The terms *wawana* or *bawana* are applied to a great variety of skin disorders recognized in conventional medicine as scabies, skin infections, or impetigo. Native healers commonly consider wawana as an andancia, acknowledging its contagiousness and tendency to spread from one part of the body to another and from one person to another person. The term *chá* is also applied to almost all types of skin lesions, but *wawana* is usually the preferred designation. The verb *bisóinama*—"to produce a discharge or pus"—is frequently applied to these conditions. Interestingly, skin disorders appear to be less common among the Tarahumaras than among people from other cultures in the area. This may be due to the fact that the Tarahumaras are more likely to live in isolation and widely scattered areas while the mestizos live in groups or communities. However, this may be just a partial explanation since these infections are also uncommon among Tarahumara children living in boarding schools and among those who live in urban areas. Internal wawana (*wawana patzá*) include a relatively more complex variety of conditions ranging from throat infections to vague discomfort in the abdomen, limbs, or head. No specific healing approaches are applied to these conditions but various ointments, poultices, and beverages are used. A poultice prepared with fresh *chu'ká* and healing ceremonies, such as *wikubema* and *morema*, described later, are sometimes used.

Kuchíwari

Kuchíwari, kuchinówari, sikówari, and *resagí* are terms labeling any disease caused by worms, whether observable or assumed to be present. *Kuchíwari* designates the maggots commonly present in decaying flesh. On the other hand, the term *sikóari* or *sikówari* is used to refer to tapeworms, pinworms, and other intestinal parasites.[43] *Kuchinówari* is applied to joint pain they believe is caused by worms. The Tarahumaras believe that flies, mosquitoes, and worms enter the body and cause fever and diarrhea.[44] Some informants mentioned that worms make noises inside the abdomen, cause stomachache, and are constantly "making more little ones." They observe that fish are frequently infested by long, white worms. *Resagí* is a condition that apparently results from infestation by the nematode *Trichinella spiralis*. Bennett reported that *resagí* is a "small bird or animal about the size of the end of the little finger" that can cause this disease, especially when sent with that purpose by a person of "bad character."[45] Resagí, they believe, may be driven away by throwing ashes or *chiltipiquín* in the air. Sometimes kuchíwari is considered a form of *wekarí* (venereal disease).

Parasites of wide distribution around the world such as *Enterobius vermicularis* (pinworm), *Ascarides lumbricoides* (roundworm), and *Taeniae* (tapeworm) are endemic in the area and are also designated kuchíwari or sikówari.[46] Larvae from *Musca domestica* sometimes grow in open wounds, particularly in the axe injuries common in the area are also called *kuchíwari*.[47] It is important to mention that these parasites have also mythical significance: worms are considered to be materialized evil spirits or forces. Tarahumara wisdom recommends staying away from water springs and streams in order to avoid contracting the kuchíwari. The kuchíwari is sometimes "extracted" by the maneuvers of a healer (*waníame*). He applies his mouth directly or uses a hollow cane to draw out the worms believed to be there. These infestations are also treated with herbal remedies, such as infusions of *amapolita* (a small poppy) and ritual ceremonies.[48] Some of these herbal remedies, such as *epasote* (epazote) and turpentine do have demonstrated antihelmintic efficacy.

Other Conditions

Sexually transmitted diseases are designated *wekarí* or *bikarí*. The words derive from the verb *bikamea*, "to be rotten or putrid." The terms are mainly applied to gonorrhea, but occasionally other diseases receive these names, particularly if the genital organs are affected. A common belief is that the person has wawana, or abscesses, inside. These diseases are considered shameful and serious prejudices often delay obtaining care. A generalized belief is that wekarí results from sexual contacts with mestizos, further stigmatizing these conditions. Sexually transmitted diseases are thought to be especially common among the Rarámuri of the gorges; the ones from the Sierra avoid intimacy with both the Mexicans and the Tarahumaras from the barrancas.[49] Increasing interactions with the urban populations of Parral and Chihuahua and greater migration to these cities has contributed to this problem. Although occasionally ritual treatments are applied to these problems, herbal remedies are more commonly employed.

Romima or *awakátzane ropara* is a condition usually affecting children but which also may afflict adults. It is attributed to food chunks, particularly fibrous, flaky, or membranous chunks, sticking to the walls of the stomach or bowels. This illness has similarities with what the mestizos and other Mexicans call *empacho*.[50] Commonly blamed is consuming poorly cooked beans or tortillas, "impure" foods, and exposure to cold or fright. Also contributing to empacho is the failure to follow rules, such as eating proscribed foods while menstruating or noncompliance with other ritual precepts. A common belief is that it may be caused by a mass of hair stuck in the stomach.[51] The supposed illness-inducing role of anger is commonly considered.

Although usually accompanied by diarrhea (*witabúa*), romima sufferers may also be constipated and colicky.

Avoidance of impure foods and the circumstances mentioned above are advised, as well as careful cooking of beans and tortillas. Romima is usually treated with herbal remedies or with healing massage (*sobada*). Among the many herbal treatments used to treat romima, those with a bitter flavor are thought to be the most effective. The beneficial effects of bitter potions are believed to be also effective for other conditions, such as wekarí, and respiratory disorders. Some common remedies such as *zempoal*, *rasó*, and *cáscara de granada* are also utilized.

"Cutting the empacho" is a technique that belongs in the so-called mystical surgeries. The patient's abdomen is anointed first with vegetable oil or lard; then the healer applies a sharp knife to the abdomen of the patient and shaves off the oil—cuts the empacho. The knife is usually moved in crosses and circles over the sufferer's abdomen.

The Tarahumaras frequently seek help from native healers or from conventional medical practitioners for conditions designated *mo'ochí okorá*, *busichí okorá*, and *surachí okorá*. The first, *mo'ochí okorá*, literally means "pain in the head." This would suggest that headache is the main complaint. However, the situation is more complex. When the Rarámuri asks the healer, "Is my head hurting, perhaps?" observers not part of the culture may find the statement puzzling. However, it is important to realize that actually the person is experiencing general malaise or a general feeling of sickness that he believes originates in the head. The term therefore does not refer just to a symptom but to several subjective sensations or signs experienced as feeling very sick. This condition is particularly striking when the sick person is a newborn child; the parents may bring the child and insist that the infant is suffering from a headache, or mo'ochí okorá. Actually, further questioning may show that they have in mind a group of manifestations such as constant crying, stuffy or runny nose, restlessness, abdominal cramping, insomnia, and perhaps fever. With older children or adults, the designation even while pointing to the head may refer to abdominal cramps and other symptoms, which if more severe may suggest to native healers conditions such as majawá or even of iwigara loss.

Busichí okorá (eye pain) is a broad term to designate all eye ailments of irritative or infectious nature. Their use of open fires to cook in houses with defective smoke ventilation makes eye irritation very common. During the dry season, dust is a pervasive eye irritant.

Surachí okorá (chest pain) is also a very common complaint. In most cases, it is a symptom of muscular source rather than of cardiovascular disease. Actual heart disease appears to have very low prevalence among Tarahumaras.[52] In addition to several herbal therapies, bathing in a hot water spring has gained popularity to treat surachí okorá.

Pasmo is a Spanish word originally applied to lesions observed on the back of horses due to abrasions caused by the friction of the saddlecloth and the saddle. The Tarahumaras and their mestizo neighbors have borrowed this word to name a disease that "falls" upon a pre-existing condition; therefore, the expression *se pasmó* is applied to a wound, a bruise, or a blow that becomes infected. An injured area that becomes swollen, bluish, and hot is said to be developing pasmo. If an abscess develops, the condition is then called *madura* or *redonda*.[53] In some instances, these lesions become infested by mites and are considered to be complicated by *kuchíwari*. Recently, the Spanish term *infección* has displaced the term *pasmo*, particularly among the more acculturated Tarahumaras and mestizos. In some areas, however, the word *pasmo* is now reserved to describe the secondary infectious or traumatic complications presented by a closed injury while *infección* is used only for the changes observed in open wounds.

Cáncer is another disease that may "fall" on a pre-existing condition. A lesion caused by pasmo may also be known as *tumor*, which patients and healers use interchangeably with the term *cáncer*. In regard to the conditions just described, it is important to note that it is very common for the persons suffering from these problems to believe they are the result of the evil influence of a hidden enemy wishing harm to the person. It is a common belief that sexual overindulgence may cause pasmo. Sexual abstinence is considered very important to avoid pasmo in a traumatized patient. Certain foods, such as pork and salty items, may also be blamed for pasmo. Chile (*korí*) is to be avoided although this does not have a rational basis. Although there is not reliable support for this, the hot and spicy quality of chili is said to produce pain on the affected area and to slow the healing process. A variety of herbal treatments are applied to traumatic lesions, ranging from animal fat and turpentine to poultices of *hierba del indio, hierba del cancer,* or *cha'gusí.* Madura calls for the use of *parches* (patches), a practice discussed in a later chapter.

Curiously, I have not found a Tarahumara word to describe convulsive seizures; instead, the archaic Spanish words *alferesía* or *ataques* are commonly used.[54] Seizures are thought to be the result of witchcraft or failure to observe incest taboos. They simply report seizures as falls, shaking, and the like. This contrasts with the mestizos, who have several words and descriptive ways to present this phenomenon. *Taranta* is another condition also named using archaic Spanish terminology, designating a state of somnolence or confusion, sometimes associated with other signs, such as fever or the appearance of serious illness.

Mental Disorders

Mental disorders are considered rare among the Tarahumaras and are only casually mentioned by scholars.[55] However, there is a Rarámuri word for mental disorders or becoming insane: *lowimea.* The Tarahumaras do consider mental illness an important condition having a profound effect in their lives. Within the context of their daily life, when mental issues are addressed they refer primarily to behavior that is socially disruptive. A person is called *lowíame* (madman) if he displays disrespectful or aggressive behavior. Little reference is made to delusional, perceptual, or hallucinatory manifestations. The relative lack of emphasis given to these problems is probably related to the way the Tarahumaras consider themselves as members of their social group. A cornerstone of their social behavior is the profound respect for the actions of others, no matter how deviant their behavior appears to be, providing it is not disrespectful or aggressive. The lowíame, withdrawn person, or depressed person is assigned activities and roles, which the group feels the person may be able to perform. The word *lowíame* often is applied to individuals who show physical agitation. As a matter of fact, the verb *lowimea* means "to shake or to be shaken in a physical sense." The relative form, *lowema,* is "to make agitated." Interestingly, this term is also applied to the effects of plants containing neurotoxic poisons, which they pour into pools of water to stupefy fish. The fish exposed to these poisons act agitated before dying. A deluded or hallucinating person also may be called *lowíame*; however, the term is more frequently used to address the loose, idiosyncratic individual displaying behaviors that appear to be irresponsible. The lowimea state, or madness, is thought to be caused by a variety of factors, among them, "offending" the ritual herbs with hallucinatory effects, such as jíkuri (peyote). Certain rules and deference or respect to them are required by their traditions. This also applies to other plants known to induce hallucinatory states, such as *tikúwari* (jimsonweed) and bakánowa. Agitation however remains a prominent feature. An informant reported, "Jíkuri will drive the one it hates crazy and make him go running with a big headache, especially when it has been consumed with batari." Taking jíkuri under direct sunlight or during the summer months is also outside the Tarahumara cultural context and will cause madness. Bakánowa consumption is said to induce silly behaviors, aggressiveness, hostility, and suicidal actions.[56] According to stories, bakánowa may cause insanity because it starts making the person feel well but then will become rabid and it may even lead him go become a bandit or a murderer. It is also mentioned that the person will hallucinate and become paranoid due to long-term use of bakánowa.[57]

The Tarahumara verb *lowetzama* indicates a type of alienation that makes one wander about without goal or direction. Other expressions, such as *lowírima* or *lowíwama*, mean "to induce craziness." *Wichuwá*, a synonym for *lowíame*, is reserved for male madmen while *wichulá* is used solely for insane females.[58]

The fruit of a plant resembling *jaltomate* is called *wichíwari* because eating it may cause a transitory state of madness.[59] Alcohol intoxication is referred to as if the individual was a true lowíame. While agitated states are considered madness, depressed conditions are thought of as devitalization or partial death, as in the loss of the iwigara. In a society such as the Tarahumara, characterized by peaceful and respectful relationships among people, poor control or overt expression of anger is considered a serious moral vice, many times considered mental illness. *Oparúame* is a term used to describe anyone from people with a quick temper to a person possessed by wrath. Such a person will be ostracized by the community. An important indicator of the intolerance of expressions of anger is that while the Tarahumara ideal norm of behavior gives the appearance of being passive and contemplative, Spaniards and Mexicans (*ichápochi*), because they behave more aggressively, are considered evil individuals who originated in the underworld where the devil dwells. In fact, I have heard several times that they are actually sons of the devil.

EIGHT

Rarámuri Healers

Tarahumara healing activities take place at different levels and in different settings. When ill or diseased, the family or the community may call an individual clearly defined as a healer within their culture, such as a shaman, or owirúame. Often, they resort to those among themselves known to have some knowledge of healing practices or capability for healing. At any rate, the first level before reaching for outside help is the individual. Self-healing skills are common; anyone who has trekked with the Rarámuri through the mountains and has spent the night outdoors is likely to have observed a Tarahumara sitting on a rock warming himself close to an open fire and pulling a handful of herbs, without moving from the place he is sitting, and placing them in a crock with water to boil. He may have pulled the herbs from his *moruka* (bundle) to prepare a medicinal tea. He may do this without telling whether the drink is a healing potion or just a pleasing beverage. Most Tarahumaras seem to be familiar with several curative herbs. However, if asked what the herbs are for the answer may be "to take them to Bakochi (Chihuahua City)." Tarahumaras are resourceful herbalists who readily share their remedies with their mestizo or *ichápochi* neighbors.[1]

The next level for healing is the family. In practically all dwellings, one may find a collection of herbs carefully tied in bunches. Families have their own remedies and seldom seek outside advice. An inventory of these herbal remedies is described in another section of the book. Another resource at the community level is the wise man, a person considered knowledgeable about illnesses, healing techniques, and herbs, but not necessary a healer. The wise person often is the one who advises whether a native healer or a conventional medical practitioner should be called. This person therefore is appreciated as a resource in the community. In those instances in which a traditional healer or a conventional physician is called, upon arrival they will find that a wise man is already there. The mature age of the wise man is a factor that may increase the respect paid him and his credibility; however, in many places relatively young people enjoy a high standing. Trust is greater if this individual is a family member, an owirúame apprentice or if he has witnessed childbirths, sicknesses, or deaths. The number of wise men in a hamlet varies, depending on the size of the community. The wise men usually collaborate with each other and rarely exhibit jealousy or competitiveness. This is only part of the answer to the question formulated by Mull and Mull about who makes a decision to use scientific health care.[2] Other reasons include level of acculturation, familiarity with the nearby clinic, hospital, or doctor through prior care, and so forth.

The Shamans

Shamans are a group of particular interest. They play the role of spiritual leaders and healers. Earlier in history they also had considerable political authority and assumed important leadership roles. As mentioned in our historical review, these men led the Tarahumara rebellions in the 1600s. The shamans have maintained their high social status and considerable power even after they renounced along with their people using belligerence as a political tool. They have remained influential, leading ritual and propitiatory ceremonies, dances, and functions intended to avert evil. They therefore also play a priestly or ministerial role. Without their shamans, the Tarahumaras would feel lost, both in this life and in what they expect after death.[3] The status of shamans in the community is higher than that of any official or the wealthiest man.[4] They enjoy many privileges, courtesy, and deference.

Si'páame

The *si'páame*, or Jíkuri shamans, are considered the most resourceful with the most effective healing skills. They have the privilege of using the hallucinogenic cactus jíkuri within the context of the Jíkuri ritual. The powerful healing ceremony performed by the si'páame is known as the *si'pimea* ("rasping") and cannot be performed by other shamans. The si'páame conducts the ceremony, as one of the healing ceremonies. The activities of these shamans will be presented in more detail in the description of the Jíkuri ceremony in another section of this book.[5]

Owirúame and Dream Healing

The owirúame ("the one who heals") is considered to be the good shaman; this in contrast with the evil shaman, or *sukurúame*. One of his important roles is healing; however, his ministrations are also aimed at preventing illness. For instance, he "cures" children to prevent them from getting sick. Moreover, his healing is not limited to humans; he may direct his ministrations to animals, to increase the productivity of fields, and to insure the soundness of celestial bodies—the sun and the moon—when they seem to decline and "require help." For instance, if their brightness seems to be fading, such as during eclipses. The owirúame are always men.[6] Most are mature; only occasionally are they relatively young or elderly. The owirúame's reputation increases among people if they feel that he is obtaining favorable outcomes. Owirúames sometimes appear to compete with each other trying to gain the interest of the community.[7]

Among the owirúame's healing methods, the use of dreams is particularly striking; owirúames claim to have the power of dreaming at will to examine their dreams' content. Through this, they can determine the whereabouts of a lost soul in instances of iwigara loss. They claim to have the power to compel the soul to return to its owner. Occasionally, they prepare herbal potions to assist in this process using commonly known herbs or some herbs considered new and magic. They may use the dream to later "guide" the ill person to a particular site where plants with healing properties may be found. Incidentally, some of the preparations, such as jíkuri, are hallucinogenic and therefore induce dreamlike states.[8] After ingestion of certain plants, the shaman is supposed to see and "travel" to locate the evils responsible for certain illnesses.[9] The owirúame are considered knowledgeable herbalists. Their collection of herbs may include hundreds or just those that are favorites. Some owirúame resort to healing devices such as the sweat bath or singing—*wikaráame*—to heal the sick. The cross is their distinctive symbol and amulet.

Owirúames are highly respected in the community and when called to assist, no effort is spared to make their stay pleasant, fulfilling any requests they might make. There is a negative side to this. If an owirúame is displeased or offended, it is believed he may turn his powers against the people and become a sukurúame, or sorcerer. This may be part of a dualistic conception of the world: the one who can heal others can also cause them disease and death.

Waníame

The *waníame* is another class of shaman noted by the use of a healing technique consisting of sucking the affected area using a hollow cane or directly placing his mouth on the affected part of the body.[10] Through this process, they extract rocks, worms (*kuchíwari*), *ru'síwari* (the Jíkuri stone), or other noxious objects from the body of the afflicted person.[11] These objects are thought to have been inserted in the body by some mysterious entity to cause harm. This therapeutic technique is similar to that practiced by the *techichinaliztli* of other Mesoamerican groups. The waníame may scan the whole body or apply his cane following an established ritual guide that identifies healing points. Once the waníame succeeds localizing and extracting these foreign bodies, he will display them to those present and burn them in a flame.[12] Many realize that the waníame may have had these objects hidden elsewhere, claiming later to having found them in the body of the ailing person. Awareness of this deception does not seem to affect the popularity of this healing maneuver.[13]

Towita

The *towita* is a healer who limits his activities to lifting depressed fontanels through the healing maneuvers described earlier. The techniques they practice vary and range from lifting the baby by its ankles and suspending it upside down to pushing the palate of the infant with two fingers, trying to elevate the depressed sunken fontanel. The towita healer will also prepare herbal potions.

Sukurúame

Sukurúame is the name applied to a class of shamans believed to possess witchcraft powers. It is claimed that they may even cause iwigára loss. Their role is the counterpart or antithesis of the owirúame. The sukurúame is considered to be in communication with or possessed by the devil. The allegation that someone is a sukurúame implies that the person is using evil powers to harm another individual and may therefore be subject to punishment. As an aside, it should be noted that the practice of witchcraft is also attributed in some instances to anyone known to be a shaman. Sukurúame are thought to be capable of transforming themselves into beasts, like the myth of the *nagual* (person transformed into an animal) in other Mesoamerican cultures. The sukurúame are usually male, aged, and supposedly operate in a limited area. Some of them are considered more powerful than others. As a payment for their wicked work, they request animals, maize, or money. They are thought to possess talismans that make them more powerful; among these objects are *sukí* or *ru'síwari* or plants such as peyote. The sukurúame who use peyote are very much feared.[14] Paradoxically, a cross, the talisman that supposedly empowers the owirúame, also enhances the malevolent strength of the sukurúame.[15] It is claimed that sorcerers, to make others ill, may resort to eating the puffball fungus *pata de perro* (*Lycoperdon* sp.) to be able to approach people without being noticed.[16] The attributed evil power is not restricted to affecting people; it is also used in animals and inanimate objects, such as homes, fields, and even celestial bodies.

It is of interest to contrast the mestizo healers with those of the Tarahumaras; this illustrates an important cultural divide between people who live close to each other. The mestizo healers are the curanderas, *sobadoras* (masseurs), and *comadronas* (midwives), most of them female and designated with the Spanish words just mentioned. In the western section and canyon country, the influence of the Sonoran and Sinaloan curandera traditions are common. Some of these native healers have become rather notorious, such as la Señora de la Piedrita, in Urique, whose healing powers were attributed to a rock shaped like a hand, and the famed sobadora Señora de Cochibampo in Sonora or the curandero from Los Capomos in Sinaloa.[17] In spite of their visibility and location in the land of the Tarahumaras, their role and attributions therefore differ considerably from those of the Tarahumara healers.[18]

NINE

The Jíkuri Ceremonial Complex

Jíkuri is an important and complex ritual deserving detailed description given its religious and healing significance. The name of the ceremony comes from the name of jíkuri, or peyote (*Lophophora williamsii*), a cactus with hallucinatory properties that is native to northern Mexico and the southwestern United States. This plant, described later in more detail, plays an important role in the ceremonies. The descriptions of the rituals presented here derive from my direct observations. The information was collected in the area of Norogachi. Four of the events, those which took place in Romichi, Bakasórare, Turuséachi, and Santa Cruz, appeared to be primarily funeral ceremonies, while others, in Wakarichi, had apparent healing purposes. Also incorporated in the narratives is information on twelve other events obtained from several Tarahumara informants living in the Pawichiki-Ruruséachi area. This section also examines the anthropological reports made by the ethnologist Claus Deimel, who studied Jíkuri ceremonies in Narárachi. Data from Lumholtz, Bennett, Pennington, and other scholars who covered Jíkuri celebrations in the upper Río Conchos and writings of Father David Brambila, S.J., are also summarized here.

During the winter of 1975–76 I learned about the activities of Jíkuri shamans, or si'páame, moving within the Norogachi-Pawichiki-Choguita area. These shamans came from the Okochichi-Sarabichi-Narárachi region. The Indians claim this area as "the land of witchcraft." In the Guachochi area, informants reported ceremonies performed in their neighborhood during the same winter by si'páame coming from Bakéachi.[1] The rituals also appeared to have been celebrated in the Creel-Sisoguichi area that winter. In western Tarahumara

Country, the ceremonies are not common, but the people seem to be familiar with elements and beliefs associated with them.[2] It should be noted that there is strong evidence indicating that the Jíkuri ceremony is a pre-Columbian custom. The ritual was frequently mentioned in the early records of the early Jesuit missionaries and the native origin is further corroborated by archeological data. The *si'píraka*, a musical instrument used in the ceremonies, has been found in old burial caves.[3]

Search for and Collection of the Jíkuri Cacti

The cactus grows in land considered sacred, located outside of the Tarahumaras' territory, in an area named La Ramada between the towns of Camargo and Jiménez in the state of Chihuahua. Tarahumaras organize ritual journeys to collect the jíkuri used in their ceremonies. This mission is carried out by a designated group of tribe elders, which include a si'páame—the Jíkuri shaman—and from six to ten men. The details of the trip are handled as a tribal secret. In advance of the travel, only those selected for the mission know who is designated, their departure time, and the time to return. The total time spent in this journey is close to a full month.[4] Prior to the trip, the men undergo careful preparation.[5] Only a few will be allowed the privilege of actually touching and collecting the cacti. The task is done with great care, ensuring that the roots of the succulent are not damaged. The buds are cut off with a flat wooden stick. At noontime, the men gather and consume a fragment of the plant. Upon returning home, usually at night, they appear intoxicated, as if they have been drinking. Their arrival is celebrated with a "curing" event. A small cross is erected and tutuguri, described before, is performed. A large bonfire is set up and the community members dance around the bonfire for two full nights.

The Jíkuri Ritual Ceremonies

The total number of persons who attend a Jíkuri ritual varies widely, depending on the social visibility of the sponsors, their wealth, and the quantity of the corn beer, or batari, available. As many as two hundred persons may be present in the event or as few as thirty-two. The celebration consists of a sequence of events that for descriptive purposes here have been labeled preamble, preparation, rasping ceremony, Jíkuri dance, communion, and epilogue.

The preamble initiates the ceremonies but it is not considered an essential part. On the other hand, if this or any of the other elements are omitted, the sponsors feel compelled to offer apologies for the omission. The preamble includes a tutuguri dance, the matachín or pariseo dance, the *nutéa* or *nutékima* (food offering), the curing of batari (described below), a consent request addressed to the regional authorities, and a head count.

The tutuguri dance is an invocation intended to raise to heaven the soul of a deceased person. This part is omitted when a funeral celebration is not the purpose. The matachín and pariseo dances described earlier in this book are performed to honor the spirit of Jíkuri and the person named in the homily recited during the ceremony. The matachín is performed if the event is carried out during the end-of-the-year cycle and the pariseo, if it takes place during Lent.

The nutea is a food offering to the dead. During this part of the ceremony, also omitted when a funeral is not the purpose, the shaman is believed to be able to communicate with the dead person, whose soul is assumed to be present. He encourages all to support the message included in the homilies celebrating the deceased. Throughout, Onorúame, or God, the transcendental being "who gave Jíkuri to the Tarahumaras," is honored. Drinking batari is an important part of the Jíkuri ritual. The beverage is "cured" before the ceremony simultaneously with the nutéa offering. A crock containing the batari is covered with a piece of cloth, which is removed by the shaman while he draws in the air a cross with a flaming *chopéke*, or stick of resinous pine wood.

The Tarahumaras meet a requirement they assume is expected by the Mexican authorities. They notify the authorities when a public healing activity takes place. Consent for the celebration is requested in writing, specifying the date and place of the event. The shamans themselves insist on compliance with providing this notification, believing that this will prevent them from being accused of sorcery. The process has certain peculiar features. The letter is written mixing Spanish and Tarahumara words. It is not clearly specified if the request is for a healing event or a religious ceremony. The document is written with pencil on ordinary paper. In one of the documents I examined, the petitioners informed the authorities that a "doctor" and his brother were in the area, summoned by sufferers from "headache and heart pain," and that this doctor knew how to cure these ills. The community in which the event would take place was mentioned in the document, which was formalized with a rubber-stamp seal from county authorities. Copies of these documents are usually not filed by the authorities. It is apparent that they consent to the procedure without understanding its nature or significance.

The shaman and the organizers of the event expect the presence of tribal authorities to help maintain order during the ceremony. Without this, objectionable behavior would be likely to occur as often is the case in events in which batari is consumed.[6] The presence of the siríame, the native governor of the Tarahumaras, is acknowledged with great deference and respect. All those attending line up in a row to salute the siríame or gobernador. After the welcoming, the siríame stands in the west side of the *yúmari* (courtyard or patio) to deliver a nawésari, or homily. This address is formal and not unlike those delivered in Norogachi after Sunday Mass. The maintenance of appropriate behavior is emphasized throughout. Individuals considered unsuitable to attend the ceremony are prevented from attending. Persons might be considered unsuitable if they are thought to carry a mysterious "bad element," which implies that they could be sorcerers. More vaguely, it is thought the person might have evil intentions for the use of the power of jíkuri. The unwelcome individuals are often notified in writing. In one of these events, I was asked in my role of owirúame, or healer, as a physician to examine a man who was considered objectionable in order to determine whether he was fit to participate. As is the usual practice, I was not asked to examine the subject to diagnose an illness, but to determine his or her hidden intentions.

After a long wait, the ritual proceeds after the nawésari delivered by the gobernador. This is followed by a supper consisting of *tónari* (boiled meat), tortillas, and *pozole*, or boiled corn. The ceremony progresses very slowly, interrupted by long waiting periods—a feature common in all Tarahumara events. The si'páame, "the one who rasps," usually has arrived to the area well in advance but he does not appear at the site until sunset. Only those close to him and the people who provide meals and batari for him know his precise location. Special significance is given to a head count; officiants need to know how many people are attending, both those who are participating in the ritual and those who are waiting outside the sacred place.

During the early stages of the ceremony, the shaman participates in the conviviality encouraged by the drinking of the first gourds of batari by playing the violin, telling sexually explicit jokes, and reminiscing about earlier times. At about nine P.M. the shaman goes to the courtyard while matachines perform three dances outside and later inside of the dwelling of the host. All of them then proceed to walk in a procession led by the si'páame, who plays the violin for the matachín dance. The group then reaches the courtyard, or patio (yúmari), where another group is dancing tutuguri.

A small wooden cross known as *presencia* is erected at one of the corners of the rectangular yard, usually to the left of the tutuguri crosses, and expressions of respect are displayed similar to those on the main cross. The singing and dancing are interrupted by one more nawésari, which is delivered at this point. The procession then continues toward the Jíkuri courtyard.

The Jíkuri courtyard is usually located in a ravine, near a small stream. A circular space approximately five meters in diameter is cleared from vegetation and rocks, and a large circle is drawn defining the area. In the center of the circle, firewood is piled up and lighted.[7] A large log is brought to the courtyard set in the west side to be used as a seat by the shaman and his main associates. A wooden cross and the offerings are then carefully set on the east side and hanged on a tree stem with several branches. This stem, according to informants, represents a tree of life, a common symbol among Indian tribes. The offerings (*mo'ibuma*) consist of meat. These include the rear quarter of an ox or a whole goat. The visceral block with the heart and lungs of the sacrificed animal is also suspended from the stem.[8] The *sacristanos* hang the meat offerings, *wári* (baskets) with *remé* (tortillas) and tónari, depositing them near the cross. The fragment of a broken pot is laid down close to the offerings to be used later for burning incense. A pewter cup containing *mé*, a watery infusion of agave, is placed close by, as well as the si'páame's *moruka*, which consists of a bag with his utensils. There they place also a large can containing water and a large jar with batari. For some of the ceremonies, the participants also set several lighted candles with the offerings. In one of them, candles are lighted one at a time; when the new candle is lighted, before placing it with the offerings, it is taken to the si'páame who interrupts his chanting to bless the candle and pray. A small pot containing batari is placed close to the place where the shaman will sit. In some instances, the presencia and the tónari are brought later to the courtyard.

So the presencia and offerings are introduced into the sacred circle, presented to the si'páame and to the main cross, and placed with the other offerings in the east side of the courtyard. As the si'páame enters the sacred circle, there is much joking, small talk, and considerable batari drinking. The siríame again recites a brief nawésari. Occasionally violin players or drummers play repeatedly the same melodies within the circle.

The shaman's assistants proceed to dig small holes in the ground, one for each two individuals celebrating within the circle and one larger and deeper in front of the place occupied by the si'páame. This hole is sufficiently large to contain a small basket holding Jíkuri, the sacred cactus. A much larger hole is dug on the opposite side, close to the cross. A *matáka* (grinding stone) is placed next to it. Following, a line is drawn in the ground encircling the patio, including in it the holes, the large log, the cross, and all the set of offerings, with the bonfire occupying the center of the circle.

The participants to the ceremony individually ask for permission to enter the ritual circle. The number of people allowed within the sacred circle ranges from nine to twelve men, women, and children. The number of participants in the circle is obtained ahead of time to estimate the amount of Jíkuri that will be need for the communion. Once in the sacred ground, the men are placed to the left of the si'páame and women on his right and the children with their mothers. During the ceremony, no one is allowed to cross the circle line except to *ishíma* (urinate), *witamea* (defecate), or *upésima* (vomit); that is, to relieve their physical needs. Leaving and reentering the circle requires the performance of special ritual motions. To leave the circle, the person walks in a counterclockwise direction, doing body turns and drawing a cross in the air in front of the si'páame. To reenter, the person walks in a clockwise direction. Similar ritual reverence is demonstrated in front of the cross and the si'páame.[9]

The si'pimea, or rasping ceremony, is the central part of the of the Jíkuri ritual. The name refers to the strumming movements required to play the si'píraka, a one-meter-long musical instrument made of brazilwood (*sitagapi*) shaped like a sword. A series of notches are carved along its main axis. The instrument is played by scraping it with a cylindrical stick of the same material. Only the si'páame is allowed to play this instrument. One instrument examined by me had forty-seven notches on the anterior surface with crosses carved on the extremity by which the instrument is held; the posterior face is without ornament. Other si'pírakas I examined had three groups of notches separated from each other by carved crosses following the pattern of ten notches, one cross, thirty notches, one cross, ten notches, and then one cross carved on the front (the rasping side). A specimen described by Lumholtz in 1902 was

146 ▾ *Tarahumara Medicine*

Jíkuri (peyote) ceremony. The *sipáame* is giving the rasping stick (*sipíraka*) and Jíkuri a sip of *batari* (*tesgüino*). Wetting the instrument with batari makes the instrument produce a different sound than when it is dry.

arranged in the pattern of a cross, twenty-two notches, a cross, twenty-one notches, a cross, and finally thirty-four carved notches.

At a given point in the ceremony, a sacristano, showing great reverence, unties the *si'páame moruka* pulling out the si'píraka and a wooden bowl (*batea*) and presents these objects to the cross and to the si'páame who receives them. The si'páame places the bowl upside down putting one end of the si'píraka on the bowl, which will act as a sound box. The instrument is lightly soaked with batari, representing ritual nourishment. Functionally, it is a way to get the desired rasping sound from the instrument as the wet instrument produces a different sound than the dry one. When played, the instrument emits a loud low-pitched sound. Scraping of the instrument is done rhythmically for periods lasting from one to three minutes and occasionally longer. The initial scrapings are brief and the sound elicited is softer. Later, the si'páame begins singing, accompanied by the instrument. Usually, sets of three melodies or a multiple of this number is played. A man playing with a crudely built violin accompanies this performance. Other instruments, such as drums, are played depending on the season. If the ceremony takes place during the end-of-the-year season, a violin is used; during Lent, drums and flutes are played.

As a next step, the sacristanos bring the Jíkuri cactus, moving slowly and demonstrating great reverence. The cactus is then lifted, following a circular pattern, presenting it to the cross as an offering in a way similar to what is done with *weróma* offerings of batari. Then the jíkuri is brought to the si'páame, who loudly greets it while at the same time he pleads for forgiveness and favors, following which he deposits it in the hole reserved for it as he continues chanting and rasping.

It is difficult to decode the meaning of the chanting even after carefully listening to sound recordings. However, praises to God and Jíkuri can be discerned. The phrases of the songs come in series of three. In between, the si'páame delivers brief nawésari homilies, which usually are addressed to "our Father and

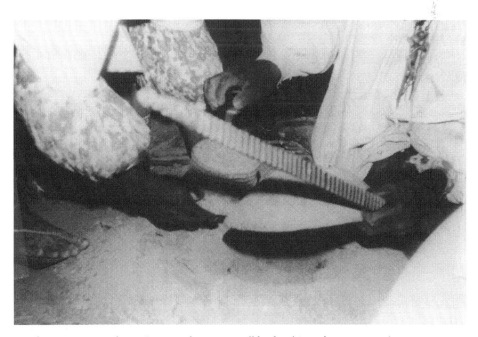

Jíkuri ceremony. The sipáame is placing a small basket (*émuri*) containing the sacred Jíkuri in a hole, dug for that purpose, where it will stay until the time of the communion. The hole will be covered throughout the ceremony by the *batea* seen in the picture, which will act as a resonance box for the sipíraka.

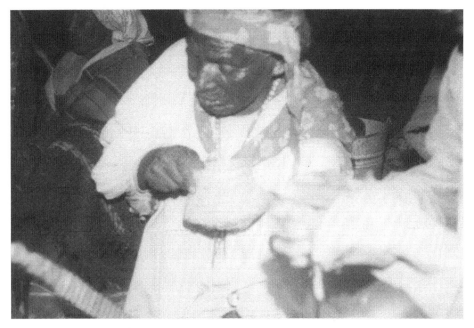

Jíkuri ceremony. The sipáame is talking to the Jíkuri in the basket. Usually he praises the cactus but also asks for its favor and presents some petitions to the spirit that lives in it.

Mother" and always end with the phrases "*Matétera ba, riosi matétera sinéame rarámuri, riosi matétera ba!*" after which the people answer, "*Matétera arí ko ba, we iwéraga semé kíria*" ("Thanks, all the Tarahumaras, thank God, thanks to God!" and "Thanks to you, very well [strongly] you played!")

At regular intervals, the bowl covering the hole containing jíkuri is removed and the contents of it are incensed. The fragment of crockery used as a censer (*incensario*) is ritually moved three times in circles over the opening of the hole. Jíkuri, the main cross, and the presencia are similarly incensed. During the phase just described, the conversations among those present make references to early times when game abounded and a great variety of animals could be observed, many of which do not exist now. Those were the long-ago days of *chérame rowí* and *chérame chumarí*; that is, the Old Rabbit and the Old Deer. Changes in the perceived position and brightness of the moon are also topics of conversation throughout the ritual with everybody repeatedly asking, "At what time will the sun rise?"

In some of the funeral Jíkuri ceremonies, a mirror is placed reflecting the image of the flames of the central fire. In one instance, the community was asked to help find the proper angle of reflection, a task that all those present found amusing. Informants mentioned that the purpose of the mirror was to keep a constant watch of the flames to get a glimpse of the dead man's soul ascending with the flames. A Tarahumara informant from Rotorichi stated that, on one occasion, participants saw in the mirror flames that instead of going up went down, falling side by side with the Jíkuri dancers of whom, after a few days, some died.

At another point in the course of the rituals, the si'páame signals to begin the Jíkuri dance. Like most events in the course of a Jíkuri, the actual timing seems to be capriciously chosen, guided only by the si'páame's own intuition or perhaps guided by some astral observation.

The dancers hang their sandals (*aká*) on the offering tree and perform barefoot, which is a symbolic gesture indicating they are stepping on sacred ground. Although briefer than other Tarahumara dances, the Jíkuri dance is not only strenuous but also dancing barefoot around the central fire implies stepping on the burning coals that have spread all over the dancing space. The number of dancers varies according to the number of people present within the circle. If a person is too old, too drunk, or too sick to dance he is expected to restrain from participating. An important dance movement is an eight-shaped evolution, repeated many times. This is the only part of the ceremony where one may identify the symbol eight that Artaud thought was so significant.[10] The dance is accompanied by the si'páame's chanting and has a rhythm as fast as that of the pascola mentioned earlier. The dancers do not wear special attire; they hang on their shoulders chains of rattles in a harness-like fashion or handle to shake them. The last man in the row clinks a metallic bell. The series of evolutions are repeated throughout the course of the night from three to twelve times, always in series of three.

The Communion

Once Jíkuri has been honored with si'píraka scraping, incensing, chanting, and dancing, one of the Tarahumara women, generally the youngest within the circle, is asked to grind the jíkuri in the matáka stone placed near the hole set for this purpose. Efforts are made to prepare the strictly necessary amount for the ceremony. Jíkuri is ground, while one of the sacristanos carefully deposits the cactus buttons on the matáka and blesses them, drawing a cross in the air while another woman pours water to the mixture. A sacristano collects the draining liquid from the inclined flat surface of the matáka into a *lochi* (gourd). The whole scene is illuminated with a *chopéke*, or torch made of resinous pine sticks.

Once the jíkuri mixture is ready, the communicants are called one by one. A cross is drawn in front of them and a cup with a few measures of the mixture is offered to drink.[11] The taste of this infusion is like that of stale water that has been kept for some time in a wooden container. After ingesting it, the communicants thank the sacristano, saying *Matétera ba!*, followed by drinking water three times using the hollow of the hand to bring the water to the mouth or drinking it from a gourd. Men consume the jíkuri mixture first, followed by women. The last receiving it is the woman who ground the jíkuri, the sacristano, and finally the si'páame.

It is difficult to obtain descriptions of the subjective experiences of participants who consume the jíkuri. Most of them avoid talking about it or their accounts are in general terms, such as, "We saw pretty things with beautiful colors." A participant however declared, "I saw a big cross of fire coming toward me and I heard very pleasant voices telling me to be happy and relax; I heard something like music coming from heaven." Interestingly, there is not a Tarahumara word for hallucinations. Dreams are called *rimúa* and this term is also used to describe the perceptual experiences induced by jíkuri. The term also means "deep thinking, to meditate, and to reflect."[12] Particularly significant is the belief that through this experience they receive special revelations from a superior power, which enables them to gain hidden meanings not otherwise accessible.

At the end of this part of the rituals, the si'páame, addresses the men present, asking them whether the holes made for the occasion have been covered with dirt. They reply that it has been done. He then walks to the women's side, where they are kneeling, and asks the same question, which is answered in the affirmative. He returns to his place, sets the si'píraka on the batea, scraping it and singing as he has done throughout the night. After assuring that all the holes have been covered, he removes the bowl, placing it upside down in the ground. Taking the remaining jíkuri, he shakes the bag to rid it of dust and places it on the bowl, drawing a cross in the air. The scraping stick and the si'píraka are placed crosswise. Finally, lifting the bowl, he hands it to a sacristano, advising him not drop its holy content, while keeping the si'píraka and jíkuri in place with the scraping stick pressed against the bowl. Turning a full circle, he advances toward the cross and then makes three full turns.

The assistants proceed to reassemble the si'páame's *moruka*, placing the objects inside and carefully depositing the empty container, which held water throughout the celebration. The smaller bag with *mo*riwá* (incense), the wooden bowl, a pewter plate, and the remaining peyote also are therefore placed there. All of this is done with great care because the Tarahumaras believe that if jíkuri escapes from their custody, it could cause great harm. The matáka is buried using the hole dug for this purpose at the beginning of the ceremony and is never used again. After this phase of the ceremony, the Jíkuri dance continues. In the case of funeral and healing celebrations, the dance may continue until dawn. In closing, the si'páame grants permission to enter to those who have been kept outside the sacred courtyard.

Closing Events

These include the morema, ba'wí ówima, si'panema, wejazo, mé or méke, ritual cleaning of the patio Jíkuri, rarajípari, and najarápuri.

MOREMA   A fragment of an old broken *sekorí* (big clay pot) is used as an incense (*copal* or mo*riwá) burner. A burning coal from the fire is picked up with two branched twigs and placed on the pot, adding incense producing a dense smoke, moving the clay in circles over the heads of those present, including the persons who were kept outside the sacred circle during the ceremony. A sacristano holds the pot close to the nose of each one of them and has them inhale the smoke, using a blanket to prevent the smoke from escaping. The clay with the incense is withdrawn, drawing in the air a ritual circle over the head. Another sacristano, using a fresh pine tree branch, sprinkles cold water on the participants' faces, all of whom reply *Matétera ba!*

BA'WÍ ÓWIMA   The ba'wí ówima, which follows in this sequence, appears to be a purifying ceremony for to those who were exposed to jíkuri. The sacristano, using a cloth soaked with water, cleanses the head, face, trunk, and the four extremities of those exposed while in the sacred circle and even of all those who were in the courtyard outside it. Since the Jíkuri ceremonies usually take place during winter, it is not uncommon to find a sheet of ice has formed in the water. Interestingly, Ba'wí ówima always elicits a mirthful response from the group.

SI'PANEMA   Another healing maneuver, si'panema, follows; the term literally means "rasping or scraping." During this process, the si'páame uses his notched stick to symbolically remove or pull out diseases from those affected, applying a rasping motion of the instrument over parts of the body thought to be diseased and over certain ritual points. Before performing these healing maneuvers, the si'páame asks the individual about those parts of the body that are ailing and then proceeds rasping three times on each of the affected areas. The rasping is applied then to the forehead, ears, nape, shoulders, chest, back, abdomen, testicles, thighs, limbs, and heels in this order.

WEJAZO   The *wejazo* maneuver finishes these purification rituals and consists of splashing a gourd full of cold water on the face of each of those present.[13]

MÉ OR MÉKE   At this end stage of the Jíkuri ritual, a sacristano administers the *owáame* (remedy) to all those present. This is a watery infusion of agave (*mé* or *meke*). With the healing mixture in a spoon, he draws crosses and circles over the head of the recipients and then places the spoon in their mouth. Men are given three spoonfuls while women receive four. Then he touches with the spoon other parts of the body, saying, *Moŏra okorá, surara okorá, sonorá okorá ma, ku sawí, ma ga nire eyena*. This means, "Headache, heartache, chest pain too, heal! Now be healthy and go!" After the mé administration, the receivers give thanks again.

As mentioned, all the holes have been filled with dirt, the matáka and its handle buried, and the main cross put away. The food offering is distributed among the people except for a part claimed by the shaman as compensation, usually including meat, tortillas, and some batari. Frequently, the si'páame leaves at this point without letting anyone present know his destination. The sacristanos remain to give food to the matachín dancers.

RARAJÍPARI AND NAJARÁPURI   At the end of the ceremony, a rarajípari—kickball race—and a *najarápuri*—wrestling contest, are organized. In the race, a man and a woman from the sacred circle run against the outsiders on the court. During the celebration the *gomáka* (ball) and *rowera* (cloth and grass ring used in female races) are kept as a separate offering site along with a pot of batari. It is said that a race is carried out only when the deceased was himself a runner but in some of the ceremonies observed, the race was carried for a deceased person who was not a runner. After some ceremonies, a najarápuri wrestling competition is held among the matachines. Throwing their kowera, or headbands, high into the air signals the end of the ceremony. However, even after this closing many remain until all the batari is consumed.

Some ritual prescriptions follow the ceremonies. Jíkuri participants are to avoid sexual relations for three days following the events and are supposed to remain in a solitary dwelling or cave. Sunlight is to be avoided, believing that it may cause madness. The ceremonies described are therefore held in winter when the periods of darkness are longer. Consumption of salt is restricted and touching persons who are sick is avoided.

The Jíkuri Ceremony as an Expression of Tarahumara Mysticism

One of the main themes of the Jíkuri ritual appears to be the belief that jíkuri is a powerful spirit or entity capable of causing great harm. The ceremony would be a means to appease this entity. Jíkuri is also considered to be a virtuous spirit; Lumholtz observed that the cactus is not kept in the dwellings because the sight of anything immodest may offend it.[14] Jíkuri supposedly has the power to restore health and give a long life and to bring benefits to those who believe and honor it. Jíkuri supposedly is continuously producing a noise—psst, psst. The Jíkuri's liking for noise supposedly is a reason why rasping the si'píraka is done—to please Jíkuri. An important function of Jíkuri is to take care of the souls of the dead, particularly if the deceased have honored it during their lifetime. In order to enter *rewegachi* (heaven), the soul of someone who has participated in life within the scared circle in a Jíkuri ceremony has to be liberated by having his family summoning and participating in a new Jíkuri celebration. Otherwise the spirit of Jíkuri will keep the soul captive, not allowing it to continue its journey into heaven. Those who have participated within the sacred circle must attend two other celebrations. Communicants who have participated in three Jíkuri celebrations are believed to gain the spiritual standing to become si'páame. On the negative side, loss of the iwigara, believed a cause of disease, might result from touching jíkuri or from failing to follow the prescriptions of the ritual.

Other beliefs closely related to Jíkuri involve the *ru'síwari* or *sukíki* stone, which is also considered to be living and is respected and feared. Everyone avoids touching it. Ru'síwari supposedly becomes at night a bird that eats children's souls.

Jíkuri also is supposed to protect the corn crops and protect from lightning. That is why the shaman asks from jíkuri protection against these phenomena. In passing should be mentioned that the peyote-deer-corn complex is a set of beliefs common among other Indian groups. Along these lines, during Jíkuri celebrations, frequent references are made to Old Deer and Old Rabbit and the times when they wandered freely over the land.[15] Also related is an assumed relationship of jíkuri with snakes. In addition to be considered a remedy for snakebites, it is also believed that it has the power to blind the snake, preventing it from seeing as men throw hot rocks into its mouth.[16]

It should be noted that the bakánowa plant has attributions that are similar to those of jíkuri. An old man, formerly a noted runner, suffered from intense joint pains. He and his friends thought that he might have inadvertently offended bakánowa and this was causing the ailment. A si'páame therefore was called to celebrate a Jíkuri to please the cactus and compel bakánowa to leave alone the distressed old man. The cure therefore required bakánowa as well as Jíkuri celebrations. Bakánowa incidentally is believed to be "more courageous" than jíkuri.

TEN

Compendium of Tarahumara Herbal Remedies and Healing Practices

This compendium presents in alphabetical order the remedies or medicinal plants commonly used by the Tarahumaras. The common term for each plant or remedy is listed, followed by the scientific names and the various names applied to them by the Tarahumaras, designated as TAR. Next, Spanish identifying terms are included, designated as SP. Sometimes the Náhuatl and Northern Tepehuan terms are included, designated as NAH. or N.T. English names are included in the entries if used in the literature. The significant botanical features of these plants are presented as well as a description of their preparation and administration as remedies. It should be noted that the medicinal claims are part of the local lore. If scientific validation is available, it is noted. The ceremonial contexts within which the remedies or plants are used are described when appropriate. References are also made regarding noncurative or practical use. Pharmacological information when available is briefly summarized. It should be noted that some plants, although not native to the area, are presented because of their common use or social significance to the Tarahumaras and other cultural groups in the area.

Aborí[1]

Juniperus pachyphlaea Torr., *Juniperus deppeana* Steud.,[2] *Juniperus durangensis* Martínez,[3] *Cupressus arizonica* Greene

TAR. *aborí, aorí, ahualí, awerí, kawarí, koarí, wakarí, awarí, gayorí, kaorí, waorí, oyorí, orwí, péchuri,* pechúruwa,[4] *waǎ, wa'yá*[5]

SP. *tázcate* (or *táscate*),⁶ *enebro, cedro, sabina, tuya, grojo, junípero, cada, ciprés, tázcate sabino, sabino, sabino colorado*

N.T. *gáyi, gágayi*

The list of synonyms listed for the genus Juniperus and the Tarahumara designations refer to several gymnosperm conifer trees of the family Cupressaceae. In rare instances, these terms are applied to a tree of the families Pinacea or Taxodiacea due to resemblances between these trees and junipers. The Tarahumara term *aborí* has considerable presence in their culture: for example, Aboréachi, the name of an important village, derives from it. The names *péchuri* and *péchuruwa*, of course, do not correspond to the series. Their meaning is discussed below. *Tázcate*, also written *táscate*, is the word used by mestizos to designate a juniper. In Tarahumara Country, *tázcate* is a synonym for *aborí*. *Tázcate sabino*, or simply *sabino*, refers only to Tarahumara's *waǎ*.⁷ Mestizos would occasionally call a *tázcate sabina* (savine) or *cedro* (cedar), but never *enebro* or *juniper*. *Waǎ* is the Tarahumara name of the cypress *sabino*,⁸ whose characteristics and uses overlap with those of *aborí*. Finally, the word *tuya* is occasionally used to refer to small tázcates.⁹

Regarding the healing properties attributed to the plant, the Tarahumaras and mestizos from Cerocahui claim that a decoction of the leaves and bark of a tázcate has a relieving effect on missing menses or amenorrhea.¹⁰ Evidently, it is conceivable that abortive effects are also sought. In Norogachi, the decoction is used to prevent hemorrhage after delivery. Incidentally, an interesting remedy utilized for the same purposes is a *tierra de tucera* decoction. This consists of sand and other particles that moles deposit at the mouth of their hole. The dust contains dirt, mole's waste, and particles of aborí and *rojá*.

Some informants suggest that a poultice made from aborí leaves may heal impetigo but *chu'ká* (*Senecio*), or pine leaves, are preferred since they are considered less irritating to the skin. A decoction prepared from leaves of aborí is also used as an analgesic or antitussive.¹¹ A watery decoction made from the leaves of wa'á, awarí, or péchuri is used for coughs or colds or as a lotion to ease body aching, indigestion, stomachaches, flatulence, respiratory disorders, and sore throats.¹² Informants readily report the uses mentioned for aborí, adding that wa'á could be used instead but that aborí's leaves are more "fleshy" and therefore a more effective remedy. Other observers have noted that the tázcate sabino decoction is used as an antimalarial agent.¹³ Aborí is also used as a decoction to bathe the legs of runners for three days before a race.¹⁴ The use of a tea made from aborí for the treatment of muscular pain also has been reported.[15] Finally, the Tarahumaras attribute mystical powers to these plants: decoctions of wa'á (*Cupressus*) are believed to protect and cure men and animals from the feared water spirit, while those of the aborí (*Juniperus*) are used to "cure" the fields and protect them from lightning and hail.[16]

Wikubema

Wikubema is a healing and purification ritual using aborí performed by itself or as part of more complex ceremonies. For this purpose, aborí leaves and branches are burned and placed on burning coals or hot rocks to produce "healing" fumes to which persons, animals, or inanimate objects are exposed.[17] The ritual is usually applied to sick individuals with fever, body aches, or other conditions. It is also used to purify persons who have touched the dead or have been in contact with other harmful, proscribed, or taboo objects, such as human bones or Jíkuri, including the utensils used to handle them. Wikubema is sometimes part of the fire baptism, the naming ceremony following the symbolic cutting of the invisible thread linking the child to a tree. Wikubema may also be part of a funeral ceremony taking place in the cemetery.[18] Persons subject to the wikubema ceremony are rubbed with bone marrow and skunk or rattlesnake fat prior to the ceremony. My informants claim that the fat has to be from a cholugo but ordinary lard also is used. Inhalation of the fumes is particularly important during the wikubema as it is in the morema or Jíkuri ceremonies. During a sweat-bath session, sometimes part of the ritual, the person may be exposed to the volatized essential oils of juniper and cypress leaves and their therapeutic action; however, the maneuver mainly carries a mystical significance.

In Samachique, usually the tutuguri courtyard adjacent to the shaman's house is used, with space provided for prayers and an area where the individual places his clothes. A wooden cross is placed in the area where the shaman prays. The sweat-bath format performed by a shaman is particularly common in some areas, such as Samachique. This is carried out in a large circular or square pit in the ground approximately five feet wide and in diameter and four feet deep. The pit is lined with rocks in the bottom. The rocks are heated with oak firewood and aborí branches are placed on the glowing coals to produce a cloud of thick smoke. Water is poured over the hot rocks to generate steam. Juniper branches are tightly wedged between the pit's walls with a crude brush lining separating the hot rocks from the communicant, who is then placed in the pit, protected from the direct heat by the branches.[19] The individual is partially covered with blankets, preventing to some extent the smoke from escaping.

Wikubema has changed little since it was first described in the anthropological literature.[20] The ceremony is used to purify the relatives before they attend a burial, which takes place in caves or in a cemetery. It is also considered important to purify the house of the deceased, burning cedar boughs in the fireplace of the dwelling. The floor is carefully swept and all of the deceased's belongings are removed. This ritual gains importance when the deceased is not the head of the family and there is an interest in keeping the dwelling. If the dwelling is abandoned, it is usually torn down. The remaining stones and other building materials are sometimes purified before being used to build other houses. Occasionally, it is decided to use a cave that previously held human remains to store hay or to lodge sheep. In these cases, long poles are used to collect the bones, and the remains are thrown away, usually into a stream to let them be carried away by the water. All of this is done since contact with human remains (*chu'í*) is believed to bring disease, both in man and animals (*engalgamiento*). Smoking the cave with aborí fumes is done to purify it.

In the case of a naming ceremony, the baby is passed several times through the smoke of burning leaves and its body moved, drawing circles and crosses. Ritual formulas are uttered softly and unintelligibly during the procedure. This ceremony may be performed one or more times for the baby.[21] After the wikubema, a celebration may follow, during which meat dishes and batari are consumed. A tutuguri dance may or may not be part of the event.

Some comments are in order to place the Tarahumara remedies within the framework of history and conventional medical knowledge. Many cultures have used the leaves and fruits of cedars, junipers, thujas, and other related trees for medicinal purposes. For example, the Sabines—an ethnic group neighboring the early Romans—used a decoction of *Juniperus sabina* L. as an abortive.[22] The diuretic properties of the berries of *Juniperus communis* L. have been recognized almost for as long as their usefulness to make gin has been known. Oil of cade from *Juniperus oxycedrus* L. was considered a dermatological panacea. The leaves and other parts of *Juniperus* sp. have essential oils shared by many gymnosperms that may be responsible for physiological effects, inducing skin flushing and easing pain. Savine oil, a derivative of *J. sabina* L. contains sabinene and sabinol, which were thought to induce abortion.[23] For a long time, juniper berries and the oil contained in them were attributed medicinal effects, such as diuretic and stimulant effects. The terpenoid volatile oil, pinene, may explain their expectorant qualities.[24]

The hard, durable, and pleasant-looking reddish wood of the trees of the family Cupressaceae makes it suitable for practical uses such as to make certain parts of the Tarahumara plow. Musical instruments, particularly matracas and the handles and bridges of violins and guitars, also required the hardness of aborí wood. Small pieces of this wood are used to build ingenious locks for doors and other miscellaneous artifacts. Aborí wood is considered the material of choice to carve crosses either to wear or to hang on the walls of houses and temples. These wooden crosses have been used at least since the seventeenth century and some authors believe that they are pre-Hispanic. Stems and twigs of waá and aborí are used to build the arches characteristic of the Holy Week celebrations and aborí branches substitute occasionally to make brooms (*pichira*) when the traditional *otate* broom is not available.

Children in Norogachi eat raw juniper berries, which they consider a delicacy. Pennington also extends this observation to the eastern foothill Tarahumaras and relates that this use had been documented since the eighteenth century.[25] Curiously, Zingg denies the fact that the Tarahumaras eat juniper berries.[26] Pennington confirmed that Tarahumaras do prepare a drink from tázcate berries.[27] Some Tarahumaras prepare a refreshing beverage with the leaves of aborí.

Akáame

Potentilla thurberi A. Gray
    TAR. *akáame*
    SP. *fresa cimarrona, hierba de San Antonio*

This wild strawberry grows in shades and well-watered areas near Norogachi and Nararachi. A tea is made from the roots of *Potentilla thurberi* and is used for coughs, diarrhea, sore throats, and other gastrointestinal conditions. The raw root is applied to an aching tooth to numb the pain.[28]

The pharmacological actions of *Potentilla* are attributed to the tannins and saponins contained in its roots.[29]

Amapola

Papaver somniferum L., *Papaver rhoeas* L.
　TAR. *chutama, chotama*
　SP. *amapola, adormidera*

Papaver somniferum, the opium poppy, is not native to the Tarahumara region. It is mentioned here because the plant plays a role in the social dynamics of the area. *Amapola* or *adormidera* (sleep inducer) are the popular names given in Spanish to the plant and especially to its beautiful flowers. The name is related to the flower's narcotic properties that have been known in the region, especially by the mestizos. Adormidera was cultivated for ornamental purposes in the old mining towns and villages of the region. A concoction of the capsules of the flowers was used by some mestizos as a remedy for arthritic pain and as a general analgesic in Uruachi and Papigochi.

For reasons alien to both Tarahumaras and mestizos, their land became in the early 1960s an area of production of raw opium for illicit purposes. The isolated and rough character of many locations within Tarahumara Country attracted growers called *chutameros* to the region. The Mexican army has conducted operations against planters since the 1970s—called then Operacíon Condor—and continues destroying crop fields and maintaining detachments in the region until the present day. As a consequence, poppy growers and traffickers have retreated to the most inaccessible places of the sierra to carry out their activities. Both the plant and the methods of cultivation and collection of opium gum are known by Tarahumaras. However, few of them participate in the production of or trade in the narcotic in spite of the financial enticements. The Tarahumaras seem to perceive a cultural dissonance with these activities and often report when the plant is grown and provide specimens to the military patrols searching the area. Occasionally Tarahumaras have been charged and imprisoned for cultivation done on their land against their wishes by the chutameros. Although amapola is not native to the area, wild specimens may be found derived from careless handling of the seed in places close to clandestine air strips and plantations.

Mestizos and acculturated Tarahumaras do cultivate other plants of the family Papaveracea for ornamental and medicinal purposes in Cerocahui, Norogachi, and Cusárare. Several varieties of *Papaver rhoeas* are cherished by the villagers, and attributed medicinal purposes belonging to *P. somniferum*, particularly against arthritis pain, swelling, and diarrhea.

Anará or Anaráka
Mascagnia macroptera (Moc. et Sessé ex DC.) Nied.[30]

TAR. *anará, anaráka, matenene*[31]

Tarahumaras from the western canyons use scrap shavings from the bark of *anaráka*, mixed with animal fat as a poultice for toothaches or to ease the pain of sprains or fractured bones. This plant has been used as a remedy to treat malaria.[32]

Ápago
Ptelea trifoliata L.

TAR. *ápago, upaga, opóa*[33]

The Tarahumaras boil the leaves of *ápago*, the hop tree or wafer ash, and use the decoction to bathe the body or face to treat joint pain.[34]

The root of this plant contains the alkaloid berberine.[35] The plant's easy-to-carve wood is used to make crosses and beads.[36]

Aposí
Erythrina flabelliformis Kearney, *Erythrina coralloides* DC., *Erythrina* sp., *Erythrina americana* Mill.

TAR. *aposí, aposhí, batosí, kaposí, a'poshí*[37]

SP. *colorín, chilicote, frijolillo, frijol quemador, frijol patol, zompancle*[38]

Scholars have identified the Tarahumara's *aposí* as *Erythrina flabelliformis*. It should be mentioned that most Mesoamerican ethnobotanists identify *colorín* with the species *Erythrina americana*.[39] Knowledge about its poisonous properties appears to have been common in Mesoamerica. The tree is characterized by its thorny branches resembling those of orange trees and its bright red seeds and flowers. Aposí grows mainly in the canyon country, but its habitat extends to the upper Urique River basin, reaching the Sikirichi-Bakasórare-Romichi area.

A widespread legend involving this plant is that of the giant Ganó. According to the story, the giant enslaved and killed Tarahumaras and took their wives to serve him. The Rarámuri eventually killed him, poisoning him with aposí beans mixed with the ordinary beans that the giant's wives were cooking for him.[40] This story illustrates awareness of the toxic nature of aposí as well as the dualistic belief that there are beans that nourish and beans that kill. The story of Ganó and the Rarámuri contains many parallels with other traditions, such as the legend of Nararachi, according to which the Tarahumaras defeated and killed Apache invaders.[41]

Regarding the remedial uses of the plant, seeds of aposí, toasted and ground, are mixed with warm water. The decoction is consumed to relieve abdominal distress.[42] Although reported effective, it must be in fact a dangerous remedy. Shamans have used aposí beans in the past; the seeds have been discovered in a Basketmaker burial site in the Tarahumara region.[43] Being an element of an important Tarahumara myth, it is not surprising that although poisonous, aposí beans were thought to possess strong healing powers. A tea made from toasted and ground aposí seeds has been used as an emetic and as a cure for toothache.[44] The seeds are also used in poultices to treat wounds and for eye infections.[45] Aposí has been considered a hallucinogen.[46]

Aposí beans, due to their bright red color, are a favorite material used as beads for necklaces and to adorn the tutuguri crosses. From the bark of aposí, a yellow dye is made, considered to have commercial value in the area.[47]

Extracts of the plant contain the alkaloids erysotrine, erysodine, erysovine, and eryspine. Some of these alkaloids have curare-like neuroblocking and respiratory-depressing effects responsible for the plant's toxicity and therefore are not used in conventional pharmacotherapy.[48]

A*rí

Coccus axin Llave, *Llaveia axin*, *Carteria mexicana* Comstock

TAR. a*rí, arí, a*warí, o'lí[49]

SP. *gomilla, gomía, gomilla de zamo,* or *samo, zamo*

A*rí is one of the most remarkable substances used by the Tarahumaras and mestizos.[50] It is a lacquer gum or wax produced by a homopterous insect, *Carteria* Mexicana,[51] *Coccus axin*,[52] or *Llaveia axin*. The male of this species is winged, with a second pair of wings, which are atrophic. The a*rí wax is produced by the female, which after metamorphosis loses its legs giving it the appearance of a worm. The female insect has a proboscis that is inserted into trees, keeping it attached to the tree and used to feed from the sap. The female's body swells as eggs develop in it; the wax is secreted and stored in the body of the insect where the larvae continue to grow after the female dies. The male, lacking a proboscis, is unable to feed and dies after fulfilling his reproductive function.[53] These insects live in trees such as the *jobo* (*Spondias mombin,* a native wild plum), *piñón* or *jumete* (*Jatropha curcas*), Palo mulato (*Bursera grandifolia*), and *ciruelo* (*Spondias* sp., another native plum).[54] The *samo* tree (*Coursetia glandulosa* Gray or *Willardia mexicana* Rose) also supports the insect.[55] This coincides with my observations in the Urique Canyon area and Guachochi.[56]

The Rarámuri harvest the wax during the rainy season.[57] A*rí is processed by rolling it by hand into thick sticks to store for the winter.[58] This resin has been identified with the lacquer used in the Mexican states of Michoacán, Chiapas, Veracruz, and Yucatán as a fine varnish to treat their beautiful wooden handicrafts.[59]

A*rí is highly appreciated by Tarahumaras and mestizos when used as a relish or condiment boiled in water. Its taste is sweetish, acidic, but very refreshing. The gum becomes sticky and elastic in boiling water before it dissolves to form a red-colored emulsion or sauce. This sauce is considered a delicacy and is sometimes additionally flavored with *isíburi* peppers. A*rí is expensive; therefore, sediments are saved and reused. A*rí is sold by traveling canyon dwellers. It may also be found in general stores in Guachochi and Batopilas. It is offered in the traditional way, prepared as small cakes wrapped in plastic or paper bags. Canyon Indians may bring it still adhered to samo or jumete sticks.

A*rí plays a medicinal role; it is rubbed on the skin and the abdomen as a local analgesic. It is also used as a remedy for gout.[60] Its astringent effects have been recognized since early times when was used to clean wounds, to treat headaches, and to relieve fever and constipation.[61] The a*rí emulsion is considered an effective remedy for empacho, dysentery, and to clean the kidneys but it especially is the favorite remedy for hangover throughout the Tarahumara region from Bakiríachi to Samachique.

Arisí

Cucurbita foetidissima H.B.K.

TAR. *arisí, garisí, karisí*[62]

SP. *calabacilla*,[63] *calabacilla apestosa* or *hedionda, sanacoche, calabaza silvestre, chichicoyote, calabaza de coyote*[64]

Arisí is a cucurbitaceous plant scattered throughout the middle elevation sierra terrain. It grows in long, crawling vines through open spaces, within plowed lands, or alongside the margins of roads and streams. It has typical yellow flowers that evolve into small pumpkins. The roots of arisí contain saponin and are used as soap to wash clothes.[65] The long vine stalks have several uses, such as animal feed and to make string twine.[66] The dried shell is used to make *sáuraka*, the typical rattle. Boys use the pumpkin as a substitute for the wooden ball (*gomákari*) used by adults in kickball races.[67]

In regard to native medicinal uses, the roots, leaves, and fruits of arisí are used as laxatives. The plant contains saponins, steroids, tannins, and organic acids, which may explain its usefulness for such purpose.[68] Cucurbitaceae also contain cucurbitacins and terpenoid glycosides in their roots and fruits; *Cucurbita foetidissima* is unusual in that it contains these glycosides also in its leaves. Studies indicate that the root extract increases the tone of the muscle in the intestinal wall of rats and mice and perhaps has antipyretic and analgesic effects.[69] An additional use, not reported before, is as a toothache remedy: the Tarahumaras place a bit of a paste, made from the fruit, on the diseased tooth.

Arrayanes (Aceite de)
Myrtus communis L.
>TAR. *siyóname wi'í, siyóname ba'wira*
>SP. *aceite de arrayanes, aceite verde*

The oil made from myrtle plays an important role among present-day Tarahumaras' remedies. Its use was introduced to the region by merchants and through other intercultural contact and it is rapidly replacing some native remedies. The oil is used to massage the kickball race runners and also to treat sprains and muscular aches. Mestizo masseurs use it as a rubbing oil while reducing luxations or treating injured bones and joints. The root, bark, and leaves of *Myrtus communis* contain a large amount of galic and tannic acids, a yellow dye, a resin, and an essential oil of agreeable flavor. In addition, the plant contains a glycoside, calcium and potassium sulfates, and phosphates.[70]

Babárachi
Caesalpinia mexicana A. Gray
>TAR. *babárachi*

The dried and ground roots of *babárachi* are sprinkled on wounds, sores, and ulcers to promote cicatrization.[71]

Bachí
Cucurbita sp., *Cucurbita pepo* L., *Cucurbita mixta* Pang., *Cucurbita ficifolia* Bouché
>TAR. *bachí, sibóbachi*

SP. *calabaza, chilacayote*

Squashes and pumpkins are among the four basic elements of the diet of American Indians. Cucurbitaceae have been well known by American Indians since ancient times. The Tarahumara summer squash is *Cucurbita pepo*. A winter squash, *Cucurbita mixta*, grows in the western canyon slopes and lowlands. The fig-leafed, yellow-striped fruit, or Malabar gourd, is *Cucurbita ficifolia*, *chilacayote*, or *sibóbachi*.[72]

Knowledge about the anthelmintic properties of the seeds is very old among the Indians, and the active principles of these seeds were later incorporated into therapeutic compounds used by western medicine. These properties are particularly effective against *Taenia* (tapeworm). This is the main medicinal use with which the Tarahumaras are well familiar.

The plants contain cucurbitacins and terpenoid glycosides in their roots and fruits. They may also contain saponins, steroids, tannins, and organic acids.[73] The leaves of chilacayote are diuretic, while its macerated seeds are used against febrile illnesses.[74]

Bacierera

Ribes neglectum Rose

SP. *bacierera*[75]

Application of the leaves of this plant on the forehead or wrapping the leaves in a cloth tied around the head are common headache remedies among the uplands Rarámuri.[76]

Bakánowa

Scirpus acutus Muehl.,[77] *Scirpus* sp.

TAR. *bakánowa, bakánoa, bakana, bakánori, bakánawa, bakánowi, chimari, tá kabora*[78]

Bakánowa was thought to be an *Echinocactus*.[79] However, the meaning of the word *bakánowa* suggests otherwise. *Bakú* (or *baká*) designates a reed and *nowa* denotes a root. Bakánowa has finally been identified as a plant of the family Cyperacea, *Scirpus* sp., and specifically as *Scirpus acutus*.[80] The plant is taxonomically related to the papyrus plant *Scirpus papirus* L. and *chufas*— *Scirpus sculentus* L. tiger nut. The root is a complex tubercle resembling a string of pellets; for this reason, bakánowa is also designated *tá kabora*. In Norogachi, bakánowa *is* often called *chimari*, a term that signifies "weed" or "grass"; this is done to conceal from those alien to the culture the fact that they are referring to a plant sacred to the Tarahumaras.

Bakánowa or *bakánori* ranks second only after peyote as the most important ceremonial plant. Its magical attributions and behavioral effects support complex beliefs accepted with reverence and awe. The plant is used with demonstrations of great respect accompanied by ceremonial singing. Ingestion is expected to put those who take it to sleep and make them dream so that they may "travel" to visit with deceased relatives.[81] Bakánowa, like jíkuri, is said to talk, sing, and make noises. Believers state that the plant travels and converses with the dead. If the person who takes it is irreverent, the plant may induce that individual to throw him- or herself into a fire.[82] Due to its property of inducing hallucinations, it is considered instrumental to communicating with God and other preternatural beings. As a hallucinogen, some claim is more powerful than jíkuri.[83]

Bakánowa, like jíkuri, is more than just a healing agent; it has a mystic significance within the Rarámuri worldview, being attributed with the power to cause and to cure disease. Bakánowa has a reputation of effectiveness in the treatment of all types of painful conditions of the joints. It is used to relieve muscle spasms and pain and is a remedy for digestive distress.[84] The whole root is stirred in boiling water and used as a drink. It is also applied externally to the back for respiratory illnesses. Bakánowa is chewed or ground and applied over wounds to help healing.[85] The plant is considered one of the best remedies to relieve pain.

The tuber is also used to cure insanity. On the other hand, it is contended that bakánowa actually drives people mad.[86] The root is chewed by the shamans and then applied locally to ease pains. Bakánowa is carried on long trips as it is considered the best remedy for relieving pain.[87] As a poultice, it is used for scratches and superficial infections. The plant is believed to give strength to runners. They apply it to themselves three days before a race.[88]

Bakúi

Phragmites communis (L.) Trin., *Arundo donax* L., *Mülhenbergia porteri* Scribn., *Eclipta alba* (L.) Hass,[89] *Panicum bulbosum* H.B.K.

TAR. *bakúi, bakú, baká,*[90] *bakubi, bakuwi,*[91] *bakasó, baké, bakiré, bakéchuri, bakusí, awá* or *awáka, waká, i'kí, witabórachi*[92]

SP. *otate, zacate común, carrizo, caña, chiwite*

Baká or *waká* means "reed, or an aquatic plant." *Bakúi*, on the other hand, is used in Norogachi to designate "an herb brought from the canyons and used to fish." Others argue that *baká* refers to a hollow reed while *bakúi* refers to a solid plant (*otate*). *Bakasó* is used to designate a variety of reeds of low height that grow near Norogachi. *Baká* designates different reeds, including *Phragmites communis*.[93] Other investigators have stated that the term *baká* refers to *Arundo donax* and *Phragmites communis* and identified *bakú* as *Mülhenbergia porteri*, a common grass.[94] *Bakuki* has been recognized as *jara*, the shaft to make arrows, and *bakúmuri* as *otatillo*, a thin reed also used to manufacture arrows.[95] *Baké*, *bakiré*, and *bakéchuri* have been identified as *Eclipta alba*.[96] Other related terms are *wasábona*, which applies to *jarilla de río*, and *awáka*, which applies to *chiwite*. Brambila identified *awé* or *wawé* as *chiwite*, or *amole*, a plant used by Tarahumaras as soap.[97] *Witabórachi* is a plant of the family Gramineae, *Panicum bulbosum*.[98] Tarahumaras reportedly also use the stems of *Phragmites communis* to make arrows.[99]

In an anonymous Jesuit transcript (ca. 1940), bakúi (or otate) was mentioned as being a remedy to strengthen the contractions during childbirth. In Norogachi, a tea from this plant is given to women experiencing a difficult labor. The stem of otate is decocted into a beverage used to relieve urinary retention.[100] It has been reported that the root of the grass *Mühlenbergia porteri* is boiled, mashed, and used as a poultice for relieving backache and articular pain.[101] The fruit of *baká* (*Phragmites*) has been used to treat diabetes and gastrointestinal ailments.[102] The inflorescences of *awá*, or chiwite, are eaten to treat gastritis and emesis; in powdered form it is applied for skin infections and pimples.[103] *Witabórachi*, eaten raw, is attributed to control vomiting and, prepared into a tea, is used for sore throat.[104]

Bakusí

Equisetum laevigatum A. Br. *Equisetum hyemale* L. var. *affine* (Engelm.) Eat.
   TAR. *bakusí, kawasíora, gaowachire, gawarísoa*,[105] *pakuchara*
   SP. *cola de caballo*[106]

Equisetum laevigatum grows along small streams in the southeastern Tarahumara region. The long stems of this plant are used in the preparation of a tea taken for chest ailments.[107] A decoction is also used as a wash in the treatment of wounds. Another species, such as *pakuchara, E. hyemale*, is used as a tea for urinary ailments.[108]

Bakusí contains thiaminase, flavone glycosides, saponins, and silica. These constituents may explain some of the therapeutic as well as the toxic properties of *Equisetum*.[109]

Banagá

Conferva sp.
> TAR. *banagá, banewá, banewátama, banawá*[110]
> SP. *lama de río*

This bright green river alga is known by the Rarámuri as *banagá* or *banewá*. It is used for its assumed antiseptic effects and possible cicatrizing qualities. Some recommend its topical application for wounds and burns; others restrict its use to burns. An informant native healer recommends it topically or taken in a tea for general medicinal purposes. I have observed the application of banagá alone or mixed with turpentine for burns by Tarahumara healers.

Bariguchi

Eriogonum atrorubens *Engelm.*, Eriogonium undulatum *Bentham*[111]
> TAR. *bariguchi, mariguchi, maribuchi*
> SP. *hierba colorada*

These plants of the family Polygonaceae—*Eriogonum atrorubens* and *Eriogonium undulatum*—grow alongside streams and in plowed fields of southeastern Tarahumara Country.[112] They are also abundant along the upper course of the Urique River. The root of the plant is extracted using a steel bar or a pick, cut in small pieces and added, fresh or dried, to nixtamal (corn dough) ground on the matáka. The herb is an important flavoring substance used as an ingredient in making *siyóname remé* (rather tasty blue tortillas). Children seek its red flowers to sip their sweet nectar.

The root of this plant is mashed and cooked or made into tea used as a remedy for the common cold (chawiste). In general, it is used to relieve coughs and other respiratory symptoms.[113] Another use reported is as an astringent and cure for toothache.[114] Mestizos use it for diarrhea, coughs, and gastric distress.[115] *Bariguchi* also is smoked as a tobacco substitute.[116]

Basigó

Tagetes lucida Cav., *Tagetes micrantha.* Cav.
> TAR. *basigó,*[117] *pasigó, basikó,*[118] *wasígokochi, yeyésowa, e'yesa, yeso*[119]

SP. *hierba anís, yerbanís, anís,*[120] *anisillo, anís silvestre, santarosa,*[121] *coronilla*[122]

NAH. *yauhtli*[123]

Basigó is an herb widely grown throughout Mexico, mainly in mountainous terrains. In Tarahumara Country, *Tagetes lucida* is the predominant species. Basigó grows in wide patches, its bright yellow flowers filling the environment with a characteristic aniseed fragrance. The herb grows in meadows, on hillslopes and mesas, but also in middle-elevation valleys in forest clearings. Basigó flowers blossom during the rainy season—June, July and August—and as late as October in some areas. Mesa de Pilares, Rikusáchi, Rikéachi, and Cusárare are but a few of the places where one may enjoy the sight and smell of these plants. The widely present locative word *basigochi* indicates that the range of *Tagetes* covers the entire sierra country. Basigó is gathered by children and women by simply pulling the plant and uprooting it. These plants are then dried and stored hanging upside down from the beams of houses.

Basigó, or in Spanish *yerbanís*, is a very important item in the Tarahumara trade with Mexicans living in central Chihuahua. *Hierba anís* is highly regarded among Chihuahua City inhabitants. The delicate tea prepared with it shares popularity with oriental black and green teas, chamomile, *hierbabuena* (mint), and the aromatic laurel.[124]

Tagetes lucida (Náhuatl *yauhtli*), has been used from antiquity in Mesoamerica alongside *estafiate*. It was used to treat those diseases attributed to Tláloc, the Toltec rain god.[125] The Tarahumaras' medicinal use of this plant has been known since the eighteenth century.[126] Tarahumaras and mestizos use it in a tea prepared with boiled flowered plants as an intestinal astringent and as a remedy for the symptoms that accompany the common cold and influenza. Basigó users appreciate its calorific, diuretic, and diaphoretic effects. It is a remedy for gastric distress, empacho, and other gastrointestinal ailments. The vapor inhaled while sipping the infusion aids in the healing process. At the feast of Santa Rosa, spraying with basigó is part of the ceremonial.[127]

Lewis and Lewis report that the genus *Tagetes* contains lactones and sesquiterpenes, agents known to induce contact dermatitis.[128] These authors report also that an extract of *Tagetes minuta*, a common variety growing in subtropical America, holds promise in cancer therapy. *Tagetes multifida* DC. is remarkable for its diuretic properties.

Chemical analyses of *Tagetes lucida* show the presence of organic acids, tannin, an essential oil, and an alkaloid.[129] *T. lucida* contains the flavones quercitagrin, whose glycosides are diuretic, and kaempferitin, with laxative glycosides.[130]

Batagá
Coix lacryma-jobi L.
    TAR. *batagá*

The sturdy, bright gray seeds of *batagá* are used by the Indians to make elaborate necklaces and other adornments. Batagá is cultivated in the western canyons and sold throughout Tarahumara Country.[131] Batagá has ceremonial and medicinal uses. In the past, the shamans were never seen without wearing such a necklace when officiating at a ceremony.[132] The seeds are believed to possess many medicinal qualities and for this reason children often are made to wear them.

Be'techókuri
Alternanthera repens (L.) Kuntze, *Alternanthera achyrantha* (L.) R. Br.,[133] *Plumbago pulchella* Boiss.
    TAR. *be'techókuri, be'techókori, garichókori, witichókare, na'chururusi, na'chúkurusi, na'chúrusi, nachúturi, nachúresi*[134]
    SP. *tianguis*. hierba del alacrán, hierba gama,[135] *caricholi*
    NAH. *tianguis, tianguispepetla*,[136] *chalcuitlatl*[137]
    N.T. *kijoso*

Alternanthera repens has an ubiquitous presence and use throughout Mesoamerica. Characteristically it grows covering the ground like the carpet (*petate*) that is used to display merchandise in Indian markets. *Be'techókuri* is the name commonly used for the plant in areas such as Norogachi and Samachique. According to several informants, it means "the one who always lives in or near the house." This refers to the fact that it is commonly found in Tarahumara dwellings or in the ruins of old houses.[138] *Be'techókuri*, or *na'chúkurusi*, has been identified as *Alternanthera repens*, an herb of wide distribution on the American continent, from the southeastern United States to Argentina.[139] *Alternanthera* is characteristically found in rocky or sandy terrains.

The medicinal properties of the plant have been known in Mesoamerica from pre-Columbian times. Descriptions of the plant and its uses date back to the early colonial period and are included in the classical studies of Hernández (1570–1577). The juice of be'techókuri has astringent properties; therefore; it has been found useful treating hemorrhoids, swollen gums, dysuria, and dysentery.[140] More recently, it has been attributed to control fever and to act as a diuretic.[141] The Franciscan "Relación de Guazapares" reports that be'techókuri was mixed with turpentine or lard in poultices and applied to bone fractures.[142] In Kwechi and Norogachi, be'techókuri is the remedy of choice for emesis. It is also used to relieve the sensation of dryness in the throat. Topically, the decoction is used as a poultice for skin infections and joint problems. In Guachochi, the tea is used to relieve headache. *Alternanthera repens* leaves are used in a tea for digestive disorders, and the long root is pulverized and infused in a drink taken for urinary retention.[143] A tea prepared from its roots and leaves is used to alleviate fever, headache, and sore throat. The decoction of be'techókuri has been used to treat typhoid fever and the ground root is used as a poultice on dermatitis and measles.[144]

The plant reportedly contains a soft resin, chlorophyll, oxalates, nitrates of sodium and potassium, and an unidentified glycoside.[145] Researchers at IMEPLAM (Instituto Mexicano para el Estudio de las Plantas Medicinales) have studied the effects of an aqueous extract of *tianguis* on the aorta and atrium of rats, which are similar to the effects of adrenaline and isoproterenol. The extract has an adrenergic effect inducing isoproterenol-like effects in the rat's isolated auricle.[146]

Bichinaba
Unidentified
   TAR. *bichinaba*
The use of the plant called *bichinaba* as a purgative has been reported.[147] The Rarámuri verb *bi'chinama* means "to peel, or remove the bark"; therefore, the term may suggest that it is a bark and not a root.

Boldo
Peumus boldus L., *Boldea fragans* Jussieu
   SP. *boldo*
This is a medicinal plant known from the American Southwest to Chile as a digestive and remedy for hepatic disorders; *boldo* is quite popular in northern

Mexico. The dried leaves of boldo are obtained from Mexican merchants and pharmacists and used in a tea for empacho and mollera caída by mestizos and acculturated Tarahumaras.[148]

Borraja

Borago officinalis L., *Amsinckia* sp.
    TAR. *borraja, sarábora, sewáchari*[149]
    SP. *borraja, morraja*

Borraja is used as a medicinal plant by mestizos and by some Tarahumaras in Samachique, Cieneguita, and Cerocahui. There are no reports of *Borago officinalis* growing in the Tarahumara area; however, some local varieties may exist given that the plant grows well in in cold mountain areas. Borage is described by Samachique and Cerocahui informants as "a plant covered by hairs with blue flowers." This matches the characteristics of the family Boraginacea, particularly the forget-me-nots.[150] A plant of the family Boraginacea—*Amsinckia* sp. (*sarábora* or *sewáchari*)—has been observed in Tarahumara Country. A commercial borage flower is available in central Chihuahua and sold in the sierra by mestizo merchants. Infusions of *borraja* are a favorite remedy used throughout the sierra for the treatment of dysentery. Bye reports that sarábora is used by the Tarahumaras as a general medicinal tea.[151]

Potassium nitrate is a predominant salt in the extract of borage. The extract also contains mucilage, albumin, and mineral salts, which may be responsible for diuretic and diaphoretic effects. It is found useful in febrile and eruptive diseases.[152]

Brasilillo

Calliandra eriophylla Benth.
    TAR. *sitagapi*[153]
    SP. *brasilillo*

The identification of this plant has been difficult because it has characteristics and uses similar to those of *sitagapi*, or brazilwood (*Haematoxilon brassiletto*). It is reported that the plant is the source of a dye for wool. Regarding medicinal uses, decoctions of the entire plant, strained and set aside for several days, are used to treat gonorrhea.[154]

Cacalosúchil
Plumeria acutifolia Poir., *Plumeria rubra* L., *Plumeria* sp.
  TAR. *rochiló*[155]
  SP. *cacalosúchil*

Tarahumara informants claim to be familiar with the effects of *cacalosúchil*; however, they seem to confuse this plant with jumete. The latex, bark, and root of a *rochiló* (*Plumeria acutifolia*), or Temple Flower, is used as a purgative. Such use has been reported for at least the past fifty years.[156] The gum of the wood of *cacalosúchil*, or Mexican frangipani, is also known to have been used as a purgative by pre-Columbian Indians in central and southern Mexico.[157] The plant is also applied as a poultice to wounds, cuts, and scratches.

Lewis and Lewis point out that the milky or yellowish juice from *Plumeria* sp. may be responsible for causing contact dermatitis.[158] Some plants of the family Apocynaceae, particularly *Plumeria multiflora* Standl., contain the antibiotic plumericine, which is active against fungi.[159] An alkaloid in the latex of *Plumeria* may kill microorganisms and larvae in infected wounds and those caused by puncture from thorns of spiny plants. An alkaloid in the latex glycoside, plumierid, is also found in lesser concentration in its leaves and roots but not at all in the latex.[160]

Calahuala[161]
Notholaena candida (Mart. & Gold) Hook, *Notholaena aurea* (Poir.) Desv., *Notholaena limitanea* Maxon var. *limitanea*
  TAR. *kalawala, kanawala*[162]
  SP. *calahuala*
  N.T. *úpasai*

The canyon Tarahumaras use a decoction of the *calahuala* (*Notholaena candida*), a fern, as a remedy to control fever.[163] Also reported is the use of a tea prepared with this plant for congested chest, heart and kidney ailments, and cough and respiratory problems.[164]

Calomeca de Los Bajos
Unidentified
  SP. *calomeca de los bajos*

According to a mestizo informant, the root of *calomeca* is used in a decoction to alleviate disorders of the kidneys and the bladder. The Aztec-like name of this herb resembles *cocolmeca*, to which it may be related.

Canela
Cinnamomum zeylanicus Nees[165]
   SP. *canela*

Cinnamon is, of course, not of local origin; however, its use has become common as a Tarahumara and mestizo remedy. The plant is used powdered or in sticks. *Cinnamomum zeylanicus* is brought from plants grown in Chiapas and Veracruz or from traditional sources in the Far East to stores throughout Mexico. In addition to its culinary uses, Mexican and Latin American folk healers believe that cinnamon facilitates digestion and controls flatulence. Abdominal pain is commonly treated by mestizo housewives and by some Tarahumaras with a hot cinnamon tea.

Cascalote
Caesalpinia cacalaco H. & B.
   SP. *cascalote*

The term *cascalote* is applied to a diversity of shrubs from which tannin can be obtained for tanning animal skins and to prepare dyes. Tarahumaras and mestizos know other tanning plants that grow in the region and are discussed later. Particularly in Tónachi, the term *curar al cascalote* refers to the process of tanning using the bark of several trees, particularly oaks. Several herbs or barks known as *cascalote* are used by Tarahumaras and mestizos to treat *mal de orín* (urinary tract infections). Bye points out that in addition to tannins, the bark of *Caesalpinia* contains also saponins.[166]

Cha'gusi or Cha'gúnari
Baccharis glutinosa Pers.,[167] *Baccharis* sp.,[168] *Purshia tridentata* (Pursh) DC., *Franseria cordifolia* A. Gray[169]
   TAR. *cha'gusi, cha'guna,*[170] *cha'gúnari, ropagónowa*[171]
   SP. *pasmo, hierba del pasmo, vara dulce, jarilla de río*

Cha'guna, cha'gusi, and *cha'gúnari* are variants of a word applied to *yerba del pasmo* (*Baccharis* sp.).[172] *Cha'gusi* usually designates the species *Baccharis glutinosa*; the term *cha'gúnari* is used for *Purshia tridentata* of the family Rosacea and *chaguna,* for *Franseria cordifolia*.[173] All of these terms derive from the verb *cha'gunama*, "to suppurate or produce purulent discharge." The name *ropagónowa, ropagórema,* or *ropagókoma* means "to have a stomachache."

The Tarahumaras recognize a large and a small variety of cha'gúnari, both noted for their whitish flowers. Pennington claims that both *Baccharis* (cha'gusi) and *Purshia* (cha'gúnari) are used to kill fish by dissolving them in ponds, but in Norogachi neither is employed for this purpose.[174]

Since the time of the *Relaciones* in 1777, *hierba del pasmo* was considered an effective medicinal plant used to heal pasmo; that is, infections or abscesses of the skin or the subcutaneous cellular tissue. The Tarahumaras use it to treat suppurative lesions (*cha'gúname* or *bisóiname*), applied as a hot poultice to reduce swelling from wounds and sores. The young leaves of this herb are also bruised and applied as a poultice to heal skin infections.[175] Tarahumara informants report its effectiveness in wawana, ripíwarii, and buyánari, conditions described earlier, and for different forms of infectious dermatitis. Some recommend boiling the herb and bathing with the decoction. However, I observed more commonly the use of the leaves applied as a poultice fastened to the affected part and kept on for several days. The leaves of *chagúnari* (*Purshia tridentate*) are mashed and used as a poultice to relieve boils.[176] The Tarahumaras drink a decoction of the leaves and stems as remedies for severe coughs and colds.[177] Bye[178] reports the use of the leaves of *ropagónowa* (*Baccharis* sp.) in a tea for gastrointestinal disorders.

Almost all members of the Asteraceae family contain sesquiterpene lactones, many of which are cytotoxic; particularly *Baccharis pteronoides* is toxic to sheep and cattle.[179]

Chawiró or Chawé

Lemairocereus thurberi (Engelmann) Britton and Rose, *Cephalocereus leucocephalus* (Poselger) Britton & Rose, *Pachycereus pecten-aboriginum* (Engelm.) Britton & Rose, *Echinocereus* sp.

TAR. *chawé, chawiró, chawirochi,*[180] *wichowáka,*[181] *bitaya mawalí,*[182] *wichurí,*[183] *napísala,*[184] *na'písora,*[185] *na'písuri, na'píchuri*

SP. *pitahaya, hecho, cardón*[186]

These giant cacti are characteristically found in the Sonoran Desert, coastal plains, and the bottom of the great Chihuahuan canyons. In canyon country transitional zones, they are found standing up alone or in groves known as *pitahayales*. The impressive cacti called *chawiró*, however, do not reach the height of the giant saguaro. It has been described thus: "Like a mezquite it has small leaves, like a *guamúchil* it carries the leaves on the top and like *chawiró* bears its fruit high."[187]

The different species of *pitahaya* or *chawiró* have been grouped together by some authors as *Cereus giganteus* Engelm.,[188] or more recently, *Carnegiea gigantea* (Engelmann) Britton and Rose. The *pitahaya dulce* of México has been identified as *Lemairocereus thurberi*.[189] The terms *chawiró* or *pitahaya* are also applied in the region to *Pachycereus pecten-aboriginum*. In fact, Pennington contends that these two species indistinctly receive the local designations *chawé, pitahaya, hecho,* or *cardón*.[190] Several varieties of pitahayas are found in the low Tarahumara lands, including *pitahaya echabógame, barbona* (bearded), *bitaya mahualí, pitahaya hecho,* red pitahaya, and *napísola pitahaya ceniza* or *marismeña*.[191] According to Bye, pitahaya echabógame is *Cephalocereus leucocephalus*; bitaya mahualí is *Pachycereus pecten-aboriginum*; the red pitahaya is *Lemairocereus thurberi*; and napisóla is *Echinocereus* sp. *Na'písora* has been identified as *Cephalocereus leucocephalus*, designated with the Spanish name of *pitahaya barbona* or *pitahaya vieja*.[192] Finally, it is important to point out that the name *pitahaya* is given not only to cacti but more commonly to their fruits. Incidentally, pitahaya fruit brought to the area comes from several species, including *Carnegiea gigantea* or *Cereus alamosensis* Coulter, from the Mexican states of Chihuahua, Sonora, and Sinaloa.

The giant cacti's yellow, red, or purplish fruits (*tunas* or *pitahayas*) are highly appreciated by the Tarahumaras as a delicacy. The Franciscan *Relaciones* of 1777 already mentioned that the fruits of *pitallas, cardones,* and *hechos* were considered important to the Rarámuri.[193] The sweetness of the fruit, especially when fresh, led Lumholtz to comment, "It has the best wild fruit growing in the north-western part of México."[194] The fruit is available from June to August and is obtained as has been done traditionally, using a long pole to reach the fruit high in the plant.

The Tarahumaras have multiple uses for the fruit. Commonly, it is eaten raw but it is also fermented to make a drink. The fruit seeds are toasted, ground, and added to kobisi in the same way as is done with squash and *eráka*.[195] The bristly covering of *chawé* is used as a comb. The wood of the cactus is used to make chairs, tables, and doors.[196] In canyon country, dry pitahaya wood is sometimes used as a substitute for chopé, the resinous pine wood used to make torches or as quick firewood to get a fire started.

In regard to curative uses, the sap of pitahayas, particularly that of pitahaya barbona, is used in the Tarahumara lowlands to treat burns. I studied its use by Tarahumaras and mestizos of the Urique Canyon who claimed that the sap could only protect the burned areas from exposure to dust. However, it appears to function also as a cicatrizing agent.[197] The remedy seems to be effective for the treatment of burns, even of second and third degrees. The sap may have antibiotic or bactericidal actions. In the canyon of Batopilas, the sap is also used on cuts, bruises, and sores. A hot tea made with stems of *Pachycereus pecten-aboriginum* is used as a purgative and to relieve pain.[198] The bark also is used to treat dysentery.[199] Stems of na'písora are used, after burning its spines, as a plaster cast in bone fractures. In the western slopes of the sierra and in Sinaloa, native masseurs commonly use it to immobilize reduced fractures.

Psychoactive substances have been found in several of the species of cacti discussed here. Tarahumaras of the western canyons use the juice from young branches of chawé (*Pachycereus pecten-aboriginum*) to induce hallucinatory phenomena during native ceremonies. Branches are crushed and the strained juice is mixed with water. The mixture appears to induce dizziness and hallucinations like those reported from jíkuri.[200] The pulp of the *Pachycereus* stems contains an active principle isolated by Hey under the name of pectinine. This substance produces a physiological reaction similar to the anhaloine contained in jíkuri, reported to induce tetanic spasms.[201] Bye relates that Späth isolated an alkaloid, carnegine, from these species and later Agurell isolated 3-hydroxy-4-methoxyphenethylamine from the alkaloid fraction of cultivated plants.[202] Bruhn and Lindgren observed the same substance in specimens of *Carnegia gigantea* and *Pachycereus pecten-aboriginum*.[203] This last substance is an important alkaloid in the psychedelic family; subjects ingesting it present cross-tolerance with LSD, psilocybin, psylocyn, and dimethyl mescaline.[204] This suggests that all of these substances may share similar mechanisms of action. Also it is possible that several among the species called *chawiró* may contain hordenine, as has been reported for *Cereus* sp.[205]

Chichiquelite

Solanum nigrum L., *Solanum verbascifolium* L.
TAR. *chichiquelite*,[206] *chichikalite*[207]
SP. *chichiquelite, quelite amargo, trompillo negro*
Chichiquelite (*Solanum nigrum*) used as an edible green is considered as an

emergency or last-resort food. It is consumed roasted, boiled, or fried.[208] The plant is collected from June to August, which is also the time to harvest other related plants. It has a bitter flavor; therefore, native herbalists have assumed that it was useful to treat gastrointestinal disorders. The tea has been used as a purgative.[209] The remedy has also been thought to be a urinary antiseptic.[210] More commonly, chichiquelite is used in poultices to treat wounds, scratches, and skin lesions. The entire branch is cooked and after cooling overnight, the remedy is applied on the chest and back to treat pulmonary conditions.[211] It also is used on aching joints to relieve pain.[212] *Solanum verbascifolium* is similarly used, heating its leaves and applying them to sores.[213] Sometimes the leaves are heated overnight and applied wrapped in a cloth.[214] The poultice may be applied to relieve headaches, pain, or wounds.[215]

S. nigrum contains in its green shoots, leaves, and fruits, steroidal glycoalkaloids—solanine, demissine, and others.[216] Apparently, the concentration of the toxic alkaloid solanine contained in the plants varies, depending on the growth stage of the plant. The more pleasant and distinctive the smell, the more poisonous the plant seems to be. In some varieties, the concentration of the alkaloid is so low that it is possible to eat them without apparent consequences.[217] *Solanum nigrum* is rich in iron, with appreciable amounts of calcium, phosphorus, vitamin C, and to a lesser extent, niacin, thiamine, and riboflavin.[218]

Chicura

Franseria sp., *Franseria ambrosioides* Cav., *Ambrosia ambrosioides* (Cav.) Payne, *Ambrosia cordifolia* (Gray) Payne

TAR. *chicura, chikura, chikuri*

SP. *chicura*[219]

This white ragweed, with characteristic irregularly dentate and lanceolate leaves, grows on the river banks of the canyon's bottom. Bye identified *chikuri* as *Ambrosia ambrosioides* and as *A. cordifolia*.[220] The herb is very bitter and apparently distasteful to grazing animals. Some herders claim that goats will only eat it if other kinds are not available. This plant is sometimes severely toxic to cattle.[221]

The leaves of chikuri (*Ambrosia ambrosioides*) are used as a poultice for inflammatory skin reactions.[222] Chicura (*Franseria ambrosioides*) leaves are crushed and placed on the head to relieve headache, while another plant, also called *chicura*, is prepared as a tea taken for headaches and stomach ailments.²²³ The bitter tea of chicura leaves is occasionally used in Morelos, Chihuahua, to treat gastrointestinal symptoms and venereal disease.

Chiká

Cercocarpus montanus Raf.
   TAR. *chiká*
   SP. *palo duro*

   Chiká (*Cercocarpus montanus*), a mountain mahogany, is a common shrub found on exposed upland slopes.²²⁴ It was formerly used to make arrows. This rosaceous plant is known to be cyanogenic under certain conditions, but its leaves are considered safe food for animals.²²⁵ Nonetheless, poisoning of cattle has been reported. Small amounts apparently do not harm humans. The leaves of this shrub are used in preparing a tea taken for the relief of colds, constipation, or dysentery, while its roots are ground upon the matáka and used as a poultice for any kind of sores.²²⁶ The leaves and bark of chiká are boiled into a decoction used for constipation, stomach ailments, and dysentery.²²⁷

Chikí

Unidentified
   TAR. *chikí*
   SP. *hojasén*

   Chikí, or *hojasén*, is a bush with very green leaves that Tarahumaras use as a strong purgative.²²⁸

Chikuri Nakara

Hieracium fendleri Sch. Bip., *Dichondra argentea* Humb. & Bonpl., *Crotalaria ovalis* Pursh, *Dichondra repens* Forst & Forst
   TAR. *chikuri nakara, chikuri*,²²⁹ *chanírusi, rawichi sitákame*,²³⁰ *raragochi*,²³¹ *bahíname, bajiname*,²³² *bajuísuri*²³³
   SP. *oreja (orejuela) de ratón, no me olvides*²³⁴

   This plant has been identified as *Dichondra argentea*, which was considered an economically important herb for the Tarahumaras because in addition to their local uses, they gathered and sold the herb in interregional trade.²³⁵ *Chikuri nakara*, or mouse's ear, was identified as a plant of the family Compositae: *Hieracium fendleri*.²³⁶ It is a tiny plant found in the uplands, especially in the western pine-clad country. On the other hand, *chikuri nakara* was thought to be *Crotalaria ovalis*, which is plainly called *chikuri*.[237] *Rawichí sitákame*, an herb with the medicinal uses reported for *Hieracium* sp., has been known for a long time.[238] Bye identified *orejuela de ratón* as *Dichondra argentea*, but he also recorded *Hieracium fendleri* as *chikuri nakára*.[239]

Dichondra argentea is reported to be used to prepare a tea to alleviate urinary disorders.[240] This tea is also used to treat diarrhea, particularly when attributed to empacho.[241] The fresh or dried leaves of *chikuri nakara* or *rawichí sitákame* (*Hieracium fendleri*) are bruised and used as a poultice for sores and wounds.[242] The powdered leaves of *raragóchi* (*Hieracium* sp.) are used by Tarahumaras in the treatment of boils, sores, and ulcers.[243] Bye includes a *Hieracium* sp., *bahíname*, as among the Tarahumara remedies for ear problems, tonsillitis, and headache.[244] Another plant, which belongs also to the family Convulvulacea, *bajuísuri* (*Dichondra repens*), has been reported being used as a hot decoction to bathe swollen legs.[245]

Cho'péinari or Cho'pénara

Hedeoma dentatum Torr., *Geranium* sp.

TAR. *cho'péinari, cho'pénara, cho'péina*,[246] *chopéinari, chopénara*

SP. *hierba del catarro*

This tiny plant *cho'pénara* (*Hedeoma dentatum*) is used to treat colds, particularly by the upland people.[247] The roots and leaves are crushed in the palm of the hand and cupped over the nose and mouth.[248] The little yellow flowers are boiled to prepare a tea to relieve colds. Another herb, also called *cho'péina*, is used to treat diarrhea,[249] although some informants question this. A mouthwash prepared from *cho'péinari* is used to relieve toothaches.[250] Leaves and roots of *Geranium* sp. are used to relieve cough.[251]

Cho'rí

Matelea sp., *Martynia annua* L., *Martynia fragens*, *Proboscidea parviflora* Wooton & Standley, *Carnavalia villosa* Benth.

TAR. *cho'rí, chorí, cho'ríkari, kachorosi*,[252] *kapochí*,[253] *gajpochí*,[254] *suréchi*[255]

SP. *Talayote, garambullo, espuelilla, espuela de caballero*,[256] *toritos, cuerno de toro*,[257] *lupino*[258]

Cho'rí is a term designating *Metelea* sp., although Tarahumaras also call the edible *quelites* (*Asclepias glaucescens* and *A. brachystephana*) *chorí* or *kachorosi*.[259] According to some authors, *cho'ríkari* or *chorí* is *M. fragens*. *Cho'rí* also has been identified as *Hieracium mexicanum* and cho'ríkari as *Martynia fragens (garambullo)*.[260]

A. euphorbiaefolia Engelm. is included in Bye's list of edible *chorí[ki]s*, while *Gajpochí a'chigali* is identified as *Carnavalia villosa*, and *gajpochí sogapi* as *Matelea petiolaris*.[261] Bye applies the observations of Palmer and Pennington regarding *Martynia fragens* or *M. annua* to *Proboscidea parviflora*.

Regarding the curative powers attributed to these plants, it has been reported that the canyon Tarahumaras crush the roots of *chorí Matelea* sp. to prepare a poultice for boils.[262] The crushed root is used in a poultice to treat burns.[263] It is also reported that the fruits of *Martynia annua* are used in preparing a warm decoction drunk to alleviate chest pain attributed to heart problems.[264] A tea prepared from *M. annua* is used in Batopilas to treat burning abdominal pain and empacho.[265]

Chúcha or Chúchaka
Unidentified
TAR. *chúcha* or *chúchaka*[266]
SP. *hierba de la muela*

Chúchaka is a plant with tiny little flowers grouped in corymbs.[267] It is one among the many remedies used for toothache. The crushed root is pushed forcefully in the cavity of the affected tooth to relieve the pain.[268]

Chu'ká
Senecio candidissimus Greene
TAR. *chu'ká, chuká* (or *chukáka*), *chuǎ, chu'yá, chiká, chikwá*,[269] *chukuá*,[270] *choká*,[271] *chukákari*
SP. *lechuguilla, hierba ceniza, chachacoma*[272]

The tendency in the Tarahumara language to change the vowel sounds within a word makes the name of this plant a source of confusion, whether one is trying to identify the plant designated by such a name or inquiring about its uses and applications. However, it is clear that *chu'ká* (or *chukáka*) is for the Rarámuri and mestizos the herb *Senecio candidissimus*. The word *chiká* (or *chikáka*) is used referring to it or, more likely, to *Cercocarpus montanus*. The forms *chuǎ, chu'ká*, and *chu'yá* may also refer to *Chenopodium album*, or even to *Chenopodium ambrosioides*.[273]

Dense patches of *Senecio candidissimus* (*chu'ká* or *lechuguilla*) are commonly observed on the mesas of Tarahumara Country in places where the pine forest has been cleared. The plant is also found in the open valleys of the pine-oak forest of the central and northern sierras of western Chihuahua.[274] Chu'ká plants contribute to the characteristic appearance of the Tarahumara uplands landscape. The plant is a grayish green, with radially arranged spatulate leaves. The leaves, if cut across, show a central wet zone with a velvety cover. The plant's tendency to spread rapidly is attributed to rhizomatous growth and the absence of grazing animals.[275]

Like other plants of the *Senecio* genus, chu'ká is rich in tannins, which explains its astringent qualities. According to several informants, the plant has diuretic effects. However, little is known about the chemical substances in the plant, which I suspect may have antibiotic properties.[276]

I first became acquainted with the uses for this plant during a trip with the children of Norogachi's boarding school from Creel to Norogachi. A nun, a Tarahumara Indian herself, traveling in the same vehicle noted a patch of chu'ká, asked the driver to stop, and had the children and other passengers collect a large amount of the plant. She later used it to prepare a decoction to bathe children in the boarding school who had scabies, wawana, or other skin lesions. Observing that the remedy seemed effective with the children, we tried it later in the clinic with hospitalized patients suffering from scabies and other skin conditions who also seemed to be helped. Native informants mentioned having used the dried plant crushed and mixed with some water to relieve skin inflammation.[277] Also reported is the use of a warm water bath containing leaves or use as a poultice to treat sores, cuts, and skin infections.[278]

A common use of chu'ká is in the treatment of venereal diseases. In Norogachi, this plant is thought to help chancres to heal. Such a remedy is sold to mestizo prostitutes, who believe it is an effective remedy. Informants from Samachique believe that the plant is effective against infectious skin lesions, whether they are deep-seated abscesses, superficial, or external lesions. This supports the common belief that *wekarí* and wawana are the same illness, one manifested in the skin and the other internally. It is claimed that a tea taken every morning or used as bath water is beneficial. In the region of Pinos Altos, it is used to treat bladder and kidney infections. In Pawichiki, an owirúame recommends chukákari to treat kidney problems. Patients with emphysema, pulmonary tuberculosis, and asthma consistently report relief. The root of chu'ká placed in a tooth with caries reportedly alleviates pain. An additional attribution that deserves further inquiry is its use to relieve nervousness, suggesting that the herb has anxiolytic effects.[279] Fever control, diabetes, and cough relief are among the several effects also attributed to chu'ká.[280]

Cientos

Achillea millefolium L.

SP. *cientos, ciento en rama*,[281] *mil en rama, aquilea*

Achillea millefolium (yarrow or milfoil) is a plant of Eurasian origin.[282] Its local use therefore is the result of transcultural exchanges. Membership of *cientos* in the family Asteraceae is readily suggested by its characteristic white exuberant corymbs. The Spanish name of *mil en rama* ("a thousand in a branch," or milfoil, in English) is an allusion to its rich inflorescence. Botanical sources refer to it as a wild herb that grows in cleared forests and lawns, but in Norogachi cientos is usually cultivated in gardens and considered an ornamental flower. The plant seems to help other plants growing close to resist parasites.

American Indian groups are familiar with it and have several uses for the varieties *Achillea millefolium* and *Achillea lanulosa*.[283] According to Lewis and Lewis, their fresh leaves are chewed to relieve toothache.[284]

In Norogachi, a tea prepared with the inflorescence is a favorite remedy for the treatment of colic. European references also attribute healing properties to the plant. The Spanish *Enciclopedia de las Hierbas* refers to its use as an ointment, pointing out that the infusion is taken as a tonic and to relieve fever, while the dried plant is used to treat baldness. The same source claims that *Achillea millefolium* has antiseptic and general remedial properties.[285]

Ciruelo del Campo

Thyrallis glauca (Cav.) Kuntze, *Galphimia Glauca* Cav.

SP. *ciruelo del campo*

Tarahumaras living in the western canyons gather the fruit of *ciruelo del campo*, a plant of the family Malpighiaceae. The leaves of the plant reportedly are used to prepare a tea to treat diarrhea. It is also used to clean wounds.[286]

Cocolmeca

Smilax sp.,[287] *Smilax mexicana* Griseb., *S. cordifolia* Humb. & Bonpl.,[288] *S. rotundifolia* L.,[289] *Piper* sp.[290]

TAR. *kokolmíka*,[291] *kurúvia*[292]

SP., NAH. *cocolmeca*[293]

Cocolmeca is a climbing vine with strong tendrils that often become lignified into a knotty, twisted stick. Canes made from cocolmeca are common in the Tarahumara region. All originate from southern Mexico. The importance of cocolmeca lies not just in the remedial properties attributed to it by the Tarahumaras. Their use suggests a link with Mesoamerican cultures. The name is Náhuatl, meaning "twisted vine."

Kokolmíka simmered in water is used to bathe and relieve joint pain. Canes made from the plant are offered by native healers to their patients.[294] It is widely believed among Indian groups from Mesoamerica that joint pain may be helped by walking leaning on canes or batons made of cocolmeca.

Contrahierba

Psoralea sp., *Psoralea pentaphyla* L., *Psoralea palmeri* Ock., *Psoralea trinervata* (Rydberg) Standley

TAR. *békochi*

SP. *contrahierba, contrayerba*

Mexican scholars commonly refer to *Psoralea* sp. as *contrahierba*.[295] In particular, *Psoralea pentaphyla* is *contrahierba blanca*. This plant grows in the warmer zones of Tarahumara Country. It may also be found near Norogachi and other places of the central highlands. The plant is assumed to have emetic and laxative effects, probably due to the alkaloid psoraline. This substance is quickly absorbed by both oral and subcutaneous routes. It is rapidly eliminated from the body. In infectious states, it reduces the fever as salicylates do.[296] Decoctions of contrahierba blanca appear to be an effective way to treat malaria. Like most snakeroots, putative antidotes for snakebite, contrahierba is a diaphoretic agent.[297] The plant has been reported since remote times to be an important medicinal plant among the Tarahumaras.[298] The roots of *Psoralea palmeri* are used in a tea for chest congestion and those of *P. trinervata*, for fever. In Norogachi, mestizos and Tarahumaras use a tea prepared from contrahierba to alleviate the common cold, tonsillitis, febrile illness, and emesis.[299] For vomiting, usually a squirt of lime is added to the tea, which improves its efficacy.

Copalquín

Coutarea pterosperma (S. Watson) Standl., *Hintonia latiflora* (Sessé & Moc. ex DC.) Bullock, *Coutarea latiflora* Sessé & Moc. ex DC., *Hintonia standleyana* Bullock, *Croton fragilis* H.B.K., *Croton niveus* Jacq.

TAR. *iwíchuri*,[300] *iwígiri*,[301] *batári*,[302] *atóari*,[303] *kayá*,[304] *chikókawi*, *sikókowi*,[305] *compakini*, *conpakini*

SP., NAH. *copalquín*,[306] *copalchi*, *palo de almizcle*, *copal*, *vera blanca*[307]

N.T. *chivúkari*

The name *copalquín* is of Náhuatl origin designating the trees *Coutarea* (or *Hintonia*) *latiflora* and *C. pterosperma*.[308] These subtropical deciduous trees are common in the canyon country. Interestingly, removal of the bark does not seem to harm the tree. The tree branches and trunks are used in construction as posts.[309] Pennington identified copalquín, or *atóari*, as *Coutarea latiflora* and *kayá*, or *batári*, as *Coutarea pterosperma*.[310] For Bye, copalquín is also *Hintonia latiflora*. He reported the use by Tarahumaras of two plants belonging to the family Euphorbiacea—*Croton fragilis* and *Croton niveus*, called *chikókawi* and *sikókowi* respectively.[311] *Copalchi* is also *Coutarea latiflora* for Cabrera, but he pointed out that the denomination *copalchi* was used by the Indians to designate different species sharing the balsamic properties of copales.[312] Pennington reports the ceremonial burning of the bark of *Coutarea latiflora* with a characteristic pungent odor.[313]

Copalquín, as well as other local medicinal plants, such as *chuchupaste*, *hierba de la vibora*, and *matariki*, have been for the Tarahumaras important products of commercial exchange. The herbs are taken by the Indians to Chihuahuan towns and cities and sold door to door or in quantity to traders in remedial herbs. This type of commercial activity dates to very early times.[314]

It should be noted that the antimalarial properties of plants of the families Euphorbiaceae and Rubiaceae known as *copalchi* have been used throughout the North American continent. This use has justification given that the bark contains the alkaloids quinine and quinidine.[315] The bark of *Coutarea latiflora* is therefore an alternative or apparent substitute for quinine from other species.[316] In the past, malaria was a major health problem in the area and cases of this disease are still endemic in the canyons; conventional medications such as quinine or chloroquine, which can be more reliably administered, have replaced the use of copalquín to control the disease.

The long experience using the bark in febrile diseases has perpetuated its use to treat recurrent fevers. The Tarahumaras have used copalquín for medicinal purposes since remote times.[317] Hrdlička reported that *kopalkin* [sic] was used to alleviate chest pains.[318] Zingg noted that in addition to the use of the bark for fever, the powdered leaves were applied as a poultice to treat gall sores suffered by animals.[319] Bye noted similar use in humans.[320] Pennington described the use of *kayá* (*Coutarea pterosperma*) scraped and boiled into a mixture used to control fever or relieve gastric acidity.[321] In Norogachi, a copalquín decoction is used to treat the symptomatic complex called *bilis*, which is attributed to repressed anger and frustration and causes digestive problems and general discomfort. The remedy is also utilized for venereal diseases and flatulence. The tea prepared from the bark is very bitter and is administered in small amounts. Some Tarahumaras recommend the tea to prevent conception, suggesting that it has abortive effect. In Samachique and Humariza, a weak tea is used to relieve stomachaches. Zingg and Pennington report that the pounded bark of *Croton niveus* is boiled in water and drunk as a diuretic.[322] Bye reports the use of the bark, leaves, and resin of *chikókawi* (*Croton fragilis*) for toothaches, adding that it is prepared into a tea as remedy for coughs and sore throat.[323] Currently, copalquín is more important for Mexicans as a remedy than for the Rarámuri, its traditional suppliers.

Copalquín reportedly is used by the Tarahumaras to sweeten tesgüino if it is too bitter after fermentation. In the Urique Canyon, the bark of several plants of the family Rubiaceae—*Coutarea pterosperma*, *Randia laevigata*, *Randia Watsoni*, and *Randia echinocarpa*, collectively known as batari—are used in the preparation of tesgüino.[324]

The bark of *Croton fragilis* contains tannins and saponins that could explain its expectorant and analgesic effects. Salmón summarizes that the croton oil contained in the seeds, leaves, and stems of *C. fragilis* consists of glycerides and a mixture of terpenoid principles known as phorbols.[325] He emphasizes the potential toxicity of this species. The bactericidal properties of copalquin suggest the presence of a substance, which could be selovicin, also found in croton oil.[326] The bark also contains essential oils, organic acids, resins, a dye, and a copalchine crystallizable principle that seems to be of alkaloid nature; quinine and its more toxic ∂-isomer, quinidine, have been isolated from the bark of copalquín, *Hintonia latiflora*.[327]

Coronilla

Berlandiera lyrata Benth. var. *macrophylla* Gray
  SP. *coronilla*
This plant of the family Compositae *Berlandiera lyrata* var. *macrophylla* grows both on the western and eastern Sierra Madre and along small streams in southwestern Tarahumara Country.[328]

The roots of *coronilla*, boiled in water, are used as a purgative throughout the area. It is also used for any condition that according to the native healer may require an "intestinal cleansing." The *Relaciones* of 1777 already had mentioned the purgative effects of the plant. This plant may be the same coronilla reported in colonial times as an antidote for the arrow poison used by the Tarahumaras. González Rodríguez quotes a 1697 letter written by General Retana describing the use of the herb: after promoting bleeding from a wound, a poultice of coronilla was applied to it with life-saving results.

According to informants, in Norogachi different uses are mentioned depending on whether the informant is Tarahumara or mestizo. Tarahumaras find it useful—similar to copalquín—to treat wekaríki (gonorrhea, or venereal disease). Mestizos, on the other hand, advocate its usefulness for backaches and lumbar pain. A mestizo informant from Guachochi claims that coronilla, applied as a poultice, is as useful as *hierba del cancer* to accelerate the resolution of hematomata. Finally, Bye reports on its use by the Tarahumaras in a cleansing wash for fever.[329] Deimel argues that only mestizos prefer coronilla as a remedy.[330]

Epasote

Chenopodium ambrosioides L., *Teloxys ambrosioides* (L.) Weber
  TAR. *chuǎ, chu'yá, chu'ká, chuá*[331]
  SP., NAH. *epasote*,[332] *epazote, ipasote, basota,*[333] *basote,*[334] *pasote*[335]

Epasote (*Chenopodium ambrosioides*) is considered by the Tarahumaras a quelite; that is, an edible herb and especially a condiment for bean and meat dishes. The plant is one of three species of the family Chenopodiaceae—*Ch. album, Ch. Ambrosioides,* and *Ch. Graveolens*—designated by the Tarahumaras *chuǎ*.[336] *Ch. ambrosioides* is more often called *epasote* (or *ipasote*) by Tarahumaras and mestizos. *Júpachi,* or *hierba del zorrillo*, designates *Ch. Graveolens*, which is characterized by its offensive smell and its tiny leaves. Some native informants indicate that the word *chuǎ* is also used to designate a plant known in

Spanish as *quelite cenizo, bledo,* or *chual* (*Ch. album., Ch. Ambrosioides,* and *Ch. Graveolens*) growing in cleared spots used for cultivation or human dwellings in the uplands. In Cusárare, Samachique, Pawichiki, Choguita, and Norogachi, these herbs are allowed to grow within the cornfields because of their usefulness as condiments and medicinal properties. *Ch. ambrosioides* grows alongside the creeks of the lower sierras and upper barrancas. From pre-colonial times, epasote leaves were used by the Tarahumaras as a condiment. The leaves of the plant may be eaten raw, cooked, or fried. When fried, epasote does not seem to have significant toxic effects.

Like other Indian groups and the Mexican population in general, the Rarámuri are familiar with the medicinal properties of *Chenopodium ambrosioides.* The current medicinal use is the same as that recorded by the Franciscan friars writing in the eighteenth-century *Relaciones.* Scholars visiting Norogachi in the late nineteenth century noted its use as a febrifuge and anthelmintic.[337] A decoction of epasote leaves is administered to children suspected of having intestinal worms. The remedy is also administered to horses, cattle, and pigs. The anthelmintic, irritative, and carminative actions of *Chenopodium* explain its use to treat various gastrointestinal ailments. It is also considered an effective febrifuge and pain reliever.[338] When *Ch. Ambrosioides,* as well as *Ch. Berlandieri* and *Ch. Album,* are used as a food condiment, frequent change of the cooking water seems to lower the plant's toxicity. Epasote in a tea is taken to relieve fever and stomachaches. It is also claimed that it relieves headaches and joint pain.[339] This remedy is used to facilitate childbirth and to alleviate skin outbreaks and itching. In Kwechi, epasote decoctions are given to relieve coughs.[340] Informants recommend it for rikibuma (helping the baby descend the birth canal). Large amounts of the infusion are used to induce abortion but the effect is not reliable. Bye claims that a stronger decoction is used externally to kill worms in infested wounds and ears.[341] Mestizos use epasote as a remedy for menstrual pain.[342]

Regarding active principles found in epasote, an essential oil is produced and collected in saclike hairy structures in the plant's flowers, leaves, and young stems.[343] The oil fractions contain limonene and ascaridole. Velásquez highlights an important fact: while ascaridole is very toxic, chenopod oil is less so, and the entire plant is still much less toxic; however, the anthelmintic properties are present in the three, and paradoxically, oil of chenopod appears to be more effective than ascaridole itself.[344] In addition to ascaridole, terpinene, pinocarvone, aritasone, saponin, sapogenine, an alkaloid chenopodine, tannins, and a water-insoluble triterpenoid have been found in chenopodium oil.[345] The therapeutic dose range of chenopodium oil is quite narrow, and toxicity, consisting of neurotoxic, central nervous system effects, including depression of the respiratory center, increases rapidly with higher doses. Intoxication may cause convulsions, cardiac and respiratory-center depression, and death.[346] The treatment of infestations by *Ascaris lumbricoides* (roundworms), has been the primary use of epasote and chenopod oil. The anthelmintic spectrum also covers *Ancylostoma duodenale*, *Necator americanus*, and some parasites of the family Cestodeae. The oil also carries a bacteriostatic action against a diversity of microorganisms, including *Staphylococcus aureus*, *Escherichia coli*, *Pseudonomonas aureoginosa*, and *Candida albicans*.[347] Amebicidal effects also have been reported.[348]

Erá or Eráka

Opuntia sp., *Opuntia tuna* (L.) P. Miller., *Opuntia hernandezii* DC.,[349] *Opuntia ficus-indica* (L.) Mill., *Opuntia phaeacantha* Engelm., *Opuntia arbuscula* Engelm., *Opuntia lindheimieri* Engelm., *Opuntia robusta* Wendl. var. *guerrana* (Griff.) Scheinvar, *Opuntia violacea* Engelm.

TAR. *erá* (or *eráka*), *iráka*,[350] *wechaira*, *ronéraka*, *witóri*, *péchuri*

SP. *nopal*, *nopal de Castilla*

The *Opuntia tuna*, or prickly pear, belongs to the *Cactaceae* genus. Although not as abundant in Tarahumara Country as in other parts of Mexico, these cacti play an important role in the Tarahumaras' culture and diet. In the lomerío region, clusters of *nopales* offer their stems and fruits as delicacies to Tarahumaras and mestizos. The great *nopal de Castilla* (*Opuntia ficus-indica*) although not a native of the region, grows in the Cerocahui mission grounds, suggesting that humans planted them there. *Opuntia tuna* and *Opuntia Hernandezii* grow isolated or in clusters in Sitanápuchi. Four species of *Opuntia*— *wechaira*, *ronéraka*, *witóri*, and *péchuri*—may be observed on the bare hillsides of western canyons.[351] Of the four varieties of nopal in the canyons, only two grow in the uplands. Among the canyon varieties, only one, *cara brillante* (or *duraznillo*) is wild.[352] *Péchuri* is the nopal called *tuna de conejo*, or rabbit's prickly-pear.[353] *Chirúrame erá*, the round-leaved nopal, is *Opuntia phaeacantha*.[354] Other species also observed in the area are *O. arbuscula*, *O. linheimieri*, *O. robusta* var. *guerrana*, and *O. violacea*.[355] The larger thorns (*machagí*) of some species are used to spear fish (*ro'chí wichamuko*). Three varieties of *eráka* are used to prepare batari.[356]

For medicinal uses, the Tarahumaras used to cut the leaves, peel, and then crosshatch them with a knife. The ground seeds of cilantro were sprinkled over them and the leaves were roasted and covered with ashes. This mixture was placed in a palm leaf and wrapped around the waist to treat abdominal pain.[357] The compaction of the bowels caused by consuming *napó* (prickly pear fruit) is usually relieved by eating a preparation of *erá* (nopal) leaves.[358] The stems of erá are used as a poultice for burns and to treat bites from poisonous animals.[359] Nopal stems are also used as casts to set fractured bones.[360]

Regarding the pharmacological properties of *Opuntia*, an alkaloid, cactine, has been identified in the stems of *Opuntia* sp. Cactine is a cardiotonic agent; the salts are useful to treat cystitis and urethritis in small amounts and in larger doses they may act as laxative and vermifuge.[361]

Escorcionera
Iostephane heterophyla Hensley
    TAR. *iwisíame*[362]
    SP. *escorcionera*,[363] *escorzonera*[364]

Pennington identified *escorcionera* as *Iostephane heterophyla*, a plant used in southeast Tarahumara Country to obtain a dye to stain wool.[365] For this purpose, the root is crushed and boiled for several hours. Tanned skins are also occasionally dyed with this preparation. *Iostephane* roots appear to have antipyretic and analgesic properties, which may relieve joint and back pain.[366] In the region of Cerocahui, some mestizos utilize a decoction prepared from the root to treat women's sterility and, in other places, it is used to relieve postpartum problems.[367] A weak decoction is used as a tea and a strong one, as a poultice. It is also used to treat bad breath, poisonous animal bites, bruises and wounds, and venereal diseases.[368]

Estrenina
Plumbago scandens L.
    TAR. *rochínue, rochínowe*[369]
    SP. *estrenina, corcomeca*

This plant of the family Plumbaginaceae has been favored as a remedy for a variety of disorders by mestizos and Tarahumaras. The latter also use its roots in the preparation of batari. Roots and branches of *Plumbago scandens*, common in middle portions of the western canyon slopes, are crushed and used in preparing a tea taken for intestinal disorders or articular pain.[370] An infusion made from estrenina is applied topically to treat wounds and skin parasites.[371]

Eucalipto
Eucalyptus globulus Labill., *Eucalyptus* sp.
   SP. *eucalipto*

Eucalypts are not indigenous to the New World but these trees have adapted remarkably well to a diversity of climates and have a ubiquitous presence on the continent. The active principle in eucalyptus, eucalyptol, is a terpenic volatile oil.[372] Expectorant and antitussive properties are effects attributed to this substance. It also has been used as a febrifuge and antimalarial agent.[373] The volatile oil however may be quite toxic at high doses.[374] As with other oils, it is absorbed through the skin, intestine, or lungs. A tea prepared from the leaves of *eucalipto* is highly appreciated by the mestizos of Norogachi, Cerocahui, and Guachochi and is taken as an antitussive remedy. Tarahumaras claim that the tea can be also used for the treatment of *kuchíwari* and bladder ailments.

Gobernadora
Larrea mexicana M., *Larrea tridentata* (Moc. & Sessé ex DC.) Coville[375]
   SP. *gobernadora, guamis*

Although *gobernadora*, the creosote bush, is not properly a Tarahumara plant, it is widely used by the population as a remedy.[376] The species is characteristic of the Sonoran and Chihuahuan deserts.[377] Creosote bush clusters reach the boundaries of Tarahumara Country west of the Río Florido where it joins the Río Conchos.[378] Gobernadora is considered a nuisance in the cattle-raising land of central Chihuahua due to its invasiveness and the consequent displacement of forage grass. The plant is well known in mestizo settlements in Guachochi and San Juanito.

Tarahumara and mestizo informants, even those from isolated hamlets, believe in the medicinal properties of gobernadora, or *guamis*. Preparation of gobernadora tea to be used as a remedy requires stirring briefly a branch with leaves in boiling water; a prolonged procedure may result in a toxic potion. Informants from Chihuahua City claim that guamis is effective in treating gallstones and renal stones. The most common use of gobernadora is in the treatment of peptic ulcers. Gobernadora seems to be an effective remedy applied externally against skin parasites and infections, and there are claims of good results in noninfectious dermatitides and skin cancer. Creosote was included in the *U.S. Pharmacopoeia* for more than one hundred years.[379]

Go'tó

Phaseolus metcalfei Wooton & Standley
  TAR. *go'tó, ko'tó, o'tó, we'tó*
  SP. *frijolillo, corcomeca*[380]

The crushed roots of *go'tó* are one of the most common agents used as a catalyst in the preparation of batari.[381] The roots and leaves are also cooked and used to make clay pots waterproof.[382] In southeastern Tarahumara Country, the roots are used in preparing a tea taken to alleviate stomach distress.[383]

Granada

Punica granatum L.
  TAR. *ganarsi*[384]
  SP. *granada*

The pomegranate tree, *Punica granatum*, a native of North Africa and western Asia, is cultivated in some areas of Tarahumara Country. When local stores have pomegranates available, the fruits of the plant are eagerly acquired by mestizos and Tarahumaras. The fruit skins are boiled to prepare a medicinal tea considered an efficacious remedy for dysentery. In Cerocahui, the skins are mixed with another infusion prepared from gordolobo (rasó), supposedly to increase its antidysenteric efficacy.[385] In Norogachi, mestizo informants claim that this mixture combined with *zempoal* is even more effective. A poultice prepared from the fruit is used to help heal boils on legs or feet and as a topical anesthetic.

Granada reportedly contains chlorophyll, pectic substances, starch, sugar, calcium oxalate, potassium sulfate, abundant tannic acid, and four alkaloids: pelletierine, iso-pelletierine, methyl-pelletierine, and pseudopelletierine. Pelletierine is an active principle and is carried in the fruit skin, bark, roots, and stalks of the tree. Pelletierine is used as an anthelmintic to treat *Taenia* spp. (tapeworms), *Bothriocephalus* spp. (a cestode worm), and *Ancylostoma* sp. (hookworm) infestatrions.[386] High doses may cause headaches, vertigo, and convulsions.

Guamúchil

Pithecellobium dulce (Roxb.) Benth.

TAR. *mikúchuni, mikúkuni*[387]

SP. *guamúchil, huamúchil, wamúchil*

Guamúchil is an important tree not only in the Tarahumara canyons but also throughout the coastal plains of the Mexican states of Sonora and Sinaloa. The tree is very tall, with some specimens reaching up to sixty feet in height. It grows along the Urique and Fuerte Rivers and in general is well known in all the canyon country. The fruits or arils of the tree are considered a delicacy by the barranqueños. Guamúchil has long been used as a source of food and also as a material used in tanning and in preparing a dye.[388] The leaves of guamúchil were boiled by the Tarahumaras to make a decoction used as a wash for sore eyes or for "clouded vision."[389]

Gu*rú

Nolina duranguensis Trel., *Nolina matapensis* Wiggins

TAR. *gu*rú, ku*rú, gu'rú, gurú, gu'ú, kurú*

SP. *palmilla*

Plants of the family Agavaceae (*Nolina duranguensis* and *Nolina matapensis*, both commonly known as *palmillas*) have considerable economic importance as a source of fiber to weave baskets, hats, and mats (wari and *émuri, koyachi*, and petates). Palmillas are not found in the high altitudes covered by the pine forest. The search for these plants in the lomeríos, or canyon regions, promotes social exchanges between the Tarahumaras living in these different habitats. The rash caused by exposure to nettle ivy (*bikáchari*) can be treated with ashes resulting from burning *gu*rú*, used as a skin remedy.[390] Occasionally, the plant is used to treat other skin disorders and in particular, burns. Ashes of old wari are used for ceremonial purposes.

Hierbabuena

Mentha canadensis L., *Mentha arvensis* Stew.[391]

TAR. *bawena*[392]

SP. *hierbabuena, yerbabuena*

Hierbabuena (*Mentha canadensis* or *Mentha arvensis*) is probably the most commonly used medicinal plant in Mexico. The approximately six-inch-high plant grows well in protected settings or cultivated. It is taken for medicinal purposes or as a refreshingly pleasing beverage. Tarahumaras in Cerocahui have been reported using it as early as 1777.[393] As an aromatic and sweet herb, believed to be "fresh," it has found several uses, ranging from gastric disorders to the common cold and pneumonia. Its diaphoretic properties make hierbabuena tea useful to relieve the symptoms of febrile diseases. Its use for diarrhea and dysentery suggests that the infusion may be more than just a source of fluid to compensate for the body's dehydration. Hierbabuena is used to relieve toothaches, either as a mouthwash or by placing fresh or dry leaves directly on the gums.[394] *Mentha arvensis* contains a volatile oil from which pulegone and thymol, or carvracol, have been isolated.[395]

Hierba de la Chuparrosa

Erodium cicutarium (L.) L'Hér.
  TAR. *simuchí, chumuchí*[396]
  SP. *hierba de la chuparrosa*

Infusions prepared from the *hierba de la chuparrosa* are used to alleviate sore throat, coughs, or to relieve stomachaches.[397] The crushed and dampened leaves are placed in the ear to relieve earache.

Hierba de la Flecha

Sebastiana pringlei Wats., *Sapium biloculare* (Wats.) Pax., *Sebastiana pavonia* Mull.
  TAR. *chuchí*[398]
  SP. *hierba de la flecha*

The name *hierba de la flecha* suggests its use as one of several plants formerly used to poison arrowheads. Innumerable historical references describe the lethality of this practice.[399] At least two plants of the family Euphorbiaceae—*Sapium biloculare* and *Sebastiana pringlei*—are called hierba de la flecha. These plants were commonly used to kill fish in the canyons of the Urique and Fuerte Rivers by dropping them in small ponds.[400] The plants contain latex that oozes when broken; the substance is a severe irritant to skin and eyes. The latex of *Sapium biloculare* contains an alcohol-soluble resin, which is toxic to warm-blooded animals, and a water-soluble saponin, which is toxic to fish.[401]

Regarding medicinal uses, the milky juice from the bark of *Sebastiana pringlei* is added to water and used as a purgative, a practice first observed in San

Miguel near Batopilas.[402] The milky excrescence from the bark of Sapium biloculare is used in an identical manner to *S. pringlei.* The Tarahumara *chuchí* has been identified as *Sebastiana pavonia* and is used as a cathartic.[403]

Hierba de la Golondrina

Euphorbia sp., *Asclepias brachystephana* Engelm. ex Torr.
    TAR. *so'wé*[404]
    SP. *hierba de la golondrina*

Hierba de la golondrina has been identified as *Euphorbia* sp.[405] It also has been identified as *Asclepias brachystephana.*[406] Pennington points out that records of the Tarahumaras using a decoction made from hierba de la golondrina to treat sores date back to 1777.[407] Modern Tarahumaras continue using it to treat sores, wounds, inflammation, and boils.[408] Cardenal reports its use for ophthalmic conditions and internally as an antidiarrheic.[409] Another or possibly the same plant, *Euphorbia heterophylla*, was used according to Zingg by the canyon Tarahumaras for medicinal purposes, particularly for the treatment of febrile illnesses.[410] Bye mentions that the latex from *kusí sigóname* (*Euphorbia plicata* S. Wats.) is applied to sores and wounds.[411]

The complex esters related to the diterpene phoabol found in the most irritating *Euphorbias* account for their carcinogenic and allergenic properties.

Hierba de la Víbora

Zornia diphylla (L.) Pers., *Zornia reticulata* Sm., *Zornia venosa* Muhlenbrock
    TAR. *sayawi*
    SP. *hierba de la víbora, yerba de la víbora*

Hierba de la víbora grows wild in the Tarahumara hill country near Narárachi. This is one of the herbs most commonly associated with the Tarahumaras. In Chihuahuan cities, the plant has a particular appeal to Mexican and mestizo housewives, who usually obtain it from itinerant Tarahumara herbalists. It grows well in gardens but is very vulnerable to frost. Most botanists identify it as *Zornia diphylla* but two different species have been recently described: *Z. reticulata* and *Z. venosa*.[412]

The herb's name, "snake's herb," does not imply that it is a snakeroot; instead, the name refers to the appearance of its leaflets, which look like the scales of a small rattlesnake.

The use of this plant among the Tarahumaras was first documented in the

Franciscan *Relaciones* dating from 1777.[413] Contemporary Tarahumaras use the herb for *mo'ochí okorá*; that is, headaches and gastric ailments. Febrile conditions are treated with a savory tea prepared by boiling the plant. The antifebrile properties of *Zornia* have been compared with those of *Passiflora*. The common cold is a usual target of this remedy, but it is also prescribed for diarrhea, headaches, and general malaise.[414] The tea is also thought to prevent the common cold and influenza.

Hierba de la Virgen
Unidentified
SP. *hierba de la Virgen*
N.T. *kuvái*

A gum obtained from *hierba de la Virgen*, a canyon shrub, when rubbed or drunk is said to lower fever.[415] Tepehuans use the plant as a poultice placed on fractured bones. The *hierba de la Virgen* of the Tepehuans was identified by Pennington as *Stevia stenophylla* Gray.[416]

Hierba del Burro
Unidentified
TAR. *urípachi*[417]
SP. *hierba del burro, zacate de burro*
N.T. *asñitu basógadΛ*

In the region of Humariza, *hierba del burro* is used by mestizos and Tarahumaras in a tea taken for gastric distress. It is claimed that the decoction keeps fleas away and that bruised leaves placed on wounds help to stop bleeding.[418]

Hierba del Cáncer
Acalypha phlioides Cav.
TAR. *banawá*[419]
SP. *hierba del cáncer*[420]

The plant has been identified as *Acalypha phlioides* of the family Euphorbiaceae.[421] Notwithstanding its expressive name, this herb is not used to treat cancer lesions. However, since the Indians conceive cancer as an adventitious disease that is a complication of wounds and other lesions, the name may suggest the attribution of a preventive role. The plant is used as an additive to flavor kobisi, or pinole.[422] Mestizos and Tarahumaras from Guachochi and Tónachi apply the roots of this plant, raw or boiled, as a poultice on wounds, especially if they show signs of secondary infection, or when the process known as a *madura*—that is, a deep abscess—appears to be developing, particularly after a contusion. It is claimed that applied to the skin it will bring the abscess to the surface of the skin to drain. The plant is also attributed antiseptic effects.

Mestizo informants of Norogachi favor its use to accelerate the reabsorption of *sangre molida*, or hematomas, while those from Rochéachi prefer to drink a tea prepared by boiling the root of this plant for the same purpose. In Norogachi, a decoction is drunk to alleviate general pain and heart conditions. *Acalypha phlioides* contains the alkaloid acelyphine. This alkaloid has expectorant and emetic properties similar to those of ipecac.[423] The plant also carries a cyanogenic glycoside.

Hierba del Indio
Aristolochia wrightii Seem.
  TAR. *sore, soreke*
  SP. *hierba del indio, hierba del apache*

This is one more of the plants of commercial importance to the Tarahumaras. The thick root of *hierba del indio* is taken to the cities of central Chihuahua to be sold and used as a medicinal herb. The plant grows wild near Turuséachi. This herb is mentioned by Fr. Antonio de Urbina, author of the *Relación de Cerocahui* of 1777, as an important medicinal plant and it is still used by contemporary Tarahumaras and Mexicans. In Norogachi, it is ground and mashed in water to treat wounds, boils, and wawana. Mestizos prefer to use it as a tea for stomachaches or externally to relieve swollen feet. It is also used for respiratory conditions and venereal disease. In Cerocahui, an infusion prepared with the root is given as an antidiarrheic agent.[424] In Kwechi, made into a suspension, it is instilled in the ears for the treatment of otitis.[425]

Hierba del Pastor
Plantago sp.
  SP. *hierba del pastor*

This plant is also mentioned in the Franciscan *Relaciones* of 1777, as a laxative.[426] Bye identifies it as *Plantago* sp. and reports that it is still used to prepare a tea for gastrointestinal disorders.[427]

Hierba del Piojo

Dalea wislizenii A. Gray

SP. *hierba del piojo*

This plant has been reported to be used as an topical insecticide.[428] Two other *Dalea*s have been reported among the Tarahumara medicinal plants: Pennington writes about *D. polygonoides*, an upland meadow plant, whose leaves are crushed and used in preparing a tea taken for headaches.[429] Bye recorded the leaves of a *Dalea* sp. (*serogoni* or *ronóraso*) being used for gastrointestinal ailments and diarrhea.[430]

Hierba del Toro

Carlowrightia sp.

SP. *hierba del toro*

A tea made with this plant is used as an antimalarial agent.[431]

Hierba de Piedra

Selaginella cuspidata Link, *Selaginella pilifera* var. *Pringleii*, *Castilleia* sp., *Selaginella pallescens* (Presl.) Spring[432]

TAR. *magora*

SP. *hierba de piedra, flor de peña*

Hierba de piedra, stone's herb, is a plant with elongated small leaves that grows as a soft covering on wet rocks in areas such as Samachique. The plant was formerly used as an additive in the preparation of batari.[433] Nowadays, it is boiled as a remedy for stomach ailments. The plant is also used to treat heart ailments, gastrointestinal distress, and to relieve pain.[434]

Hierba de Santa María

Aristolochia brevipes Benth.

SP. *hierba de Santa María, Santa María*

This is a common plant growing in poor and exposed soil in western and eastern Tarahumara Country.[435] In the gorges area, it is used as a tea to alleviate febrile conditions.[436] The roots and lower portions of the stems of this herb are decocted into a preparation taken for intestinal disorders.[437] In Norogachi, mestizos and some Tarahumaras use the bitter tea to treat the flu and articular pain.

Götzl and Schimmer have shown that 9-methoxytariacuripyrone, a nitroaromatic compound derived from *Aristolochia brevipes*, has a strong mutagenic activity in *Salmonella* assays.[438]

Higuerilla
Ricinus communis L.
    TAR. *oriraki*,[439] *sonual*[440]
    SP. *higuerilla, higuera*
    N.T. *mukúkuri*

Ricinus communis (castor oil plant) is probably of African origin, but it is found throughout the tropics and in warmer temperate regions.[441] Ricinus is locally called *higuerilla* and grows in Kírare on the rim of the great Barranca de Urique. Raw leaves of higuerilla have been observed wrapped in a cloth around the head to treat headaches. An ointment prepared from the leaves cooked with lard and wrapped in other leaves was used for large running sores.[442] The purgative properties of the oil apparently are unknown to the Tarahumaras.[443] Fresh beans of the castor oil plant are heated and placed on boils or bruises.[444]

Lewis and Lewis report that *Ricinus communis* is one of the plants containing lectins that can act as mitogens. They also report the presence of phytotoxins, protein molecules of high toxicity.[445]

Hinojo
Foeniculum vulgare Mill.
    SP. *hinojo*

This is an herb native to Europe, but favored throughout Mexico as a spice. In Norogachi, it is attributed to have a sedative effect on anxiety when boiled into a tea. It has been used to relieve colic, gases, vomiting, irregular menstruation, and coughs.[446]

Igualama
Vitex mollis H.B.K.
    TAR. *jarí, julia*,[447] *haríki*,[448] *jalí*[449]
    SP. *igualama, uvalama*

Igualama, or *jarí*, has been identified as *Vitex mollis*[450] However, the term *igualama* is also commonly applied to *capulín* (*Prunus* sp.). The Tarahumara word *jarí* is rarely used and the Spanish term *igualama* is always preferred, as it is throughout the Mexican states of Sonora and Sinaloa where the black, sweet fruit, which ripens in May and June, is appreciated as an occasional desert. Igualama or jarí grows sporadically in the Tarahumara western canyons. The fruit is added to goat's milk and given to babies who cannot digest their mothers' milk. Regarding its medicinal attributions, it is supposed to relieve fever when prepared as remedy, either by steeping its leaves or boiling them.[451] It had been reported previously that leaves and stems were decocted into a tea as a cure for fevers. Bye adds that the tea is given to cure majawá (fright) and also is applied externally on bites of poisonous animals.[452]

Inmortal

Asclepias tuberosa L.

SP. *inmortal*[453]

This plant was repeatedly mentioned in the Franciscan *Relaciones* of 1777, which attributed to it a positive effect in the treatment of headaches.[454] Contemporary Tarahumaras use it thus: dried, pulverized stems are held in the cupped palm of the hand and sniffed to relieve nasal obstruction.[455] Cooked and in small amounts *inmortal* is taken for stomach disorders.[456]

Cardioactive glycosides and toxic resins appear to be present in plants of the genus *Asclepias*.[457]

Iwí

Eucheria socialis, *Arbutus arizonica* (Gray) Sargent, *Arbutus glandulosa* Mart. & Gal., *Arbutus xalapensis* H.B.K.

TAR. *iwí, urúbisi, kurúbasi*[458]

SP. *gusano de madroño, bolsa de madroño, madroño*

These trees are conspicuous members of the pine-oak forest. The typical red and checkered bark of these trees, which appears to be peeling, contrasts intensely with the rest of the vegetation and land formations. The wood and other parts of *urúbisi* (strawberry tree) find a multiplicity of uses within the Rarámuri culture. *Iwí* or *gusano de madroño*, "madroño's worm," is actually the larvae of *Eucheria socialis*, a butterfly of the family Lepidopterae.[459]

Descriptions from the early twentieth century indicate that larvae extracted from the cocoon's small white sac of silky texture in the strawberry tree were cooked in pots.[460] It was confirmed later that these insects make a pocket-like cocoon in the *madroño* about one inch long that the Tarahumaras boiled or roasted for eating.[461] However, this custom seems to have been lost.[462]

The root of the thick-barked urúbisi is used in western Tarahumara Country boiled with salt as a remedy for diarrhea.[463] Tarahumaras used the "pocket-like cocoon" as a bandage.[464] Iwí is reported to have a hemostastic effect.[465] Bye also reports on the belief that eating the larvae, iwí proper, helps the eaters to have long-lasting teeth.[466]

Jíkuri

Lophophora williamsii (Lem.) Coulter, *Ariocarpus fissuratus* (Engelm.) Schum., *Epithelanta micromeris* (Engelm.) Weber, *Mammillaria* sp.

TAR. *jíkuri, santo póleo*[467]

SP. *peyote*

Lophophora williamsii, peyote, or in Tarahumara *jíkuri*, is a small, soft, bluish-green, spineless cactus, native to northern Mexico and the southern United States. This cactus plays an important role in ritual ceremonies of Indian tribes belonging to Kroeber's Ute-Aztec linguistic family of which the Tarahumaras are a member.[468] The plant is incorporated in their ritual, healing, and funeral ceremonies. Interestingly, this cactus contains mescaline and many other alkaloids with hallucinatory properties. The ceremonies centered on jíkuri are called by the same name. At least two other plants occasionally are used, including *Ariocarpus fissuratus* and *Epithelanta micromeres*.[469] It should be noted that *L. williamsii* grows in the Chihuahuan desert away from the land populated by the Tarahumaras.[470]

Throughout Mexico, peyote is thought to relieve joint pain. Earlier chronicles indicated that it was used to treat placental retention.[471] The Tarahumaras have used powdered peyote to treat wounds.[472] It also has been a remedy in the treatment of fractures and orchitis.[473] Peyote is applied externally, after being ground on a metate or chewed, to treat bruises, snakebites, joint pain, and other afflictions.[474] Peyote has been considered an ultimate cure when all others have failed. According to Pennington, *Lophophora williamsii* and *Ariocarpus fissuratus* plants are chewed and applied as a poultice on bruises, bites, and wounds.[475] Analgesic effects are sought in rheumatic conditions using mixtures of peyote juice and water, both as a drink and as a rubbing compound. The use as treatment for headache has also been mentioned.[476] Several authors have observed its use in the kickball race as an amulet tied to the back of a runner to supposedly strengthen him and give him speed.[477]

Jíkuri plants contain diverse pharmacological agents. Several alkaloids have been isolated, including mescaline,[478] anhaloidine, peyotine, lophophorine, and anhalamine.[479] Mescaline and peyotine have morphine-like effects, while the properties of lophophorine and anhaloidine resemble those of strychnine. These phenethylamine and isoquinoline alkaloids appear to be responsible for the auditory and visual hallucinations induced after ingesting the cactus.[480] In addition to the alkaloids mentioned, more than fifty-five related compounds have been isolated from peyote extracts.[481]

There are many accounts of the therapeutic benefits attributed to the plant. In addition to analgesic properties, it has been claimed that peyote improves circulation, increases renal filtration, and produces a state of mild intoxication.[482] The alkaloids appear to cross the placental and mammary barriers. The physiological effects of *Lophophora* as described by early writers indicate that fatigue is relieved and sexual desire is suppressed. Jíkuri was reputed to have a beneficial effect on snakebites, burns, and wounds.[483] Prominent in the many reports on the effects of the plant are descriptions of perceptual and behavioral phenomena, including colorful kaleidoscopic visions and states of depersonalization and dissociation.[484] Psychotherapeutic usefulness also has been claimed.[485]

Juanita
Unidentified
SP. *juanita*

This is an herb or perhaps one of several sharing the same name, characterized by its small, rounded leaflets. Tarahumaras from Samachique recommend its analgesic and antispasmodic properties in the management of stomach ailments and indigestion.

Jube
Tauschia tarahumara[486]
TAR. *jube, sepé*[487]

Constance and Bye only recently have classified this plant, which was described as a relatively uncommon herb from the humid upland meadows of central western Chihuahua. The leaves of *jube* have been used as a quelite, or wild edible plant, in the past. The deep rhizomes are used to treat toothache by placing a small piece of the rootstock in the dental cavity.[488]

Jumete
Jatropha curcas L.
>SP. *jumete*

This is one of the plants in which a*rí, mentioned earlier, is deposited. Stirring a branch of *jumete* in a container of water gives the liquid strong laxative properties, making it too irritating for practical use. There are many local anecdotes about its mischievous use. For instance, during a wedding in Urique a prankster mixed some jumete with the main dish for the wedding feast causing several of the guests leave the celebration with colic and diarrhea.

The purgative properties of *J. curcas* were known to the ancient Mexicans who knew it as *cuahuayouachtli* and used it also as an antispasmodic and analgesic.[489]

Zingg and Pennington claim that the milky juice of *J. curcas* is used as a remedy for "sore eyes."[490]

Lewis and Lewis report that the seeds and fruits of *J. curcas* contain a purgative oil and phytotoxin curcin. They also report that a number of cytotoxic diterpenes, such as jatropham, are isolated from *Jatropha* sp.[491] Orellana claims that the leaves and twigs of *J. curcas* contain α-amyrin (a triterpene), a mixture of β-sitosterol, stigmasterol, campesterol, 7-keto-β-sitosterol, stigmast-5-ene-3β, 7α-diol and stigmast-5-ene and 7β-diol.[492] Four antitumor compounds, including jatropham and jatrophone, are reported from other species of *Jatropha*.[493]

Júpachi
Chenopodium graveolens. Willd.,[494] *Chenopodium foetidum* Schrad.
>TAR. *júpachi, pasuchi*
>SP. *hierba del zorrillo, epasote de zorrillo, quelite apestoso*

As a member of the family Chenopodiaceae, *júpachi*, or *hierba del zorrillo*, shares many of the medicinal properties of *Chenopodium ambrosioides*, or epasote, mentioned earlier, Although also considered a quelite, or edible herb, júpachi is most often considered throughout Mexico as a medicinal plant. Some Tarahumaras and mestizos take a tea prepared from júpachi as a substitute for coffee.

Cabrera lists among the chemicals contained in *C. foetidum* an essential oil, chlorophyll, a pectic matter, an acid resin, calcium bicarbonate, potassium chloride and sulphate, and potassium nitrate.[495] I was particularly interested in the effects of this chenopod in facilitating the menstrual flow in usual doses. On the other hand, in higher doses it causes pelvic congestion, uterine contractions, and may induce an abortion.[496]

In Tarahumara Country, some favor júpachi over epasote for stomachaches. The two species are believed to facilitate rikibuma; that is, helping a child descend through the birth canal. Bye reports on its use in a tea to control diarrhea.[497]

Kachana

Pionocarpus madrensis (S. Watson) Blake, *Thalictrum fendleri* Engelm., *Thalictrum pinnatum* Wats., *Centaurea americana* Nutt.

TAR. *kachana, kachano*,[498] *kachénora, bisiki, wichowéaka, wichiwéaka*,[499] *rabini*[500]

Bye reports that the roots of *kachana* (*Pionocarpus*) are used either as a tea or as a wash for joint afflictions. He also suggests that the root has antipyretic and analgesic properties. It is also reported that it is used to relieve the pain of decayed teeth.[501] *Thalictrum* sp. is also known as *kachana* or *kachano*. When Tarahumaras of the western canyons are overcome by heat prostration, they are given a tea made from the leaves of *Thalictrum fendleri*.[502] A decoction made with these leaves may also be used as a rubbing lotion for the same indications. Roots of *bisiki*, another variety of *Thalictrum* are used to relieve cough.[503] *Wichiwéaka* (*Thalictrum* sp.) is used as a remedy applied to burns. *Kachana* (*Centaurea*) is used by mestizos as a nerve remedy while the Tarahumaras use it to treat sores.[504]

Kasará or Kasaráka

Fimbristylis sp.

TAR. *kasará, kasaláka*

Kasará, or *kasaláka*, a sedge from the uplands, is used as a decoction for colds and pneumonia.[505] Several branches are ground, including the roots, and strained to obtain a drink. The plant is also ground and added to tesgüino to prevent colds.

Keyóchuri
Zexmenia podocephala Gray, *Lantana involucrata* L., *Lantana camara* L.

TAR. *keyóchuri, keŏchuri,*[506] *geyóchuri, geóchuri,*[507] *kebóchuri, kibóchuri,*[508] *teóchuri,*[509] *teŏchori,*[510] *geyóchara, reyóchari*[511]

SP. *peonía, peonía del país*[512]

Zexmenia podocephala is a perennial herb of the family Compositae that grows in the Urique Canyon and in the neighborhoods of Norogachi and Narárachi. *Lantana involucrata is* a plant from the family Verbenaceae, which Mexicans call *peonía*. It is a shrub about eighteen inches high with small fragrant balls growing around a center.[513]

The Rarámuri trade the roots of *Zexmenia podocephala* to Mexicans who in turn sell them to people in Creel and Parral.[514] In Norogachi and Narárachi, the large tuberous roots and leaves of *Zexmenia podocephala* are crushed and placed in a basket, which is then immersed in the river, small streams, or ponds to kill fish.

Peonía was first recorded as a medicinal herb in the 1777 Franciscan *Relaciones* of Cerocahui, Chínipas, and Guazapares. Fr. Falcón Mariano in his *Relación de Wawachiki* (*Guaguachique*) reported also a peonía that the Indians call *tegótzoli* but claimed that the Rarámuri made little use of this plant's properties. The *Relación de Guazapares*, on the other hand, calls peonía *quobhuli* and claims the Tarahumaras used it for empacho. Present-day Tarahumaras make a beverage from *keyóchuri*, using it for a variety of gastrointestinal disorders, including nausea, constipation, and empacho. Mestizos and Northern Tepehuans also use it for empacho. In Norogachi, a peonía tea is also recommended for the treatment of venereal diseases.

Contemporary reports indicate that peonía has several healing uses; the roots are finely ground on the metate and mixed with lard to make an ointment used for a disease that "eats in from the outside," eventually reaching the lungs and causing death.[515] Women drink these roots, pounded and boiled into a decoction, during childbirth.[516]

Lantana camara contains an atropine-like alkaloid, lantanine as well as a phototoxic triterpenic derivative, lantadene-A, and an antibacterial, antifungal, and anti-inflammatory agent, umuherengin.[517]

Kichínowari
Unidentified

TAR. *kichínowari, kichinówari, kichínowa, kichínoa, kichínoari, kichínori*[518]

Kichínowari is an herb with rough leaves and a single flower. The flower consists of a little tuft of purple filaments attached to a central head.[519] In Norogachi and Kwechi, the plant is called kichinówari. It is used as a laxative in a similar way as *naká*, described later. When used for this purpose, it is usually administered after one day of fast.

Kirí
Elytraria imbrincata (Vahl) Pers., *Elytraria squamosa* (Jacq.) Lind.

TAR. *kirí, kochawasira*[520]

SP. *cordoncillo*

This plant shares its Spanish popular name, *cordoncillo*, with the important medicinal plant *Piper sanctum*. *Cordoncillo*, or *kirí*, is a plant belonging to the Acanthacea family; it is called *cordoncillo* because it resembles a string. Some sources identify it as *Elytraria squamosa* growing in the Mexican states of Sonora, Sinaloa, Colima, Coahuila, Puebla, México, Baja California, and Chihuahua.[521] Botanists familiar with the Tarahumara region identify it as *Elytraria imbrincata*.

Mestizos and Tarahumaras of Cerocahui use a decoction of kirí as a laxative. They also claim that it helps to relieve pain caused by renal calculi (kidney stones). Pennington reported that a tea prepared from kirí leaves is administered for fever or diarrhea.[522] The entire plant is also decocted in a rinse used for cleansing wounds. The plant is used to prepare a tea to treat cough, pulmonary ailments, catarrh, and sore throat, and for the alleviation of fever.[523] The plant is also used to treat venereal diseases, malaria, and fever.[524] The expectorant and mucolytic actions of the foliage of *Elytraria* are attributed to the pungent tannins and saponins it contains.[525]

Korí
Capsicum annum L., *Capsicum annum* var. *annum, Capsicum frutescens* L., *Capsicum frutescens* L. var. *baccatum, Capsicum baccatum, Capsicum annum* L. var. *minimum, Capsicum annum* L. var. *glabriusculum* (Dunal) Hieser & Pickers[526]

TAR. *korí, o'colí,*[527] *isíburi, síburi, korí síviri, síbori, chile piquín, chiltepín,*[528] *chile bolita*

SP. *chile, chile piquín, chiltepín, chile bolita*

Chile, or in Tarahumara *kori*, is, with maize, beans, and squash, an important element in the diet of the native people of the Americas. The fruit is a source of vitamins A and C and a condiment in the Rarámuri diet. It is also greatly appreciated for its medicinal and mystical significance. The Aztec name of the fruit, *chilli*, has been incorporated in practically all the languages of the world. The Tarahumara name *kori* is applied to all the varieties of chile, adding a modifier when a particular variety of chile is specified. *Capsicum baccatum* or *frutescens*, the native wild *chiltepín*, on the other hand is plainly called *isíburi*. Several varieties of chile are consumed, including local as well as other varieties that have arrived through transcultural exchanges. All the chile varieties belong to the family Solanaceae and to either one of the species *Capsicum annum* or *Capsicum frutescens*. Korí is a product of cultivation while isíburi is wild. The cultivated varieties derive from seeds or transplants. The Tarahumaras are familiar with the low germination potential of the seeds of this plant. Wild *C. frutescens* grows along the margins of streams in the western canyons; it also may be found along trails, the seeds being brought there by human defecation. Regarding the harvesting techniques, once the fruit is ripe, the plant is pulled from the ground and shaken vigorously over a skin or fiber mat to dislodge the fruit. In Morelos and Batopilas, children and women collect the fruit directly from the shrub without pulling the plant from the ground. The ease with which isíburi grows in the gorges has made of it an important crop to harvest for interregional trade. Recently, rises in the price of chiltepín has made this activity more profitable. Guachochi has become an important trading center of isíburi.[529]

Regarding medicinal uses, the Tarahumaras are guided by their concepts of disease; being *picante*, or pungent, implies that the herb possesses a quality appropriate for diseases considered having the opposite property. Chile is thought to keep away sickness. An admixture of chiltepín with *chupate* (*Ligusticum porter*) is used to treat headaches and colds.[530] Salmón reports that the fruits are used in a poultice to relieve arthritic pain in the hand. Chile has reportedly been used to treat rabies in cattle and men.[531] Korí is believed to relieve chest heart pain while isíburi is used for earache.[532] *Korí chókame*, black chilli, has been used to prepare a tea to relieve fever.[533] The Tarahumaras and mestizos use isíburi to treat hangovers. A popular remedy is a mixture of isíburi and a*rí in water for the purpose just mentioned.

Kori is incorporated in the Tarahumara mystic belief system. The gentile Tarahumaras of Kírare add a bit of ground chile to a mixture of pinole and water consumed after the purifying ceremonies carried out for houses that used to belong a deceased person. Korí is also incorporated in funeral ceremonies.[534] Bye observed people casting powdered chile piquín in the air as a "cure" to keep away evil spirits.[535] In Norogachi, chile is thought to offer protection from the harm caused by sorcery. One informant claimed that burning korí removes the sorcerer's spell from the victim's body. *Risagí*, small mythical birds believed to cause disease, are thought to dislike chile; therefore, these birds are driven away by throwing ashes or chiltipiquín in the air.[536]

The pungent principle in *Capsicum* is capsaicin while its principal coloring matter is capsanthin, a crystalline carotenoid. The oleoresin obtained from *Capsicum* sp. has been used in medicine as an irritant and carminative and contains capsaicin, which when applied to the skin causes a feeling of warmth followed by an intolerable burning. Lewis and Lewis list capsaicin as an agent that may cause contact dermatitis and is suspected to have a carcinogenic potential.[537]

Korísowa

Polygonum punctatum Ell.
   TAR. *korísowa, makarísowa*[538]
   SP. *hierba chilosa*

As its name indicates—*korí*=chile, *sowa*=root—*korísowa* is a pungent, *picante* (spicy) plant.[539] The herb belongs to the family Polygonaceae; its flowers are pink and its leaves have a characteristic pungent flavor.[540] It grows near Sitanápuchi and Norogachi and is one among the fish-stupefying herbs used by the Indians to fish. The plant is placed in ponds or still water to narcotize the fish and collecting them with a *nabóara* (basket).[541] It is also used in cooking to add piquancy to *esquiate* (*elote*, called *esquite* in other parts of Mexico, a fresh corn dish) or as a general condiment. Korísowa is reported to cause dermatitis if exposed to unprotected skin. Tarahumaras from Bakasórare claim that the herb has nasal decongestive properties. A bit of the herb is smelled, making one sneeze decongesting in this way the upper airways.

Kusarí

Prionosciadium thapsoides (DC.) Math. var. *pringley* (S Wats.) Math. & Const.[542]
   TAR. *kusarí*

SP. *hierba del oso*

Kusarí or *hierba del oso* has been identified as *Prionosciadium thapsoides* var. *pringley*.⁵⁴³ It is the same plant from the family Umbelliferae mentioned by Zingg as well as that reported in the eighteenth century "Relación de Cerocahui."⁵⁴⁴ *Kusarí* is used to prepare a bitter decoction taken to relieve fever, colds, or chest affections.⁵⁴⁵

Kusí urákame

Tecoma stans (L.) H.B.K.⁵⁴⁶

TAR. *kusí urákame, koyáwari*⁵⁴⁷

SP. *palo amarillo, tronadora*

Tecoma (or *Bigognia*) *stans* is an important medicinal plant in Mexican traditional medicine. Among the Tarahumaras, *palo amarillo* was and is a household remedy used extensively.⁵⁴⁸ The yellowish flowers of *kusí urákame* are used to prepare a tea taken to relieve colds and employed as a rubbing compound applied to the chest for "heart pain."⁵⁴⁹ This *kusí urákame* may be the same plant reported by the "Relación de Guazapares" as a remedy for sore eyes.⁵⁵⁰

Tecoma stans, incidentally, was one of the plants studied by the physicians of the Mexican Medical Institute, a scientific organization that blossomed at the turn of the century. Later, during periods of revival of the interest in herbal medicines and the chemical structure and pharmacological effects of their active principles in experimental animals and humans, the antidiabetic properties of *Tecoma stans* or *tronadora* became a subject of study. Since then, several pyrindane alkaloids have been reported, including teocomanine and tecostatine, followed later by the discovery of other secondary alkaloids and, more importantly, their components: for the pyridinic alkaloids, actinidine, and for the piperidinic ones, skitantine.⁵⁵¹ These substances, according to some authors, could serve to develop antidiabetic agents.⁵⁵²

Ku'wí

Arctostaphylos pungens H.B.K.⁵⁵³

TAR. *ku'wí, u'wí, uwí, u'wíchari, u'wíchara, ku'wíchora, ku'wíchari,*⁵⁵⁴ *ku'wíchara*

SP. *manzanilla, manzanita, pingüica, tepezquite*

NAH. *tepesquitl, tepesquisúchil*

N.T. *yóri*

Ku'wí is a shrub crowned by evergreen leaves and miniature apple-shaped fruits formerly called "bear's grapes," or *uva ursi*. The name of the plant in Spanish, *manzanilla* ("little apple"), is quite appropriate. Tarahumara children consider these fruits a delicacy; *ikánuri*, the parrots that migrate from the hot lands and the canyons, eat them as well as other fruits like piñones and aborí sprouts.

The contorted, peeling, reddish branches of these shrubs are ubiquitous elements of the Tarahumara landscape, especially in the areas of transition between the sierra and the lomerío. *Arctostaphylos pungens, Arbutus glandulosa,* and *Arbutus arizonica* grow as small shrubs only three to six feet high in poorly drained slopes flanking the uplands; only where pines are mixed with oaks do these species of *Ericacea* grow like trees.[555] Zingg located the ecological niche for *A. pungens* at an intermediate altitude level or at the upper river gorges.[556] Man has played a role in restricting the spread of ku'wí through the burning of fields for agricultural and other purposes.[557]

A feature of this plant is the display within the same branch of areas that are dry and have a dead appearance and other areas that look alive and green. This particularity has intrigued native observers, who see in this the coexistence of death and life, a primordial Indian dualism. Ku'wí is for the Tarahumaras, as well as for many other American Indian groups and Mexicans from urban areas, an important medicinal plant of economic importance.[558]

The Tarahumaras appreciate the mild astringent effect of preparations of this plant.[559] Pennington reports that a tea prepared from leaves of u'wí is drunk to relieve bronchitis or other pulmonary infections.[560] In Norogachi, the same tea is used for whooping cough. Bye reported the use of the leaves of *A. pungens* in a wash to cleanse sores.[561]

The wood of *A. pungens* is used in instruments requiring hardness and durability. It is used to make spoons, ingenious locks, and wooden necks for violins.[562] Ku'wí sticks are used as a fire starter although chopé, the resinous pine wood, is preferred because it does not emit sparks.

The berries are appreciated not only for their medicinal attributes but also as foods in time of famine. In Cerocahui, the berries are also used to make a refreshing beverage.[563] It is claimed that these also may be used to make batari and may be added to atole as a condiment.[564]

Until recently, *Arctostaphylos*'s fruits, "bear-berries," played an important role in pharmacopoeias from many countries. The diuretic effects of *A. pungens*'s fruits, leaves, and extracts have been recognized for a long time and until the advent of the diuretic chlorothiazide they had great acceptance as renal and cardiological remedies. The fruits contain tannin, a resin, gallic acid, and a glycoside, arbutine, a hydroquinone derivative to which is attributed a urinary antiseptic effect.[565] In Mexican pharmacopoeias, their diuretic effects were recommended for kidney and bladder ailments, venereal diseases, and some prostatic disorders.[566] As a urinary tract panacea, the remedy's putative efficacy was extended to the treatment of kidney stones and other conditions, such as bladder infections and urinary incontinence. Uva ursi was also thought to be astringent and tonic and was used to treat diarrhea, flu-like symptoms, and bronchitis.

Lachi

Juglans mexicana *Engelm.*, Juglans major *L.*, Carya illinoensis *(Wang.) Koch.*, Juglans rupestris *Engelm.*

TAR. *lachi*

SP. (the fruit) *nuez, nuez cimarrona*; (the tree) *nogal, nogal cimarrón*

Junglans major is one of the trees growing widely dispersed along streams in western Chihuahua excepting the uplands. American pecan trees, *Carya illinoensis*, grow in eastern Chihuahua along the Florido and Conchos Rivers and in the western side of the state, along the headwaters of the Urique River near Norogachi.[567] The fruit of these trees—pecans, both wild *nuez cimarrona* and cultivated varieties—are found in several places in Tarahumara Country, especially near older towns and settlements. The wild varieties usually produce bitter nuts, but some *nogales cimarrones* do produce edible fruits. Most of the pecans and nuts used as food or for assumed medicinal properties are brought to Tarahumara Country from locations in the adjacent Chihuahuan basin and range and plains where pecans are commercially grown.

Occasionally the Tarahumaras use the wood of the nogal cimarrón tree to make common tools. Young leaves of *Carya* and *Juglans* are used by them to narcotize fish; the foliage of *Juglans rupestris* appears at the time when food supplies are becoming scarce and fish gains importance as a resource.[568]

Concerning medicinal applications, mestizos and Tarahumaras from Cerocahui use a decoction of pecan shells as a drink to treat gynecological and obstetric bleeding. These tannin-containing remedies are possibly selected by the native healers due to their bitter quality rather than based on an intuitive awareness of their pharmacological properties.[569]

Among the pharmacological components from the bark and fruits of *Juglans mexicana* with probable medicinal value, the following have been mentioned: chlorophyll, starch, tannin, malic acid, citric acid, Phipson's nucitanic acid (nucine), and a sweet-smelling essential oil.[570]

Láchimi
Phrynosoma orbiculare
TAR. *láchimi, lachi*
SP. *sapo cornudo, camaleón*[571]

The horned lizard, *Phrynosoma orbiculare*, although obviously not an herb, has been used by the Tarahumaras in healing ministrations. The spines of the lizards are toasted and pulverized for application to the head of individuals suffering headaches who are asked to repeat, "I am sick, heal me."[572] Zingg mentioned that there are several myths involving this lizard.[573] This remedy is now rarely used, if at all.[574]

Lario
Phytolacca sp.
TAR. *lario*[575]

The roots of *Phytolacca* sp. are decocted into a tea drunk to "purify the blood" and the flowers are used in a poultice for poisonous animal bites.[576] The roots contain a glycoside, a glycoprotein, a resin, phytolaccatoxin; a water-soluble triterpene, phytolaccigenin; and a toxic alkaloid, phytolaccine.[577]

Laurel
Litsea glaucescens H.B.K., *Litsea glauscecens* H.B.K. var. *subsolitaria* (Meisn.) Hemsley[578]
TAR. *laureri*
SP. *laurel*

Laurel, the Mexican bay leaf (*Litsea glaucescens*), is a popular medicinal herb used locally. The leaves are important objects of trade sold in cities of central Chihuahua by itinerant Tarahumaras along with herbs such as *matarí, basigó, wasía,* and *hierba de la víbora*. Laurel leaves are collected on canyon slopes; their fragrance and flavor are appreciated as a condiment in local foods.

The leaves of laurel are used to prepare a medicinal tea. The plant was considered as early as 1777 to be an important medicinal plant for the Tarahumaras.[579] The treatment of headaches is the preferred use of laurel for the Tarahumaras. As a medicinal tea, laurel is used throughout Mexico for disorders of the gastrointestinal tract, indigestion, constipation, and "intestinal fevers." The remedy is also used to treat respiratory tract ailments. The tea is thought to have diuretic, diaphoretic, antipyretic, and mild analgesic properties.

Regarding the presence of possible active agents, flavonoids and triterpenes have been identified from plants of the genus *Litsea*.[580] Laurel, of course, also contains an essential oil.

Limoncillo
Pectis stenophylla Gray, *Pectis stenophylla* A. Gray var. *stenophylla*[581]
    TAR. *limoncillo*

Limoncillo, consumed in the form of a strong tea prepared from its leaves, is believed to relieve urinary difficulties. Colds are treated by inhaling its vapors.[582] A tea prepared from the whole plant is used for headaches and gastrointestinal disorders.[583]

Linaza (Aceite de)
Linum usitassimum L.
    TAR. *linasa wi'i*
    SP. *aceite de linaza*

Apparently flax has been cultivated for some time in the land of the Tarahumaras, given that the Indians have a basic knowledge of how to grow this plant. Local merchants, however, obtain the commercial linseed or flaxseed oil used as a remedy by contemporary mestizos and some Tarahumaras in Norogachi from sources outside the area. Linseed oil is used as a balsam on sore muscles and arthritic joints. Occasionally it is used internally, in minute amounts, as a remedy for bladder ailments. A tea is prepared from the roots of *Linum* sp. for cough and respiratory ailments.[584]

Ma'achiri
Unidentified
    TAR. *ma'achiri, ma'achiriki, me'achiri*
    SP. *ciempiés, patas-cien*

This small plant grows under the large rocks characteristic of the landscape in the margins of the upper Urique River in the vicinity of Bakasórare. The plant has small fleshy leaflets with a delicate velvety cover. The leaflets grow from the sides of a central stem resembling a centipede, (ma'achiri). Typically ma'achiri grows on the wet and shady sides of the rock. The plant is boiled by the Tarahumaras in a decoction for gastric bleeding.

Ma'chogá

Cheilantes tomentosa Link, *Cheilanthes pyramidalis* Fée, *Dryopteris normalis* (Sweet) G. Don., *Woodsia mexicana* Fée, *Adiantum capilus-veneris* L.,[585] *Asplenium monanthes* L., *Pellaea termifolia* (Cav.) Link var. *wrightii* (Hook.) A. Tryon, *Pteridium aquilinum* (L.) Kuhn var. *pubescens* Underw.,[586] *Woodwardia spinulosa* Mart. & Gal.[587]

TAR. ma'chogá, ma'chobá, ma'chugá, machugá, mochogá, mochokwá, muchogá, wa'chigara, wachigónuri, píbora[588]

SP. helecho, zarzaparrilla

A great variety of ferns collectively known as *ma'chogá* grow in the land of the Tarahumaras and many of them are favored as remedies. The leaves of *Cheilantes tomentosa* are used to prepare a tea for urinary disorders and roots of *Dryopteris normalis* are mashed and used as a poultice for headaches or crushed and boiled to make a tea to relieve dysmenorrhea. A decoction made from leaves of *wachigónuri* (*Woodsia mexicana*) is a favorite analgesic. Leaves of this plant are sometimes pulverized, moistened, and held on the skin of joints to relieve articular pain.[589] *Adiantum* and *Asplenium* ferns are used in teas for heart ailments while the whole plant of *Cheilanthes pyramidalis* is boiled into a tea for chest congestion, chest pains, headaches, as well as kidney, gastrointestinal, and heart ailments. It is also reported that *píbora* (*Pellaea termifolia*) is used in a hot decoction for a congested chest and chest pain.[590] Finally, *Pteridium aquilinum* is used to prepare a tea claimed to have analgesic properties.[591]

Mystical powers are attributed to some of ferns; for example, *Woodwardia* is used in a decoction to protect men and animals from the feared water spirit.[592]

Polypodiaceae non-flowering ferns have played an important role in western medicine for a long time. The male ferns—*Dryopteris filix-mas* (L.) Schott and *Dryopteris marginalis* (L.) A. Gray—have an oleoresin that has enjoyed popularity as an anthelmintic agent against *Taenia* sp.[593] The oleoresin paralyzes the muscular layer of the intestine as well as the analogous contractile tissue of the tapeworm. Fern-derived remedies have toxic effects: bracken (*Pteri-*

dium aquilinum) contains thiaminase, an enzyme that destroys thiamine and causes serious poisoning, especially in horses and cattle. It also contains shikimic acid, a carcinogenic agent.[594]

Mala Mujer

Solanum rostratum *Dunal*
    TAR. *so'iwari*
    SP. *mala mujer*

Solanum rostratum, or buffalo bur, with its peculiar thorny leaves is common in Samachique. The thorns are very painful to the touch and its name, *mala mujer*, or bad woman, derives from an analogy drawn from this characteristic. The plant is used in a tea given to sick individuals suffering from fever and vomiting and to women during menstruation.[595]

Malva

Althea officinalis L., *Malva parviflora* L., *Abutilon trisulcatum* (Jacq.) Urban, *Sida rhombifolia* L.
    TAR. *malba, marba, tuchiso*[596]
    SP. *malva, malvón*

As early as 1777, *malba* was considered one of the plants of medicinal importance among the Tarahumaras.[597] The Spanish word *malva* designates several species from the Malvaceae family. The plant is used for its ornamental and remedial properties in Mexico and many other countries. Malva plants grow in the Tarahumara region and are particularly appreciated in the canyon country. The Tarahumaras of the gorges know *Abutilon trisulcatum* and *Sida rhombifolia* only by their Spanish name *malva* and they are attributed similar medicinal properties.[598]

Mestizos in Cerocahui and those living throughout canyon country utilize an infusion of malva flowers as a gargle for throat and tonsil afflictions. Taken as a tea, malva is considered effective for the relief of stomachache and used as an enema is a remedy for constipation. In Samachique, Cusárare, and Cerocahui, malva enemas are applied before childbirth to expedite labor and ease the contractions.[599] Reportedly the Tarahumaras use the mucilaginous properties of the juice of *Abutilon trisulcatum* and *Sida rhombifolia* as a hair wash, and leaves are used as a poultice for sores and cankers.[600] The leaves of *Malva parviflora* are mashed, mixed with ground corn, and rubbed on sores.[601] Boiled in a tea, the leaves are used as a purgative. A tea made from the whole *Malva parviflora* plant is taken for headache, fever, indigestion, stomachache, and excessive gas.[602] The Tarahumaras use it for burns; mestizos prefer it for intestinal pain.[603]

Malva has been considered an emollient and inflammation-soothing agent. The pharmacologically active constituents are mucilage, pectine, sugar, tannin, asparagine, dense oil, and a yellow dye.[604]

Manzanilla

Matricaria parthenium L., *Matricaria chamomilla* L.[605]

SP. *manzanilla, manzanilla de Castilla*

Chamomile, or *matricaria*, is a plant of European origin and a member of the daisy family, with white and yellow flowers. It is known throughout Mexico as *flor de manzanilla*. The plant is available to the Tarahumaras through Mexican merchants. For the mestizos as well as for the Tarahumaras, chamomile tea is both refreshment and remedy. In Norogachi, the tea is taken for stomachaches. If *manzanilla de Castilla* is not available, *manzanilla del río* (*gordolobo*, or *rasó*), a native plant, is used. Mestizo midwives in Cerocahui and Urique use an enema prepared with a decoction of manzanilla to accelerate labor and ease the pain of childbirth.[606] Chamomile is used throughout Mexico, in hospitals as well as homes, to prepare a savory tea with multiple therapeutic attributions; especially favored is its use as a remedy for gastrointestinal ailments, an antispasmodic, and to control fever. It also is used to induce menstruation and as a vaginal douche for leucorrhoea. Its use as eyewash also has been reported.[607]

María

Unidentified

SP. *María*

A tea is prepared with *María*'s fragrant flowers. Some mestizos and Tarahumaras from Cusárare use it as an abortive or to induce menstruation.

Mariwana

Cannabis sativa L.

TAR. *mariwana*

SP. *mariguana, marihuana, mota*

Mariwana, or *Cannabis sativa* (marijuana), is a deciduous, annual tall plant, with a stiff, upright stem, divided serrated leaves, and glandular hairs.[608] The source of *Cannabis* is foreign to the Tarahumara area but some plants grow wild from seeds lost during transportation. Mariwana is well known by the Indians but recreational use by them is uncommon. The Tarahumaras view such use alien and a sign of having adopted the objectionable ways of the chabochi. Smoking it is considered to lead to aggression and crime; nevertheless, increasingly the young acknowledge having tried it, especially while visiting the cities. *Cannabis* is also called *chutama*, a term usually applied to the opium poppy.

Regarding its medicinal uses, throughout the sierra and barrancas an alcohol lotion containing a single leaf of marijuana is rubbed on aching joints and muscles to relieve arthritic and muscular pain. This remedy incidentally is well known throughout Mexico. Strangely the lotion is kept with an old leaf of mariwana, sometimes for years, believing that its effectiveness is greater than that of a lotion containing a fresh leaf.[609]

The *Cannabis* plant contains several active agents, such as Δ-9-tetrahydrocannabinol, the main psychoactive ingredient.[610] It also contains approximately sixty other cannabinoids; several of them are currently the subject of considerable research.[611]

Some researchers see in *Cannabis* components promise for a wide range of pathologic conditions ranging from mood and anxiety disorders to movement disorders like Parkinson's disease, neuropathic pain, atherosclerosis, myocardial infarction, and others.[612]

Mastuerzo

Tropaeolum majus L.

SP. *mastuerzo*

Mastuerzo, or cress, is not native to Tarahumara Country, but for many generations its flowers have embellished the gardens of the old towns in the sierra. Throughout Mexico, the leaves and stems of mastuerzo are macerated in water to prepare a remedy used for ailments such as swelling of the mucosa of the mouth, aphtae, tonsillitis, and gastric ailments.[613] In Norogachi, the flowers are used by mestizos to treat stomachaches. The Tarahumaras do not use this remedy unless suggested by a mestizo healer.

The leaves and seeds of mastuerzo contain chlorophyll, a yellow glycoside called tropoleine, an astringent acid resin, an allyl sulfurated essence, calcium acetate and sulfate, and potassium oxalate.[614]

Matarí

Cacalia decomposita Muhl., *Odontotrichum decompositum* (Gray) Rydb.[615]
    TAR. *matarí, matariki, materí,*[616] *matirí,*[617] *pichawi*[618]
    SP. *matarique*

Matarí, identified as *Cacalia decomposita*, also has the scientific name *Odontotrichum decompositum*.[619] The root of this plant, used by the Tarahumaras as a remedy, has a bundle of long, fleshy, tuberous projections that converge on the stem. The plant has considerable capacity to regenerate from the severed roots left underground. This feature may have helped it to survive in spite of the intensive harvesting done by itinerant Tarahumara herbal traders who collect matarí, along with copalquín, yerbanís, hierba de la víbora, and laurel, to sell in communities outside their land. Matarí is highly appreciated throughout Mexico as a remedy for diabetes. However, diabetes is uncommon among the Tarahumaras; therefore, they have other remedial uses for it.[620] For instance, decoctions of the root are applied to ailing joints. A poultice made from crushed roots is used on wounds to accelerate healing.[621] Brambila reported the following Indian saying—"*Matariki wabé chipúame ju; sayawi i'kisa, matarí ma'choga uchabo ba*"—which means matarí is a very bitter herb that is applied, crushed, to those bitten by rattlesnake.[622] For this use, the roots of matarí are boiled for several hours, the mixture is strained, and small gourdfuls of it are drunk each morning for three or four days.[623] Pounded and boiled for few minutes, it is used as a remedy for colds.[624] Occasionally, a tiny piece of the root is packed into a decayed tooth to relieve toothache. It has been said that a tea made from matarí leaves and roots is of value to treat fever in individuals with malaria.[625] Other applications are as a treatment for body aches, colds, fever, and gastrointestinal disorders. Deimel reports that Tarahumaras use it for burns, but mestizos prefer it to treat diabetes.[626] Also described is its use for neuralgic pain and skin infections.[627] Added to the broad list of indications mentioned are liver and kidney ailments.[628] The use of matarí as a laxative is proverbial. There are many bajíachi and bonfire stories among the people of canyon country telling how accidental or mischievous administration induces severe and continuous colicky diarrhea, with effects as severe as those of jumete.[629] Informants report how these effects of matarí may be stopped by drinking cold atole.

A traditional use of matarí is to massage the muscles of the kickball runners with a lotion prepared with the oily exudates from the roots applied for two or three consecutive days before the race.

It has been claimed that due to neurotoxic effects, the Rarámuri have used it to kill fish.[630] However, my informants and other observers have not confirmed this observation.

Regarding pharmacological information, matarí has been found to contain an aromatic essential oil, resins, tannin, fat, glucose, salts, calcium carbonate, potassium and sodium sulfates, and the alkaloid senecine. These substances may have analgesic and antispasmodic effects. Its active principles decrease glycogenesis, which could explain its use as an antidiabetic.[631] Externally used, it may have an effect by forming a protective layer over wounds. Incidentally, administration of significant amounts induces vomiting.[632]

Ma'tegochi
Unidentified
> TAR. *ma'tegochi, mategochi, rirusí*[633]
> SP. *lengua de perico*

According to observers, the leaves of *ma'tegochi* are heated and used in a poultice to treat sores. The latex from the leaves is squeezed in the eye to treat the burning and pain of styes.[634]

Mateó
Buddleia sessiflora H.B.K., *Buddleia cordata* H.B.K.
> TAR. *mateó*,[635] *mató, matowi*,[636] *batoi*[637]
> SP. *teposán, tepozán*[638]

Mateó (*Buddleia sessiflora*), or butterfly bush, has male and female varieties. It grows in gardens and on the edges of plowed lands in the middle portions of the western canyons.[639] It is known in Mexico with the Náhuatl name, which has been incorporated into Mexican Spanish, of *tepozán*. This tree has a long history in Mexican herbal medicine. It was described in the early writings of Sahagún and Ximénez. The plant is attributed diuretic properties. Regarding their use by the Tarahumaras, mateó's leaves—*B. sessiflora* as well as *B. cordata*—are heated and placed on boils and sores.[640]

The roots and leaves of mateó contain lipids, organic acids, a resin, an essential oil, tannin, glucose, pectine, mineral salts, and an alkaloid, buddleine.[641]

Mateóchiri

Unidentified

TAR. *mateóchiri*

The root of a *mateóchiri* is used by the Tarahumaras of Norogachi and Kwechi to treat febrile conditions.[642]

Matesa

Gentiana sp., *Gentiana defensa* (Rettb.) G. Don.,[643] *Stemmadenia palmeri* Rose & Standl.

TAR. *matesa*[644]

The Tarahumaras use the whole plant of *Gentiana* sp. to prepare a tea taken for chest congestion. The tea is also used as a shampoo.[645] Tarahumaras living in the uplands use the latex of another plant of the family Gentianaceae, *Stemmadenia palmeri*, as a poultice for styes. The dried latex is also pulverized and applied to boils and festering sores. Bye believes that an alkaloid in the latex physically binds the flesh and kills microorganisms and larvae.[646]

Mawirí

Fouquieria splendens Engelm., *Fouquieria fasciculata* Nash., *Fouquieria macdougalii* Nash.

TAR. *mawirí, mawarí, simuchí chuwará*[647]

SP. *ocotillo*

Mawirí (*Fouquieria splendens*), a plant characteristic of the gentle slopes where it grows in colonies as well as alongside streams in the Chihuahuan desert, can also be found in the Tarahumara canyon country. This plant, called in Spanish *ocotillo*, has white flowers that blossom in May. The mashed plant is used as soap.[648]

Scrapings of ocotillo are soaked and rubbed on children to "cure fright."[649] This remedy is administered as a wash not only to those with susto but to those believed to be ill due to exposure to ripiwiri, or whirlwind. Small bits of dry ocotillo wood are boiled overnight and taken for fever and to stimulate digestion.[650]

The bark of plants of the genus *Fouquieiria* has been shown to contain tannins and saponins.⁶⁵¹

Mé, Só, and Se*ré

mé: Agave *sp.*, Agave bovicornuta *H. S. Gentry*, Agave patonii *Trel.*, Agave schotii *Engelm.*, Agave hartmanii *S. Watson*,⁶⁵² Agave chihuahuana *Trel.*, Agave schrevei *H. S. Gentry*, Agave wocomahi *H. S. Gentry*, Agave vilmoriniana *Berger*,⁶⁵³ Agave lechuguilla *Torr.*⁶⁵⁴

TAR. *mé, meke, mesagori, oweke, chawé, chawirí, watusá, watosá, sapurí, ta'chugí, metagóchare*;⁶⁵⁵ *só*: *Yucca decipiens* Trel., *Yucca schotii* Engelm., *Agave schotii* Engelm.

A. lechuguilla Torr.; s**e*ré**: *Dasylirion simplex* Trel.; *Dasylirion duranguense* Trel.; *Dasylirion wheeleri* S. Wats.

TAR. **só**: *só, soko*; **se*ré**: *se*ré, se*reke*

SP. **mé**: *maguey, mescal*; **só**: *palmilla*; **se*ré**: *palmilla, sotol*

N.T. **mé**: *mai*

The Tarahumara terms *mé, só*, and *se*ré* encompass a group of plants from the genus *Agavaceae* of the flowering order Liliales. The Tarahumara word *mé* designates succulents with rosettes of narrow spiny leaves and tall flower spikes of varieties that include *Agave bovicornuta* and *Agave schotii* growing in the middle section of the western canyon slopes. The term *só* includes plants such as *Yucca*, or *A. lechuguilla*, found throughout the Tarahumara region except in meadows and pine-clad uplands. *Yucca decipiens* is common in northeastern Tarahumara Country where it grows along shrubs in rocky, sunny exposed areas. *Se*ré* (*Dasylirion* sp.) has stiff, swordlike leaves and spikes of white bell-shaped flowers. The trunk is sturdy with a massive base, branching profusely in old trees. This plant grows on rocky sections of the middle portion of canyon slopes and in the open slopes of canyons on the flanks of the northern longitudinal valleys.⁶⁵⁶

The study of the *Agavaceae* introduces us to a complex of important ethnobotanic information regarding Tarahumara customs, religious beliefs, healing practices, and survival strategies. From early times, the Indian peoples, including the Tarahumaras, learned the utility of plants of the family Agavaceae. It must have puzzled the Indian mind how these plants could survive in the desert and still provide them with food, beverages, healing remedies, poison to catch fish, soap, and fiber to manufacture many articles. The use of mé to make or obtain beverages probably began shortly after the arrival of the Indians on the American continent.

The incorporation of mé in Indian rituals and healing ceremonies began early.[657] Not surprisingly, plants of the Agave family gained mystic significance and became elements within their religion.[658] The agave was the plant of the moon deity and, according to tradition, the first thing that God created.[659] In the Tarahumara language, *mé* signifies agave and *mechá*, moon, or in Náhuatl, *metl* and *metzli* respectively.[660]

Many activities evolved around the collection, preparation, and consumption of beverages and potions derived from cacti and agave, including some with psychoactive action.[661] The use of mé to prepare tesgüino or batari is a pan-Sonoran cultural trait.[662] *Agave schotti, A. patoni,* and *A. lechuguilla* are used not only to elaborate a fermented beverage but are also added to batari, which is made of corn, to give it body. Such customs link the Tarahumaras with other groups, such as those from the present American Southwest. There is evidence that the use of the earth oven to cook agave and consume it as food is quite old. The hearts of *Agave patoni, A. schotti, A. lechuguilla, A. chihuahuana,* and *A. hartmanii* are baked or roasted and eaten by the Tarahumaras.[663] *Mezcal* is the common Mexican name of the roasted hearts and leaves of the agave. Incidentally, after their arrival, the Spaniards pressed these roasted hearts to obtain their juice and fermented it, distilling alcoholic beverages such as tequila or mezcal.[664]

The roots of se*ré are used as a curdling agent and food and the leaves as liners for mezcal pits. In Bakasórare, a flavorful tea is made from the leaves of mé. A special delicacy, reported from the days of Lumholtz, is *velluzas*, the inflorescence, or *quiote*, of the agave plant.[665]

Se*ré is mainly used as a source of a strong fiber (*ixtle*) to make rope, baskets (wari) and carpets. In earlier days it was used to make clothing and sandals. The rope from *só* is used to tie the poles used to make the Easter arches erected to symbolize the Stations of the Cross. From se*ré, material is obtained to make scarecrows as well as to make the ariweta that women toss while running their races. The roots of *Agave schotii, Agave lechuguilla,* and *Yucca decipiens* are used as soap to wash blankets.[666] *Dasylyrion wheeleri*, se*ré, or sotol leaves are used to make the rosettes that adorn churchyards and arches during the Tarahumara Easter celebrations.

Throughout the Americas, agaves have been a source of medicinal remedies for many different ethnic groups. The Tarahumaras shred the outer leaves of *Agave bovicornuta* to obtain a juice that is applied to festering sores.[667] *Mesagori* and *metagóchare* sap are used for eye conditions.[668] The leaves of *A. wocomahi* are used in a tea to reliever headaches.[669] Decoctions of *mesagori* are believed to be useful for kidney ailments, coughs, and rheumatic pain. The base of the leaf of this palmilla se*ré is used as a spoon to drink the mé that has been prepared into a tea and is thought to be a preventive of disease while the roots are prepared to make a liquor taken for chest afflictions.[670] The seeds of *Yucca* sp. are used as a purgative and also to treat dysentery.[671] The juice from the leaves of *Yucca schotii* is used as a shampoo.[672] Scrapings from the flower stalk (*quiote*) of *Dasylirium wheeleri* are rolled into small balls and placed on the head to relieve headache; these scrapings are held on by hand or in some instances by a cord also made of agave fiber.[673]

Mé is also the name of a cold infusion prepared from *Agave* sp. used as part of various healing ceremonies. The ritual administration of mé is included in those celebrations, frequently accompanied by the consumption of infusions of sitagapi, wasárowa, and ro'sábari, the other major medicinal and ceremonial plants. Men are given three spoonfuls and women four to be cleansed. A watery cold infusion of mé is placed as an offering in tutuguri celebrations; the participants are asked to walk, one by one, to an altar and drink mé, using a leaf of agave as a spoon.[674] However on most occasions, the liquid is administered by an owirúame, sipáame, or sacristano, who places it directly in the mouth of the communicants with a spoon. The Tarahumaras consider mé a powerful healing or cleansing remedy. Healing maneuvers with it, particularly from *Agave lechuguilla* pounded and mixed with either *Dasylirion simplex* or *D. duranguensis*, are also applied to animals and to the fields to make them more productive.[675]

Saponin glycosides are present in most plants of the family Agavacea.[676] These substances have a bitter taste and irritate mucosal membranes, destroy red blood cells, and are in general toxic, especially for cold-blooded animals. When hydrolyzed, saponin glycosides yield an aglycone or sapogenin.[677] Agave hearts, leaves, and roots may contain sapogenin, gitogenin, and tigogenin.[678] The Rarámuri have learned to detoxify the plants prior to food preparation by baking and roasting them. In addition to saponins, maguey sap contains abundant polysaccharides. Although the detergent and hemolytic properties of saponins may contribute to the antimicrobial effects of the sap, the high concentration of polysaccharides may be responsible for its bactericidal action.[679]

Mechawí
Lysiloma watsoni Rose
   TAR. *mechawí*,[680] *mechawíki*
   SP. *tepeguaje*[681]

Mechawí (*Lysiloma watsoni* Rose), a desert fern tree, is a plant that enjoys popularity as a general remedy among the Indian groups and mestizos of the Mexican states of Chihuahua, Sonora, and Sinaloa.

The wood from mechawí is used to carve tesora, the siríame's batons, symbols of authority among the Tarahumaras. As regards its curative properties, taken by itself prepared as a tea or in combination with other remedies, it is considered efficacious against gastrointestinal ailments, venereal diseases, and dental problems. Some persons consider it a panacea for a wide range of health problems, including cancer. The crushed bark of mechawí is used to prepare a plaster used to set bone fractures. The Tarahumaras and mestizos of Norogachi prepare a drink with the bark of the plant to treat gonorrhea.[682] This bark is also applied to aching teeth to relieve pain.[683]

Mechawí contains tannins, mainly in its bark, which is used as a tanning agent to process animal skins.[684] In addition to tannins, the bark of *Lysiloma* contains irritant saponins.[685]

Membrillo
Cydonia vulgaris Pers.
   SP. *membrillo*

Membrillos (*Cydonia vulgaris* Pers.), or quinces, belong to the family Rosaceae. This tree is native of western Asia and was brought to Mexico by the Spaniards during colonial times. The tree has a flexible hard wood and quinces, its delicious pear-shaped fruits, ripen in September. The fruits are cooked to make reddish gelatin-like blocks, the quince dessert known in Spanish as dulce de membrillo or cajeta de membrillo. This paste is a delicacy consumed accompanied by manchego or Mennonite Chihuahuan cheese. Membrillos are highly esteemed throughout the state of Chihuahua; the excellent quinces are typical of Villa Aldama, near Chihuahua City.

As a medicinal remedy, the Tarahumaras and mestizos of Cerocahui use the raw fruit to prepare a decoction used as a general remedy. The mucilaginous core of the quinces is eaten to treat diarrhea and dysentery.[686] The astringent effects controlling diarrhea may be due to tannin.[687]

The pulp of membrillo contains glucose, tannin, malic acid, a sulphurated substrate, pectin, and an essential volatile oil. The core of the fruit contains a substance known as cydonine, which in contact with water becomes viscous. The seeds contain amygdaline, emulsine, starch, and cydonine. The peel of the fruit contains vitamin C.[688]

Mo*riwá or Mo*riwáka

TAR. *mo*riwá, mo*riwáka, mo*rewá*,[689] *muruwaka*[690]

SP. *copal, incienso*

"Scents and fumes of *Cupressus* and *Juniperus* are used throughout the Tarahumaras' life cycle," according to Bye.[691] Nevertheless, *mo*riwá* (incense or *copal*) derived from insect larvae castings made from *Bursera* sp. is the most important. Mo*riwá is collected from caves, a laborious practice that has been only partially studied. It is said that the clusters of mo*riwá are a rocklike cocoon obtained from grubs found in certain places in the lowlands. This cocoon generates considerable smoke when thrown into a cup of glowing coals.[692] If mo*riwá is not available, it may be replaced with *choré*, dried turpentine, or *poliya*, a resinous wood eroded by termites or moths.

Mo*riwá, like the incense of many peoples around the world, is used for purification rituals in many Tarahumara ceremonies. The incense smoke is believed to cleanse impurities and disease. It is inhaled to achieve greater internal exposure and supposedly cleansing. The smoke is used to free from evil the air of newborns and it plays a propitiatory role in funeral ceremonies.[693]

Mo*riwá smoke has been used for the alleviation of illnesses, including chest pain.[694] A tea prepared with mo*riwá is taken to relieve chest congestion and also has been used for uterine disorders.[695] Sometimes it is pasted to the temporal area of the head with a piece of paper to treat earache.[696] These practices have managed to survive until the present.[697]

Mukí Chiwara

TAR. *chiwara, mukí chiwara*

SP. *leche materna, pecho*

Mukí chiwara, human milk, is an important ritual and medicinal item among the Tarahumaras. In Norogachi, dead children are sprinkled with milk directly from the mother's breast and, using a finger, three crosses of milk are traced on the baby's head and chest; a few droplets are placed in the infant's mouth. An informant explained that this is for the baby to have nourishment during its journey to the other world. Mestizos and some Tarahumaras in Cerocahui use human milk as a remedy. For instance, it is dropped in the eyes as eyewash for patients suffering from conjunctivitis.[698]

Munísowa
Rhynchosia pyramidalis (Lam.) Urban
 TAR. *munísowa*
 SP. *chante pusi, ojo de pescado*[699]
 Pennington claims that this conspicuous climbing vine is known among the Indians as *munísowa*. Its seeds are ground on a metate and mixed with any available fat to prepare a poultice applied to aching limbs or backs.[700]

Naká or Nakáka
Asclepias sp., *Acourtia* sp.
 TAR. *naká, nakáka, chumarí nakara*,[701] *chi'wáiname, chi'wáinari, chi'wánari*,[702] *wakasí cha'mérowa*[703]
 SP. *lengua de vaca*,[704] *lengua de buey, chino*[705]
 Naká, or *chino*, is a herbaceous milkweed plant (*Asclepias* sp.) known in Spanish as *lengua de vaca*.[706] Some Tarahumaras call it *wakasí cha'mérowa*, a literal translation of the Spanish *lengua de vaca* (cow's tongue), as the herb is called by the mestizos. Zingg described a plant called *lengua de buey* as a member of the *Composite* species, which probably was the same as the naká described here.[707] According to Deimel, the plant corresponds to *Acourtia* sp.[708]
 For healing purposes, the herb is burned and the ashes are used to make a poultice that is applied to the abdomen to relieve colic or headaches.[709] Pennington and Brambila record its use in a poultice for rheumatic and chest pain.[710] The Tarahumaras and mestizos use it also as a remedy to treat skin problems. Tarahumaras are aware of the laxative effects of naká but they do not use it primarily for constipation but to induce the expulsion of a supposed noxious agent from inside the body. The plant is believed to "pull out" from the abdomen illnesses or "evils," even if they are located outside the intestines. Before administering the remedy, the individual is supposed to fast for one or

two days, ingesting only gruel or broth after which a tea made from the plant is ingested.[711]

Nakáruri

Vigueria decurrens Gray[712]
 TAR. *nakáruri, nakúruri, nakárori,*[713] *nakáori*[714]
 SP. *hierba de la mula*[715]

 Nakáruri, a browse from western Chihuahua, grows in damp places along ravines in southeastern Tarahumara Country. It is used primarily as a piscicide.[716] A poultice prepared from the bulbous roots of the pungent nakáruri is used to treat boils and skin infections. This remedy is preferred if maggots have infested the boils or other skin lesions.[717] The roots of this plant are also boiled in a tea to treat gastrointestinal disorders.[718]

Na'pákori

Lippia palmeri S. Watson, *Lippia berlandieri* Schauer, *Monarda austromontana* Epl., *Prunella vulgaris* L., *Tabebuia palmeri*[719]
 TAR. *na'pákori, napákori, napákuri, napá, ma'pákori, ma'pá, mapá, aborígori*[720]
 SP. *orégano, orégano silvestre*

 Na'pákori (*Lippia palmerii*) grows abundantly near Tuceros and other points on the Mesa de Guachochi, while *Tabebuia* grows in the western canyons.[721] Na'pákori has been an important Tarahumara medicinal plant since the Spanish chronicles of 1777.[722] The plant boiled into a tea is taken for sore throat, gastrointestinal disorders, cough, backaches, and as a general remedy. It is also used in mixtures with other herbs to control fever.[723]

Nawá

Wedleia sp.
 TAR. *nawá*

 A tea prepared from the root of *nawá* is used for empacho and other stomach ailments.[724] Nawá is used as an emetic to expel from the body the *awagásini* (food chunks) causing awakátzane ropara.

Nawé

Tephrosia talpa Watts, *Tephrosia leiocarpa* Gray

TAR. *nawé, chábari*[725]

Nawé (*Tephrosia leiocarpa*), a plant from the family Leguminosae, grows alongside shady ravines between Pawichí and Narárachi and throughout the headwaters of the Conchos River by the communities of Wichabóachi and Humariza. This is one of the plants widely used as a piscicide in the eastern uplands and hill country, by placing it in *nabóara* (fishing baskets) to stupefy fish.[726] Most scholars have consistently identified nawé as *Tephrosia*. Field descriptions of the herb, however, present such a diversity of botanical characteristics that one may assume that the name is given to plants of several different species. Actually an informant noted that any plant with piscicide properties is called *nawé*.[727]

Nawé is used in humans and animals to kill fleas, lice, and ticks.[728] Informants from Humariza claim that nawé is effective against ticks and lice. The juice obtained from squeezing the roots is used to extract teeth by applying it to the diseased tooth, which it breaks, and the fragments are then easily pulled out.[729]

Two active principles seemingly with piscicide action have been isolated from *T. vogelii* (Hemsley) A Gray: tephrosin and tephrosal.[730]

Okó

Pinus sp., *Pinus ayacahuite* Ehrenb., *Pinus reflexa* Engelm., *Pinus lumholtzii* Rob. & Ferm., *Pinus cembroides* Zucc.,[731] *Pinus ponderosa* Dougl., *Pinus arizonica* Engelm., *Pinus duranguensis* Martínez, *Pinus chihuahuana* Engelm., *Pinus engelmannii* Carr., *Pinus leiophylla*. *Engelm*.

TAR. *okó, wiyó, mateó, mategó, mateyó*,[732] *okóweri, ropichiri, papachiri*,[733] *sawá*

SP. *pino*

Pine trees, or *okó*, provide a characteristic green cover to the uplands. The Tarahumaras call all species and types of the genus *Pinus* by the term *okó*; however, they also have specific names for the different types of pine trees. *Pinus ayacahuite* (*pino blanco*) is *wiyó* and *P. reflexa mategó* or *mateyó*. *P. engelmanni* is known as *okóweri* or *ropichiri* (*pino real*). Both *pino negro* (*Pinus chihuahuana*) and *pino colorado* (*P. leiophylla*) are known as *sawá*.[734]

The trees have considerable cultural significance. The Tarahumaras obtain wood, a highly valued material, to build their shelters, fences, and many other objects and structures. The cones of the tree, *okóweri* or *ropichiri*, are used as combs. The hard husk of the individual scales is removed in half of the cone to expose the hard fibers and the other half is left intact as the back or holder of the comb. Tarahumaras of the sierra chew pine needles to quench their thirst on their long treks. *Choré*, or dried turpentine, is applied as pitch on *raperi* (violin) strings.

Derivatives such as turpentine, the fluid, pale, opalescent oleoresin, have many remedial applications in spite of their potential toxicity.[735] Turpentine is absorbed through the skin, lungs, and intestinal mucosae. When ingested, it has a bitter taste that dries the mouth and later induces salivation. It has some antiseptic properties and is eliminated through the lungs; it has therefore been used in the treatment of respiratory problems. Turpentine has balsamic properties and induces diuresis and diaphoresis.[736] Lumholtz reported that Tarahumaras and Mexicans boiled the needles of *P. lumholtzi*, using the decoction as a remedy for stomach trouble; he stated that the taste resembles that of anise seed.[737] The needles from P. *lumholtzii*, *P. ayacahuite*, and *P. reflexa* are used in a drink taken to relieve coughing.[738] The pine-needle tea made from *sawá* (*P. chihuahuana*) is used to relieve headache.[739] Individuals bitten by snakes are treated by having them inhale fumes of resinous burning wood (chopé or *ocote*) or by cauterizing the snakebite with a burning stick.[740]

Turpentine obtained by notching the trunk of *Pinus reflexa* is used to prepare a drink taken for lung congestion. The resin, or choré, is also used as a rubbing compound applied on rheumatic joints. Occasionally choré or *poliya*, a decayed or termite-infested wood, is used as incense. It is also applied to burns and wounds. Pitch from the large cones of *Pinus ayacahuite* is heated and used in an ointment to cure foot infections.[741] The long needles of *P. lumholtzii* are used in a tea for indigestion, coughs, stomachaches, and flatulence.[742]

In Cusárare, the *coyón del pino*, most likely an epiphytic fungus growing on pine trees, is administered in a beverage for coughs and colds.

Olivo (Aceite de)
Olea europea L.

SP. *olivo, aceite de olivo*

Olive oil is obtained from local merchants and is used by mestizos and Tarahumaras from Cerocahui to remove insects and foreign bodies from the ear and to treat ear ailments. It is also rubbed on the abdomen of women in labor to accelerate the childbirth. Olive oil is given orally to children to prevent constipation.

Oná or Onáka
TAR. *onáka*
 SP. *sal*

Oná is common salt, sodium chloride. It is used rubbed on the skin (*friegas de sal*) for any condition associated with fever or prostration. Salt is also applied in patches (*chiquiadores*) to the forehead or temples and held in place with a handkerchief. In the past, powder from ground shells from Conchos River brought by the Rarámuri of Narárachi was saved and mixed with salt to be used as a remedy for eye troubles.[743]

Ortiga[744] or Ortiguilla
Urtica gracilis Ait., *Urtica dioica* L., *Urera caracasana* (Jacq.) Griseb., *Tragia nepetifolia* Cav., *Tragia ramosa* Torrey
 TAR. *ra'uré, ra'urí, raó*,[745] *ránuri*,[746] *ra'iwá, rijiwa, rajuri, rajíreke*[747]
 SP. *ortiga, ortiguilla*

Ortiguilla, a perennial herb, may be found in all the sierra country, including Norogachi. In October, the stems of ortiguilla have dried in Samachique but in Cusárare they are still green and keep their leaves. *Ortiguilla*, or *ra'uré* (*Urtica gracilis*), is one of the plants usually growing along streams in the upper Río Conchos region with similar distribution in eastern Tarahumara Country. *Tragia nepetaefolia* and *Tragia ramosa* are common in southeastern area.[748]

A decoction of *Tragia ramosa*, called *ra'urí*, is drunk for heart ailments, but only the leaves are boiled into a tea for this purpose.[749] Leaves of *Tragia nepetifolia* are boiled into a tea drunk for fever, and those of *T. ramosa* are applied as a poultice for headache.[750] Bye also reports that fruits and leaves of *Urera caracasana* are used to prepare a tea drunk for aching and those of *Urtica dioica* for headache, measles, and pox. It is said that the plant is rubbed on the runners at the kickball race to stimulate them to run faster, but I have not witnessed this. A decoction of *T. ramosa* is used by Tarahumaras to wash sores and skin ulcers.[751]

Ówina
Ratibida mexicana (S. Watson) Sharp
TAR. *ówina, ójina,*[752] *ónora, ónowa*[753]

Leaves or roots of *ówina* or *ójina*, a yellow-flowered perennial from the upland meadows, are used by Tarahumaras of Samachique, boiled in water, to treat stomach ailments, especially those in which vomiting is a prominent symptom. The name of the plant seems to come from the Tarahumara verbs *owama* or *ó'wima*, meaning "to cure, or to heal." The leaves and roots of *Ratibida mexicana* are pulverized and infused in hot water and taken to cure headaches or colds or used as a poultice for festering sores.[754] The tea is also used for gastrointestinal disorders.[755] Cardenal claims that it is used to treat diarrheas, especially those attributed to susto or majawá.[756]

Palo Dulce
Eysenhardtia polystachia (Ortega) Sarg.
SP. *palo dulce*
NAH. *tlapalezpatl*[757]

This plant, according to *Enciclopedia de México*, used to have extensive medicinal uses in Mexico as a diuretic and remedy for multiple afflictions of the bladder and kidney.[758] Currently, it has only limited use in western Chihuahua canyons. Its bark is crushed and decocted into a mixture drunk to relieve pain caused by internal injuries.[759]

Palo Mulato
Bursera grandifolia Engl.,[760] *Bursera* sp., *Zanthoxylum pentanome*
TAR. *iweri,*[761] *rusíware*[762]
SP. *palo mulato*

The *palo mulato* tree thrives in tropical deciduous forests, growing up to thirty feet (nine meters) tall. It has deep steel-blue-gray to green bark with rusty, papery peelings. Above its solid trunk, a mature tree spreads irregular branches of attractive form. Masses of pink flowers appear early in the rainy season before the trees leaf out.

Palo mulato is highly prized as a medicinal plant by the Mexicans of Chihuahua City and Parral, justifying the long journey that the Tarahumaras make to bring pack loads of the plant to those cities.[763] The mestizos of Cerocahui, who get the plant from the nearby canyon country, are advised not to confuse palo mulato with *norote* (*Bursera penicillata*) a plant that according to them, has no medicinal properties.[764] Palo mulato has also been identified as *Zanthoxylum pentanome*.[765] However, plants from this genus have not been reported in the area.

The bark from palo mulato is highly appreciated by Tarahumaras and mestizos, as well as by the people of central Chihuahua, for its analgesic effects on joint pain resembling the effects of salicylates. In the sierra, it is believed that a decoction from the bark has an invigorating effect on the person taking it. Tarahumaras also use palo mulato to treat dysentery, colds, headaches, and pneumonia.[766]

Papache

Randia echinocarpa Moc. & Sessé ex DC.,[767] *Randia laevigata* Standl., *Randia watsonii* Robins,[768] *Randia thurberi* S. Watson

TAR. *kakwara*,[769] *apatzo*,[770] *apachi*,[771] *kajákori*,[772] *kapuchí*,[773] *a'kaweri*,[774] *akagékori*,[775] *kakáwari, a'kágegiri*,[776] *batari*[777]

SP. *papache*

Papache is the Spanish name given to at least two species of the family Rubiaceae that grow in the canyon country. *Papache grande* (*Randia echinocarpa*) is the largest. Bye identified *Papache chiquito* as *Randia watsonii*.[778] It is found alongside the Urique River and its tributaries, especially on the pluvial floodplains such as Urique and Guapalayna. Highlander Rarámuri journey to the canyons to gather papaches, the globular fruits with heavy yellowish husks, purplish flesh that is black when dried, and flat, tiny seeds. Another species, *Randia laevigata* (*kapuchí*), grows at slightly higher elevations where the plants grow in association with oaks. As early as 1777, the Tarahumaras were reported to eat the fruits of papache, an observation confirmed in 1885.[779] Considerable attention has been given to the contemporary consumption of papaches as food.[780] Authors have remarked that, in spite of its bitterness, the fruit is considered a delicacy. Scraps of the husk of papache are used as a substitute for *basiáwari* (*Bromus* sp.) in the brewing of batari.[781] Its use to facilitate the fermentation of tesgüino is common close the Urique Canyon. Some Tarahumaras call this plant *batari*, which is also the Tarahumara name for tesgüino.[782] Mexicans of Sonora consider the fruits of *Randia* good for gastric distress, while in Sinaloa the fruit was used as a remedy against malaria.[783]

Informants from Urique claim that papaches, eaten raw or boiled into a tea, are useful against diarrhea; in Guapalayna, just two miles downstream, they are used for constipation. In both places, however, they agree about the usefulness of papache to control coughs.[784]

Parches
SP. *parches, chiquiadores, lienzos*

Throughout Mexico, patches, or in Spanish *parches* or *chiquiadores*, are popular healing devices. The patches are prepared with medicinal herbs; these devices are rather crude, consisting of pieces of cloth dampened in the healing substance and tied with a string to the forehead or temples of the sick person. Sometimes they are glued to the skin. The patches resemble those used in conventional western medicine; however, the rationale for their use is different. The traditional cold/hot concepts of disease seem to be incorporated in their use. For instance, a potato patch is considered "cool"; therefore, its application is expected to relieve fever. In this case, raw potato slices are fixed to the temples to "let the heat escape." It is thought that the devices will "collect" the noxious agent. For instance, when applied to an abscess, the patch is supposed "to gather the pus and suck it out." This concept differs from the idea of transdermal absorption of the healing substance. The remedy does not go in; the "illness" is expected to go out. In Cerocahui, beans cooked and ground are applied with a bandage or a handkerchief tied to the temples to draw out the agent causing illness and relieve headaches. Common substances such as sugar and salt are often used in these devices. A common device is a vinegar-soaked cloth applied on the forehead of the ailing person to relieve headache.[785] A particularly interesting patch is the one applied to babies thought to have a sunken fontanel in the condition of caída de la mollera. After "lifting" the fontanel through the towita ritual mentioned earlier, a patch is placed to prevent the fontanel from "sinking" back.

Perritos

Antirrhinum sp., *Antirrhinum majus* L., *Cologania angustifolia* Kunth
SP. *perritos, boca de dragón*

Perritos, or snapdragons, are commonly observed in Norogachi's gardens and as in many other parts of Mexico are cherished as ornamentals. These colorful plants of the family Scrophulariaceae bear flowers reminiscent of the features of dog's faces; from this similarity the name *perritos* derives. Characteristically, the flowers have a tubular shape with a closed liplike mouth that hides the stamens. Children like to press the upper part of the tube, making it look like a barking dog. Perritos belong to the *Antirrhinum* genus, particularly to the species *Antirrhinum majus*.[786] Mestizo women in Norogachi prepare a tea to control emesis using the perritos flowers. Hewitt reported that perritos are used to prepare a multipurpose healing tea.

Poleo

Mentha pulegium L.[787]
    TAR. *wijúpachi*
    SP. *poleo*

In Cerocahui, *wijúpachi* or *poleo*, in English "pennyroyal," a plant of the mint family, is administered in a tea to relieve colic pains. It is also is given to treat insomnia.[788] Wijúpachi is also considered useful as an antitussive[789] and to treat anxiety, flatulence, menstrual disorders, coughs, stuffed nose, and susto.[790]

Prodigiosa

Unidentified
    SP. *prodigiosa*

This herb is used in the area of Samachique for suboccipital headaches, or "dolor de cerebro." The leaves and flowers of the plant are used to make a bitter tea ingested for the purpose.

Rasó

Gnaphalium wrightii A. Gray, *Gnaphalium indicum* L., *Gnaphalium conoideum* H.B.K., *Verbascum thapsus* L.
    TAR. *rasó, chiyowi, chiyowi rasorá*,[791] *jubí, rosábori, rosábari, rosobóchame*[792]
    SP. *manzanilla del río, manzanilla del arroyo, gordolobo*,[793] *hierba lobo*[794]

The herb is called *manzanilla del río* (or *del arroyo*) to distinguish it from the more delicately flavored *manzanilla de Castilla* mentioned earlier. The medicinal uses of both plants are similar suggesting that in earlier times the Spanish settlers found in this plant the American equivalent of the European chamomiles. Manzanilla del río has only a few branches covered by a yellowish "wool."[795] It is also called *gordolobo*. It blossoms in July and August, with its flowers in a compact cluster. *Manzanilla del río* is classified among the Compositae.[796] *Gnaphalium wrightii* has been identified as *manzanilla del río*, or *rasó*.[797] Bye also has identified the Tarahumara *rosábori* and *rosobóchame* as *Gnaphalium* sp.[798]

In Norogachi, mestizos and Tarahumaras use the tea of rasó to alleviate abdominal pain. In Cerocahui, a decoction of gordolobo and *cáscara de granada* was used to treat dysentery.[799] The stem and flowers of manzanilla del río were boiled and given to children as a remedy for colds.[800] Leaves and flowers of *Gnaphalium wrightii* were used to prepare a drink to control diarrhea or coughs.[801] The glandular and woolly leaves of the herb *Gnaphalium conoideum* are used for diarrhea, stomach problems, colds, and as an anti-inflammatory.[802] Gordolobo contains chlorophyll, mucic acid, oxalate and sulfate of potassium, an alkaloid, and mucilage. The mucilage is thought to act as an expectorant and anti-inflammatory and also to decrease venous congestion.[803]

Rayó

Cardiospermum halicacabum L.

TAR. rayó

Tarahumaras from the western canyons use the heartlike pods and scrapings of stems to prepare a tea for gastrointestinal disorders.[804] After boiling it for several hours, a bit of lime juice is added to the mixture and taken as a remedy for a week.

Remolino

Unidentified

SP. *remolino*

The plant called *remolino* is used by the Tarahumaras and mestizos of Cerocahui decocted into a tea and mixed with fat of cholugo and snake plus a few drops of lime juice to treat bronchitis.[805]

Repogá

Platanus wrightii S. Wats., *Platanus glabrata* Fem., *Alnus oblongifolia* Torr.

TAR. repogá, repokwá,[806] ropogá, usako, saka,[807] usá

SP. aliso,[808] sicomoro, álamo blanco[809]

Repogá or *aliso* (*Platanus wrightii*) is a giant, slender sycamore tree that grows frequently among stands of *Achras zapota* alongside streams in the canyon country. The wood of repogá is valued by Tarahumaras and mestizos from the canyons for its use in the construction of dwellings, tools, and handicrafts. It is also used as a dyestuff and to make the balls used in rarajípari races.[810]

The bark of repogá is crushed and decocted into a drink taken as a general medication.[811] It has been claimed that the bark is used as a blood strengthener.[812] Bye reports that an infusion of bark soaked in cold water of *Alnus oblongifolia* is used as to prevent the illness attributed to the "feared water spirits."[813]

Reté Kajera
Parmelia caperata (L.) Ach., *Usnea subfusca* Stirt., *Usnea variolosa* Mot.

TAR. *reté kajera, riteawáare*[814]

In the Tarahumara language, *kajera* means "bark, peel, or skin"; therefore, *reté kajera* would be the "skin of the rocks." This name is applied to an upland lichen that grows on the surface of rocks giving them a coated appearance. Several plants of the family Usneaceae have been identified as *reté kajera*, including *Parmelia caperata, Usnea subfusca,* and *U. variolosa*. The first species is used as a remedy; the latter two, as batari flavoring and dyestuffs. *Parmelia caperata*, commonly found on rock surfaces, is dried, crushed, and dusted on wounds.[815] *Usnea* sp. is prepared into a tea and given to women during childbirth.[816]

Rikówi
Unidentified

TAR. *rikówi*

The bulb of this plant is boiled into an infusion used to expel the flatworm *Taenia*. The addition of pumpkinseeds is believed to improve this action. The same infusion is used to treat other stomach ailments.[817]

Ripichawi
Perezia thurberi Gray[818]

TAR. *ripichawi, pipichawi,*[819] *pipíchowa*[820]

This plant is attributed similar effects as jumete and naká and is therefore used as a laxative. Bye reports Tarahumaras use its root in a tea as a tonic.[821] *Ripichawi* is considered valuable in the treatment of gastric disorders.[822]

Ripura

Cosmos pringlei Robinson & Fernald, *Cosmos parviflorus* (Jacq.) H.B.K.
    TAR. *ripura, ripuri, wasarépuri, kujubi*[823]
    SP. *babisa, bábiz, hierba chilosa, matagusano*

Ripura grows on the mountainsides flanking the tributary streams of the upper Urique River, for example at Bakasórare, near Norogachi and Samachique. The plant has a tubercular multiple root of a reddish color with an almost caustic flavor; for this reason, it is also called *hierba chilosa*. Rubbing it with the fingers and smelling it makes one sneeze. The pungent flavor of ripura softens a bit when the roots are dried. The roots of ripura boiled into a tea are one of the most popular antidiarrheic remedies; mestizos and Tarahumaras on both sides of the grand canyons appreciate this effect.[824] Some informants claim that scraps from the raw root may be used instead of the tea for the same purpose. Other informants claim that the chewed or crushed raw root placed on the abdomen as a poultice is as effective as ingesting the scraps or the tea. Uses of ripura include the treatment of the symptomatic complex called empacho. It is not surprising that the plant is given for intestinal disorders, stomachaches, headaches, and chest pain. A beverage prepared from wasarépuri is used by Tarahumaras in Kwéchi for the treatment of diarrhea.[825] Similar properties are claimed for *Cosmos pringlei* and the related species *C. parviflorum* (*kujubi*), which are used in a tea for gastrointestinal disorders and as a general medicine.[826] The roots of ripura are pulverized and used to prepare a poultice applied to sores caused by maggots or worms. Its use as a topical insecticide to rid people and animals of lice has also been mentioned by my informants.

Riwérame

Calliandra humilis (Schldl.) Benth. var. *reticulata* (A. Gray) L. Benson
    TAR. *riwérame,*[827] *kochínowa, kochíawa*
    SP. *sensitiva, sin vergüenza*[828]

It has been reported that during some kickball races the dried leaves and seeds of *riwérame* are chewed by deceitful contestants who blow their breath into the face of a runner to make him slow down and collapse.[829] In Norogachi, the plant used for this purpose is known as *kochínowa* (*Calliandra* sp.). The crushed roots and leaves of a tiny herb with a purple flower, called *rarajípame* (kickball runner) are steeped and drunk as an antidote for the effects of riwé-

238 ▾ *Tarahumara Medicine*

rame. On the other hand, Tarahumaras living in western canyons eat the seeds of riwérame as a remedy for sleeplessness.[830] SEMARNAP (Secretariat of the Environment, Natural Resources and Fisheries) claims that the foliage is placed in the cradles of small children to put them to sleep. The same source claims that the smashed roots of the plant are placed on caries to relieve toothache.[831]

Ro'chíwari
Lepidium virginicum L.
    TAR. *ro'chíwari, rochíwari, so'chiri*[832]
    SP. *pata de cuervo*,[833] *quelite pata de cuervo*,[834] *quelite de invierno, quelite chilada*[835]

Ro'chíwari is a peppergrass apparently indigenous to the New World.[836] Planted in upland gardens in October, it is gathered in March or April to be eaten as a quelite.[837] *Lepidium* forms rosettes on the fields during the winter, flowering during the summer.[838]

Ro'chíwari is burned ceremonially before a cross to prevent hail from damaging the crops.[839] The crushed root is prepared into a beverage useful to control internal bleeding, coughs, and other chest ailments.[840]

Rojá
Quercus sp., *Quercus arizonica* Sarg., *Quercus oblongifolia* Torr., *Quercus crassifolia* Humb. & Bonpl., *Quercus chihuahuensis* Trel., *Quercus omissa* DC., *Quercus durifolia* Seem., *Quercus rugosa* Neé, *Quercus fulva* Liebm., *Quercus albocincta* Trel., *Quercus hypoleucoides* A. Camus., *Quercus emoryi* Torr., *Quercus viminea* Trel., *Quercus coccologifolia* Trel., *Quercus toumeyi* Sarg., *Quercus tuberculata* Liebm., *Quercus incarnata* Trel., *Quercus grisea* Liebm., *Quercus epileuca* Trel., *Quercus watsoni*
    TAR. *rojá, sipura, gumíchari rojísoa, rojísowa, rokoró, okoró, bawítoa, bawítowa, wawítari*,[841] *aó*,[842] *umíchari*,[843] *bachiera, machíchare, makóchare, péchuri, epéchari, chajawa, chaawa*[844]
    SP. *encino, encino blanco, encino rojo, encino colorado, encino chaparro, encino roble*

Rojá is the generic name used for oaks. Although Brambila claims that the term *rojá* is restricted to name the white oak, *Quercus chihuahuensis*, it is also used to designate a species of oak known in western Chihuahua as *encino rojo* or *encino colorado*, which is also called *sipura* or *gumíchari*.[845] At least two other species, *Quercus crassifolia* and *Q. oblongifolia*, are also designated *rojá*.[846] The *encino chaparro* is usually called *rojísoa* or *rojísowa*. *Bawítoa* or *bawítowa* is the oak with wood used to make plows.[847] Finally, *rokoró* or *okoró* is the name of the largest of the oak trees in the Tarahumara regions, the *encino roble*.

The hard wood of rojá is used to carve components of tools requiring hardness, such as the body and handle of plows. The most common use of oak wood both by the Tarahumaras and their mestizo counterparts is however as firewood. The best source of dry rot (*yesca* or *sorá*) to start fires is the rotted trunks of *Quercus arizonica, Q. fulva*, or *Q. epileuca*. The wood of knots of rojá was used to fashion *gomákari*, the balls used in rarajípari races.[848] Tanning of skins using the bark of rojá is an important use among Tarahumaras and mestizos. The ashes of rojá bark are used occasionally in lieu of lime to mix with nixtamal, the dough used to make tortillas.[849] The acorns of sipura and wawítari are consumed by the Tarahumaras. The big leaves of rokoró (*Quercus* sp.) are used as spoons.

The main active substance in the bark is tannin which is an astringent, diuretic, and a local and general vasoconstrictor.[850] A drink prepared by boiling a sap obtained by notching the bark of *Quercus chihuahuensis* is given to women during their pregnancies.[851] This same sap is prepared as a tea given to people thought to have heart difficulties.[852] Pennington and Brambila report that fresh bark of *Quercus arizonica* is crushed and moistened for use as a salve rubbed upon the armpits and neck when these parts of the body are inflamed or have open sores.[853] The western Tarahumaras in the canyon use a decoction of the bark of *encino sepurí* from the sierra or of *encino cusi* or *aó* from the canyons to treat diarrhea.[854] A decoction made by steeping the leaves of *Quercus viminea* is taken to relieve stomach upsets.[855] Drinking three liters of red oak bark per day for three days is used to induce abortion but the effects most likely are unreliable.[856]

Rojá Sewara

Tillandsia benthamiana Klotzsch ex Baker, *Phoradendron engelmanii* Trel.,[857] *Phoradendron* sp.[858]

TAR. *rojá sewara, rowáka sewara*,[859] *rojá kuwawira, kuwawi, ku'chaó, ku'chagó, kuchawó, ku'chayó*[860]

SP. *muérdago, toje, hierba del pájaro*[861]

Rojá sewara (*Tillandsia benthamiana*) is an epiphytic plant that grows on oak trees in the canyons.[862] These plants are known to contain raphides (oxalate crystals), proteolytic enzymes, and flavonoids. Overuse of this remedy may lead to intoxication that resembles that from ergot derivatives.[863] Boiled in water, the plants are used as a strong purgative. *Toje* is considered a remedy for dysentery, an antispasmodic, and a blood-pressure-lowering agent.[864] More recently, a tea from *Tillandsia benthamiana* has been reported used as an analgesic, an anti-inflammatory, and a wash for rheumatism.[865] *Rojá kuwawira* is taken to control the symptoms of venereal disease.[866] A decoction prepared from *ku'chaó*, which may be a different epiphytic plant, is used by the Tarahumaras to alleviate respiratory conditions. A tea prepared from *ku'chao*, (*Phoradendron* sp.) is used in the treatment of coughs, gastrointestinal ailments, and venereal disease or as a purgative.[867] Tarahumaras favor it to treat burns while mestizos use it for colds and stomachaches.[868]

Rojásowa

Helianthemum glomeratum (Lag.) Lag. ex Dunal
TAR. *rojásowa, jarába*
SP. *hierba de la gallina*[869]

Zingg reported that *rojásowa*, a plant of the family Cistaceae, is a small shrub with yellow flowers.[870] A decoction prepared from it is used to control fever and also to relieve coughs.[871] Externally, it is applied for skin infections and wounds and, internally, the raw crushed leaf is considered a urinary antiseptic.[872] In Samachique, an infusion prepared from *hierba de la gallina* is appreciated as a general medicinal tea.

Ronínowa

Stevia serrata Cav., *Stevia salicifolia* Cav., *Stevia* sp.
TAR. *ronínowa, roínowa, renoia,*[873] *ri'túnui*[874]
SP. *hierba de la mula; hierbamula, raíz de la muela; hierba del piquete*[875]

Ronínowa or *hierba de la mula* is a characteristic feature in the landscape of the sierra, particularly in Mesa de Paréwachi, Basigóchi, Pawichiki, and Rotoríchi, its white corymbs topping one- to two-feet-high plants with lanceolate leaves. It is common in areas cleared of pine and oak. In Cusárare, the plants begin to die out in October, while they still are blossoming in Samachique.

Blooming field of *Stevia serrata* (*ronínowa*) in Tarahumara Country.

Ronínowa is used to sweeten and give body to batari.[876] The stems of *Stevia salicifolia* are used as a piscicide in a similar way to nakáruri.[877]

The root of ronínowa is a strong analgesic when a bit of it is applied to an aching tooth. The whole plant is also used in a poultice to treat snakebites and trauma.[878] The leaves of *Stevia serrata* and *S. salicifolia* are sometimes crushed to plug ulcerations caused by worms on animals. Boiled into a fragrant and tasty tea, it is taken hot for colic.[879] A tea made from the leaves of *Stevia* sp. is used for colds, fever, and gastrointestinal ailments and a tea from the root is used as a laxative.[880]

Other varieties of *Stevia* or a probably related species are described as bushes with purple branches called *hierba mula* that are used in a decoction taken for rheumatism. An informant from Laguna de Aboréachi claimed that a crippled woman was healed after only a few mornings taking this tea.

Rorogochi

Plantago major L., *Plantago australis* Lam. ssp. *hirtella* (H.B.K.) Rahn,[881] *Mimulus guttatus* DC.

    TAR. *rorogochi, rorogótzari,*[882] *roró,*[883] *basaró, basarónowa*

    SP. *lantén, lantén cimarrón, llantén*[884]

Rorogochi is a wide-leaved edible plant, which has been consistently identified as *Plantago major*.[885] The also edible yellow-flowered *basaró*, or *lantén cimarrón* or wild *lantén*, has been identified as *Mimulus guttatus*. The Tarahumaras also call plants of the species *Plantago argyrea* Morris, *P. galeottiana* Decne., and *P. hirtella* Kunth by the name *rorogóchi*. These are used as additives for pinole and esquiate or eaten as pot greens. Tarahumaras from Guazapares reportedly have used lantén for medicinal purposes at least since 1777.[886] *Plantago major* is used in preparing a tea that is taken as a laxative.[887] Treatment of empacho, a condition characterized by abdominal cramps, is one of the plant's uses, as is the treatment of diarrhea and dysentery.[888] The fresh leaves of rorogochi are applied to skin sores, burns, wounds, and bites from poisonous animals. A decoction of rorogochi is used to clean lungs and bladder and as a healing substance for ear or eye infections.[889]

Ro'sábari

Artemisia mexicana Willd., *Artemisia ludoviciana* Nutt. var. *mexicana* (Willd.) Keck[890]

    TAR. *ro'sábari, rosáwari, chipuso, sipurí, chipuso chókame*[891]

    SP. *estafiate, ajenjo del país*

    NAH. *iztauhyatl*

Estafiate, known by the Tarahumaras as *ro'sábari*, is a plant considered sacred by the ancient Mesoamericans.[892] Even today, it is incorporated throughout Mexico in magic and herbalist healing practices within the context of cleansing rituals, or *limpias*.[893] Fr. Falcón Mariano in his "Relación de Guaguachique (Wawachiki)" in 1777 stated that the root of an herb called *tosáiali* (probably ro'sábari) growing in the temperate canyons was used to combat indigestion and worms.[894] The Tarahumaras use ro'sábari as a remedy for gastric disorders and abdominal pain. An infusion prepared by boiling it in water is also used for headaches, fever, and dysmenorrhea.[895] The tea is taken by women to promote menstruation.[896] A tea from its leaves is used for diarrhea, stomachaches, and other gastrointestinal disorders.[897] The active pharmacological agent in ro'sábari has been reported to be a volatile essential oil found in the blossoming plants and leaves.

An informant stated that there are two different species of ro'sábari but only one has medicinal properties. Some claim that Tarahumaras do not use ro'sábari as a remedy acknowledging that mestizos use it for chest pain and stomachache.[898] Ro'sábari is used to scare away the chu'í by placing a few branches under a pillow.

Rosákame Sikui

TAR. *rosákame sikui*

SP. *termita, hormiga blanca*

These insects—termites, not ants despite their name—are rubbed on warts to remove them.

Royal

SP. *royal, polvo de hornear*

This term *royal* derives from the brand name of a popular commercial baking powder. The name is applied in Mexico to all baking powders and some baking yeasts regardless of their actual brand name. These substances have reached the Tarahumara communities and have acquired remedial uses. For instance, some Tarahumaras from Norogachi claim that gastric distress, probably gastritis, may be relieved by eating a spoonful of these powders at bedtime and drinking lime juice in the morning.

Ruda

Ruta graveolens L., *Ruta chalepensis* L.

SP. *ruda*

Rue, a woody plant with small yellow flowers and bitter, strongly scented leaves, grows wild in the canyons and in some upland settlements. It is cultivated in flowerpots in Cusárare. The herb, originally from Europe and Asia, has been part of the European lore as a device against spells that have been cast and the works of the devil.[899] European and American pharmacopoeias have incorporated it as a remedy to stimulate menstrual flow and it also has been used to induce abortion.[900]

The Tarahumaras use *ruda* as a remedy to treat ear and stomach ailments. It has been considered an antispasmodic and a remedy for epilepsy. The Tarahumaras, as well as people throughout Mexico, believe that crushed leaves of rue placed in the ear are an effective remedy for earache. A tea prepared from rue leaves, stem, and roots is administered for stomach disorders.[901] Mestizos in Norogachi mix it with valerian to treat low blood pressure and dizzy spells.

Rue contains a starch, inulin-gum, nitrogenous substances, an essential oil, and rutic acid or rutine.[902] The essence is secreted by special glands in the plant's leaves. It is composed of a mixture of terpenes and terpenic derivatives, alcohols, aldehydes, ketones, and esters.[903] Rutine can be isolated by distillation in the presence of water as a ketone (methyl-caprinol). Rue leaves are bitter and irritate the mucosae inducing nausea.[904]

Rurikuchi

Bouvardia glaberrima Engelm.
    TAR. *rurikuchi, rurubusi, trompotia*[905]
    SP. *trompetilla*

Pennington reports that a decoction of leaves and stems of *rurikuchi* is one of the favorite Tarahumara medicines for heart problems.[906]

Rurubuchi

Eupatorium sp., *Eupatorium odorata* L.
    TAR. *rurubuchi*[907]

The genus *Eupatorium* is extensively used in Mexican popular medicine as well as by North American Indian groups. The sticky juice of *Eupatorium* sp. was used as purgative by the Tarahumaras. The raw roots are ground and drunk with water to relieve stomach afflictions.[908]

Eupatorium contains tremetol, glycosides, and sesquiterpene lactones. These substances may explain both the toxic and therapeutic actions attributed to the plant.[909]

Sábila

Aloe vera L.
    SP. *sábila, sávila, aloe, aloe vera*

Sábila is considered in many parts of Mexico a plant with mystical and therapeutic properties; it is particularly appreciated as a protector against witchcraft. Both in the uplands and canyon regions sábila is grown in flowerpots and used both as an ornamental and remedial plant. Bye reports its use by Tarahumaras to treat burns and sores.[910]

Salvia

Hyptis albida H.B.K., *Hyptis emoryi* Torr., *Hyptis suaveolens* (L.) Poit., *Hyptis* sp. *Salvia officinalis* L., *Salvia melissodora* Lag.,[911] *Salvia tiliaefolia* Vahl., *Artemisia ludoviciana* Nutt. var. *gnaphaloides* (Nutt.) T. & G.

TAR. *rosábari, rosáwori*
SP. *salvia*

The scientific name of the large, wild North American sage is *Artemisia ludoviciana* var. *gnaphaloides*; this indicates that *ro'sábari* and American sage may be variants of a single species.[912] The classic *salvia* (sage), reportedly an indigenous plant in Europe and the United States, is of course *Salvia officinalis*. However, for the Tarahumaras, a plant of the family Labiatae or Lamiaceae, *Hyptis albida*, is also known as *salvia*, or sage.[913] Even more confusing, *Hyptis emoryi* and *Hyptis* sp. are respectively called *rosábari* and *rosáwori*.[914] The presence of other varieties of salvias in the Tarahumara region also has been reported; that is, *Salvia melissadora* and *S. tiliaefolia*.[915]

The Tarahumaras cook the flowers, stems, and leaves of *Hyptis emoryi* in water and give it as a tea to women in labor "because it is very hot."[916] A tea made from *H. albida* is drunk by women to hasten the elimination of the placenta and fetal membranes. Small wads of its flowers are placed in the ears for earache, and the entire plant is rubbed on the body to relieve rheumatic pain.[917] *Hyptis suaveolens* is used as an aqueous extract to treat cataracts and gastrointestinal ailments.[918] Other uses of *Hyptis* sp. are to relieve heart ailments, headaches, fever, and to lose weight.[919] The Tarahumaras use the whole plant of *Salvia tiliaefolia*, boiled into a tea, for gastrointestinal disorders and headache.

Sangregado

Jatropha malacophylla Standl., *Jatropha cordata* (Ortega) Mill. Arg.
TAR. *ratowa*
SP. *sangregado, sangre de grado*,[920] *mata muchachos*[921]

Sangregado (*Jatropha malacophylla*), or *sangre de grado*, is one more of the Rarámuri plants mentioned in one of Alfonso Reyes Ochoa's classic poems. It is however an herb considered more a plant from the desert rather than typical of the Tarahumara region. Reyes praised the property of the herb to tighten loosened teeth. The Tarahumaras use sangregado (*Jatropha malacophylla*) to treat toothache and as a general medicine. Tarahumaras in Norogachi used the bruised leaves of *Jatropha cordata* as a poultice for sore eyes.[922] The leaves were also added to the water to bathe children, believing that this will give them strength.

Sa'parí

Sisyrinchium arizonicum Rothr.
    TAR. *sa'parí*,[923] *torimásari*[924]

This plant of the family Iridaceae grows in the pine forest of the uplands near Narárachi. Sa'paréachi, the name of an important ranchería, derives from the name of the plant. Its use as a piscicide has been reported.[925] A decoction of its leaves is taken to suppress diarrhea.[926] A poultice prepared from *sa'parí* roots is applied to the abdomen to treat stomachache. An informant observed, "It burns as if it was fire for just a while, but then one heals up." The roots or the decocted whole plant is taken to relieve gastrointestinal ailments. Another plant identified as *Sisyrinchium* sp. is used to prepare a tea to relieve toothache and as a general medicine.[927] This plant is known as *ramegónowa* (*ramé*=tooth, *okó*, *gó*=pain, *nowa*=root). The whole plant of sa'parí is crushed and chewed also for toothache.[928]

Saráame

Turnera ulmifolia L.
    TAR. *saráame*

Saráame is a common plant growing along upper margins of streams in southeastern Tarahumara Country. It is used to prepare a tea to treat diarrhea; the leaves are steeped in very hot water for this purpose.[929]

Sarabí

Prionosciadium townsendii Rose
    TAR. *sarabí, sarabiki*
    SP. *jicamilla*,[930] *hierba del oso*[931]

Tarahumaras from Cerocahui state that *sarabí* is an herb that relieves hunger, implying that more than a remedy, it is a source of nourishment, especially in times when other food is not available. On the other hand, a tea made from the herb is used to treat febrile ailments.⁹³² Roasted sarabí is used to treat sore throat and relieve the distress of flu.⁹³³

Sarépari
Unidentified
> TAR. *sarépari*

Sarépari is a foul-smelling herb whose roots are reputed to be an effective remedy for diarrhea.

Sayawi sa'para
Crotalus sp., *Crotalus atrox*, *Crotalus triseriatus*
> TAR. *sayawi sa'para*
> SP. *víbora (carne de)*

Tarahumaras hunt rattlesnakes mostly to sell their dried meat to Mexicans, who use it as a remedy for rheumatism and cancer, but among the Rarámuri it is not used it for these purposes. Occasionally, its meat is consumed as food.⁹³⁴ Rattlesnake fat is however used by the Tarahumaras as a rubbing ointment during sweat-bath ceremonies.

Sewáchari⁹³⁵
Tithonia fruticosa Canby & Rose
> TAR. *sewáchari*

The sunflower-like heads of *sewáchari*, a common plant along the borders of plowed fields on middle portions of the western canyon slopes, are crushed and steeped in a drink taken for coughs.⁹³⁶

Simonillo
Conyza sp., *Conyza filaginoides*⁹³⁷
> SP. *simonillo*

In Alfonso Reyes Ochoa's poem "Yerbas del Tarahumara," *simonillo* is one of the remedies the itinerant Rarámuri herb seller offers in the streets of Chihuahua City. Simonillo was also one of the herbs studied by the extinct Mexican National Medical Institute of the early twentieth century.⁹³⁸ Throughout México, simonillo has been used to treat catarrhal afflictions of the stomach and gallbladder. Deimel claims it is used only by mestizos for diarrhea.⁹³⁹

Sinowi Ramerá

Serjania mexicana (L.) Willd.

    TAR. *sinowi ramerá*

    SP. *diente de víbora, diente de culebra, wirote de culebra*[940]

Sinowi ramerá has been used to treat pain. The stem of the plant is cut in half, the prickles removed and placed next to the skin. It is used for rheumatism, applied to the painful areas. A fresh fragment is applied when the old one dries up. This procedure is repeated until pain relief is obtained.[941]

The stems of sinowi ramerá are also crushed and used as a poultice applied to bruises, sprains, and fractures.[942] A tea prepared by boiling this plant is supposed to promote conception. Bye claims that the stem and leaves of this plant are used in a wash for rheumatic pain.[943]

In Cusárare, a *wirote* is roasted and applied on burns. It is said to speed the healing of burns.

Sitagapi

Haematoxilon brasiletto Karst.[944]

    TAR. *sitagapi*

    SP. *palo del Brasil, brasil*

    NAH. *cuamóchitl, huitzcuáhuitl*

Sitagapi, or brazilwood, grows in the depth of the canyons; the convoluted shape of its stems and branches fascinate the Tarahumaras who have great respect for its hardwood from which ritual instruments such as sipíraka and tesora are made. The Tarahumaras use the red dye obtained from the plant and attribute a beneficial mystical power to it. After their celebrations, they drink an emulsion of sitagapi in a similar fashion as they do wasárowa. Prepared into a tea, it is used to cure spots on the skin and as a laxative. Interestingly, the tea is used for diarrhea and kidney disorders, edema, and pneumonia.[945] The young branches of sitagapi are used in the preparation of a compound that is rubbed on people suffering from jaundice, bathing the person in the red liquid.[946]

The wood of this plant contains brazilin, a dye used in the past to color wines and toothpaste, and its derivative, hematoxyline, which has been widely used as a stain in pathology laboratories.

Sitákame

TAR. *sitákame, sitá*[947]

SP. *almagre, ocre*

Sitákame ("red") is a powdered ocher rock used to paint the bodies of the pascorero dancers as well as to decorate ritual instruments like the large *norí-ruachi* drums. It is also used to paint the spears used in Guapalayna during Holy Week. Sitákame is applied as an ointment on burns. In Sarabéachi, the owirúame draws a cross within a circle with this ocher on the chest or other aching body part to relieve pain, a technique that was used earlier throughout the Tarahumara region but that in recent times has lost popularity.

Sogíwari and Chiná or Chinaka

sogíwari: *Eryngium carlinae* Delar., *Eryngium heterophyllum* Engelm., *Eryngium* sp.; **chiná** or **chinaka**: *Eryngium longifolium* Cav.

TAR. **sogíwari**: *sogíwari, soíwari, so'íwari, so'irí,*[948] *sawíwari,*[949] *saríwari, saríbari;*[950] **chiná** or **chinaka**: *chiná, chinaka*[951]

SP. **sogíwari**: *hierba del sapo, flor de San Pedro, huisapol*; **chiná** or **chinaka**: *huitzquelite, cardo lechero*[952]

Sogíwari or *soíwari* (*Eringyum carlinae*) is an herb of the family Umbelliferae, found in upland meadows and occasionally growing under the shade of pine trees. More infrequently it is found in shaded portions of streams in eastern Tarahumara Country. Soíwari is also identified as *Eryngium* sp. or *E. heterophyllum*,[953] while *chinaka* is identified as *E. longifolium*.[954] Leaves of *Eryngium carlinae* are added to *keorí*—parched and ground corn—in water or cooked as a quelite.[955]

The "Relación de Guazapares" reported that *hierba del sapo* was used to prepare a purgative and the "Relación de Chínipas" indicates that it was used to treat skin infections.[956] Currently, a tea prepared from sogíwari is used to treat dysentery.[957] The flowers of soíwari are bruised and applied as a poultice to styes.[958]

E. carlinae is used against gastrointestinal illness, chest congestion, or as a general medicine.[959] *Eryngium heterophyllum* is a source of flowers and leaves used in a tea to relieve cough.[960] The Tarahumaras make a tea with roots or the plant to use against diarrhea, chest congestion, ocular ailments, and headaches.[961] Rheumatic gout and dysentery are indications for the sogíwari tea.[962]

Eryngium longifolium (*chinaka*) is boiled into a tea used for chest pain.⁹⁶³ Cardenal claims that the cooked or macerated seeds of chinaka are used against intoxications by poisonous mushrooms, particularly *Amanita muscaria*.⁹⁶⁴

Solda

Unidentified

SP. *solda*

Tarahumaras and mestizos from Cerocahui use a decoction made from *solda* leaves as a contraconceptive, administering the potion fifteen days before menstruation.⁹⁶⁵

Sopépari

Senecio hartwegii Benth.

TAR. *sopépari*

Sopépari, a large shrub growing on humid slopes and in shadowy ravines near Norogachi, is considered one of the five most potent piscicides used by the Tarahumaras. The roots are crushed and rubbed on the skin to kill lice or ticks or made into a tea and used as a purgative.⁹⁶⁶

According to Salmón, *Senecio hartwegii* contains senecionine, senecephylline, jacobine, jaconin, a cyclic diester, and tannin.⁹⁶⁷

Sunú

Zea mays L.

TAR. *sunú, su*nú*

SP. *maíz*

Sunú, or corn, the main staple of the American Indians, has also been attributed medicinal and mystic properties.⁹⁶⁸ In particular, the corn styles are boiled in a tea and given as a diuretic and to treat urinary tract infections. Such effects were known since remote times by the Aztecs.⁹⁶⁹ Chemical analysis of the styles found a mucilage, starch, pectine-like substances, calcium bicarbonate, calcium chloride, potassium nitrate, and potassium chloride.

The burned cobs (*olote* or *o'ona*) are given pulverized to treat stomachaches and diarrhea. Raw corn tortillas are eaten to relieve inflammation and toothache.⁹⁷⁰ A curing ceremony for newborns in which the infants are rubbed with the charred end of a burning corncob has been described. The shaman marks three parallel lines lengthwise over the child's head and three across.⁹⁷¹ A similar cure done on kneeling adults also has been described.⁹⁷²

Batari, the native corn beer, also has been used in many ritual curing ceremonies.

It is given to infants along with the mother's milk, believing that it will keep the infants free of sickness. In a curing ceremony, the newborn child is sprinkled with batari by the shaman to give him strength.⁹⁷³ Kennedy describes how the therapeutic and prophylactic properties of batari extend beyond curing humans to cattle, corn, and the fields.⁹⁷⁴

Tavachín

Caesalpinia pulcherrima (L.) Sw.
TAR. *makapal*⁹⁷⁵
SP. *tavachín*

The seeds of this plant are roasted and consumed by the canyon Tarahumaras. The roots of *Caesalpinia pulcherrima* are used to prepare a drink believed to be a cure for gonorrhea.⁹⁷⁶ It is also used as a wash for fever.⁹⁷⁷ *Tavachín* is also used as a treatment for susto, as a strong laxative, and to induce menstruation.⁹⁷⁸

Tikúwari

Datura meteloides DC. ex Dunal, *Datura stramonium* L., *Datura discolor* Benth., *Datura inoxia* Mill., *Datura lanosa* Benth., *Datura quercifolia* H.B.K., *Datura ceratocaula* Ort.⁹⁷⁹
TAR. *tikúwari, rékuba,*⁹⁸⁰ *uchiri,*⁹⁸¹ *uirí,*⁹⁸² *rikuri*⁹⁸³
SP. *toloache, estramonio*

Tikúwari or *toloache* (Jimson weed) is one more of the plants with psychoactive properties attributed mystical and therapeutic properties by the Tarahumaras and by other American Indian groups.⁹⁸⁴ The Franciscan document "Relación de Guazapares" noted that Tarahumaras used tikúwari for medicinal purposes. The friar reported that toloache was used for *flucciones*.⁹⁸⁵ According to their beliefs, powerful spirits were assumed to dwell in certain plants and particularly in tikúwari. The spirits may take revenge when angered or insulted.⁹⁸⁶ Zingg reported that the plant therefore was greatly feared by the Rarámuri who seldom dare to touch it: "They believe that anyone who pulls it out will become insane and die."[987] This belief has been also reported more recently by several scholars. Informants from Norogachi, however, had difficulty recognizing the herb and endorsing the reported beliefs attributed to tikúwari.

The Tarahumaras have been observed applying leaves of tikúwari to the forehead to alleviate headaches and shortly removing them, probably to prevent the induction of toxic symptoms.[988] Incidentally, the pulverized leaves of tikúwari are used as one of the ingredients in the preparation of batari and as a poultice prepared from the heated leaves as a very common remedy for swelling and sprains.[989] The leaves are also used in preparing a decoction taken for diarrhea.[990] Leaves, seeds, and roots of tikúwari are still used by western canyon Indians to prepare a drink used as a ceremonial offering or during ceremonies that promote visions or a feeling of exhilaration. A small portion of this drink is also taken by Tarahumara healers while assessing a sick person.[991] The leaves of *Datura lanosa, Datura quercifolia,* and *Datura ceratocaula* are smoked to treat asthma.[992]

Finally, the Rarámuri share a belief present throughout Mexico, that toloache may be given concealed in food to unfaithful men intoxicating them and preventing them from leaving home.

Datura has active principles including the tropane alkaloids scopolamine and L-hyoscyamine. Depending on the dose administered, antispasmodic or on the other hand toxic, anticholinergic effects may result, including psychosis, confusion, and death. The genus also contains hydrocyanic acid, isobutyril aldehyde, and malic acid.[993] The hallucinogenic effects probably are more complex than may be explained by the action of anticholinergic post-ganglionic blockade or the intoxication with the major tropane alkaloids. Other alkaloids, also contained in *Datura*, may be present, such as tropine and scopine, and interact with scopolamine. *Datura stramonium* has been included among the plants containing lectins—plant proteins with mitogenic properties.[994]

Tila

Tilia sp.

SP. *tila*

Throughout Mexico and by the U.S. border, *tila* flowers are used in preparing a tea with tranquilizing and hypnotic effects. Tarahumaras and mestizos have adopted this use. As in other parts of Mexico, *té de tila* is commonly consumed during funeral wakes.[995]

Toronjil

Cedronella mexicana Benth.[996]

SP. *toronjil*

Toronjil grows in humid terrains; its fresh, agreeable odor resembles the one of citrus fruits. In Norogachi, mestizos, knowing the medicinal virtues of toronjil, plant it in gardens and flowerpots. Toronjil tea is used against susto and for chest pain (*surachí okorá*). Others use it for anxiety and stomachache.[997]

Among the active principles of this plant are a colorless volatile oil, tannin, a bitter resin, and a mucilage. These constituents help it to be considered a general stimulant, appetite enhancer, diaphoretic, and digestive.[998]

Trébol

Melilotus indica (L.) Allioni, *Meliotus officinalis* (L.) Lam.

SP. *trébol*[999]

The sweet-smelling *trébol* has leaves that are crushed and dampened for use as a poultice applied to the head to alleviate headache.[1000] It is also used internally for sore throat and as a general tonic.[1001]

Tu*chí

Anoda triangularis (Willd.) DC., *Anoda cristata* (L.) Schltdl., *Viola umbraticola* H.B.K., *Viola* sp.

TAR. *tu*chí, tuchí, bakarísoa, wakarísowa, wajorísowa*[1002]

SP. *tuchí, malva*

Tuchi or *tu*chí* has been over time one of the best-known Tarahumara medicinal plants. *Tu*chí* or *tuchí* is commonly translated as *malva*. *Viola umbraticola* (tu*chí) grows in the upland meadows where its fresh leaves are eaten raw or cooked.[1003] Tu*chí also has been identified as *Viola* sp.[1004] A tea prepared from the leaves of *Anoda triangularis* has been used for febrile conditions.[1005] Tarahumaras from Kwechi report that *tu*chí* is useful to treat dysentery.[1006] *Bakarísowa* (*Viola* sp.) is used against diarrhea by Tarahumaras while mestizos use it against intestinal worms.[1007]

Unto sin Sal

SP. *unto sin sal*

Unto, or suet, is the fatty tissue that surrounds the kidneys of the pig. It is usually employed externally in Cerocahui by mestizos and Tarahumaras for aching, pain, and internal soreness. It is applied even to eyes and ears. In recent times Vick's VapoRub, a commercial rubbing compound, has displaced in many homes the *unto sin sal* for all its applications.[1008] In Norogachi, siyóname wi'í, or aceite de arrayanes, is preferred for the same uses.

Uré

Fraxinus papillosa Ling., *Fraxinus velutina* Torrey
 TAR. *uré*
 SP. *fresno*

Uré is the ash tree that is found in many New World ecosystems, including the Tarahumaras' mountains. Tarahumaras use the hard wood of uré to carve spoons, ax and hoe handles, and to make certain parts of the violin. An informant claims that if sitagapi wood is not available, the ceremonial *sipíraka* might be carved from uré wood. The bark of uré is ground on the mataka, mixed with water, and placed before a cross, especially during the harvest ceremonies. The Rarámuri use the cooked tender leaves of uré as food. The buds and leaves of these plants have been used extensively by several American Indian groups as remedies for rattlesnake bites and as an astringent and febrifuge.[1009] Some authors believe that the medicinal uses for uré have been overlooked by the Tarahumaras; nevertheless, the leaves are cooked and used in ceremonial healing.[1010] The Tarahumaras consider a cold infusion from uré bark ground on the mataka as a general cure when placed in front of a ritual cross.[1011] Cardenal claims that an infusion prepared from the leaves is used to treat rheumatism and kidney ailments, while the bark is used for placental retention, fever, and intestinal parasites.[1012]

Urí

Vitis monticola Buc. *arizonica* (Eng.) Rog. and Rog., *Vitis arizonica* (Eng.) Rog. and Rog.
 TAR. *urí, orí, awí*
 SP. *uva, uva silvestre, cimarrona*

Urí, a wild grape, is a highly appreciated gift of nature for the Tarahumaras. This fruit ripens in June in western canyon country and somewhat later at higher elevations.[1013] A decoction is made with the vine and root of the plant and taken to aid childbirth.[1014]

Usabi

Prunus capuli Cav., *Prunus serotina* Ehrh.,[1015] *Prunus brachybotrya*
TAR. *usabi, u*sabi,*[1016] *kusabi, usá, ko*[1017]
SP. *capulín, igualama*

Usabi, or *capulín*, is used in small amounts as a spice to add piquant flavor to esquiate. The bark and leaves of usabi, bundled and crushed, are also one of the several piscicides used by the Tarahumaras. It is claimed that a diluted infusion is a sedative and analgesic. Zingg reported that Tarahumaras believe that eating the leaves of *Prunus brachybotrya* will kill the gnats that bite them and are such a nuisance in the gorges during the hot season.[1018] In canyon country, the fruit is mixed with water and abundant sugar to prepare syrup to treat whooping cough.[1019] A tea made with leaves of usabi is prepared to relieve the cough of bronchitis, and even of whooping cough.[1020] In general, it is supposed to be effective in chest afflictions.[1021]

In Bakasórare, usabi is used for headaches and colds. A decoction of usabi is prepared to relieve headaches, urinary tract infections, and stomachaches.[1022] Although Tarahumaras use usabi as a ritual and general remedy, mestizos seem to restrict its use to relieving coughs.[1023]

The piscicide action mentioned earlier is attributed to the cyanogenic glycosides of the plant; cyanhydric acid is formed if distilled in the presence of water. Any remedy prepared with usabi may cause serious poisoning. Just six or seven leaves in 100 grams of water could be toxic.[1024]

Valeriana

Valeriana procera (Kunth) FG. Mey., *Valeriana officinalis* L., *Valeriana edulis* Nutt.
SP. *valeriana, hierba del gato*[1025]

Valerian is used throughout Mexico where it is reputed to be second only to tila an effective sedative. *Valeriana edulis* grows in the region and its roots of are boiled into a tea used as a general remedy.[1026]

But my observations indicate that valeriana is usually obtained from mestizo merchants as an alcoholic tincture or as ground dry plant.[1027] In Norogachi, valeriana is used as a sedative and in Samachique it is used for stomach ailments also. A mixture with rue is given for low blood pressure and dizziness.[1028]

Verbena

Verbena caroliniana L., *Verbena wrightii* A. Gray, *Verbena elegans* H.B.K.[1029]
    TAR. *owáame*,[1030] *wajíchuri, anariye*[1031]
    SP. *verbena, moradilla*[1032]

The Tarahumaras have used *verbena* medicinally since colonial times.[1033] A decoction of the whole plant of *Verbena wrightii*, except the roots, was used for fever.[1034] The leaves are moistened and applied as a poultice for the relief of boils and contusions; the roots are dug up during the winter and prepared in a similar way to the leaves, and the flowers, boiled into a tea, are drunk to cure boils.[1035] Throughout the sierra, verbena poultices are used for boils and similar lesions, alternating with applications of hot water and salt. In Rochéachi, the mestizos prefer to drink a tea prepared from the leaves. In Cusárare, Tarahumaras and mestizos apply crushed leaves of verbena directly to caried teeth to relieve toothaches. Bye reports that the leaves of verbena are also boiled in a tea, used for wounds, sores, and gastrointestinal disorders. Another species, *V. elegans*, is used for similar purposes.

Bye attributes the healing action of verbena to its rich content in tannin, which forms a protein-tannate film that may help wounds to heal.[1036]

Waré

Cissus sp.
    TAR. *waré*

Waré (*Cissus* sp.) is a woody climber of the grape family. The plant is repeatedly mentioned in the eighteenth-century Franciscan *Relaciones* as used throughout the Tarahumara land: in Guazapares, it was ground into small fragments or powdered to help wound healing; at Tutuaca, a wash served the same purpose, while in Suasivo, near Chínipas, it was used to help heal sprains and fractures; finally, in Wawachiki, it was used to prepare a tea for stomach disorders.[1037]

The sticky juice from the stems of *Cissus* sp. was extracted by pounding and used as a plaster to set fractured bones.[1038] Waré continues being used to help heal sprains and fractures.

Wasárowa

Karwinskia humboldtiana (R. & S.) Zucc., *Buddleia* sp.,[1039] *Parthenium tomentosum* DC. var. *stramonium* (Greene) Rollins

TAR. *wasárowa, wasároa*,¹⁰⁴⁰ *sorowa, kusí júkame, ra'mori, amurí, kioch'wital*¹⁰⁴¹

SP. *palo apestoso, palo hediondo, kakachila*

Zingg identified *wasárowa* as *Buddleia* sp., a plant of the family Loganiaceae.¹⁰⁴² Our informants however unambiguously state that wasárowa is the same plant as *kusí júkame*, which is *Karwinskia humboldtiana* from the family Rhamnaceae. In some areas, the Tarahumaras prefer the Spanish term *palo hediondo*, the "stinking tree." *K. humboldtiana* grows in thick stands over cutover slopes near canyon settlements, and in open spaces in the lower elevations of western canyons.¹⁰⁴³ It also has been identified with *Parthenium tomentosum* var. *stramonium*, a plant of the family Compositae.¹⁰⁴⁴

Sticks of the plant were brought up from the gorges to be sold to the Indians of the highlands.¹⁰⁴⁵ The foreshaft of Tarahumara arrows was made of this palo hediondo, a wood also used to make sewing needles.¹⁰⁴⁶

The plant is of considerable ceremonial importance. It is used during ritual purifications as part of the Jíkuri ceremony and other celebrations. Similar to what was described for mé and sitagapi, at the end of a ceremony the male attendees are administered three teaspoonfuls of a cold infusion of wasárowa to clean their souls and their bodies from possible pollutants. Women require four spoonfuls. The minister stands by the main cross. He takes the *bitóriki*, or wooden dish, which contains water with wasárowa and offers it to the cross, sprinkling it with the liquid. Then he walks around the cross three times and drinks three teaspoonfuls of the liquid. Next, he draws a cross on every person's forehead, chest, shoulders, and nape, starting with the men. As he draws the crosses, he says, *Mí repá atí kemu onó, kemu eyé má, a chi ko mi tibuka.* "Above you is your Father; your Mother too will take care of you." Finally, he administers the beverage to women and children.¹⁰⁴⁷ The respect that the Tarahumaras demonstrate for wasárowa is similar to that expressed for mé, sitagapi, and the hallucinogenic plants.

Fr. Falcón Mariano, in his "Relación de Guaguachique (Wawachiki)" published in 1777 mentioned that "in the warmer canyons various shrubs called *guasálaguac* [sic] could be found, which are hot to the taste not unlike a radish. The Indians crush the wood and drink the sap to relieve gastric pain." It was reported that fragments of palo hediondo or the *chuchupate* root were wrapped in a piece of cloth and tied around children's necks, believing that the strong smell would protect them against disease.¹⁰⁴⁸ The leaves of *amurí*, identified as *Karwinskia humboldtiana*, were wrapped in cloth and tied to the head to treat headaches. *Buddleia* sp. was used by a shaman as a magic cure for children whose spirits were "eaten by the water serpent."[1049]

Currently wasárowa is used as curative remedy for majawá, susto, and loss of the iwigara. The crushed bark of *Karwinskia humboldtiana* is still used to prepare a drink taken by Tarahumaras and Tepehuans to treat fever. The sap of *wasároa* (*Parthenium*) is applied to a hurting tooth to relieve the pain and is used to prepare a wash or a poultice for joint pain and wounds.[1050]

When a person is about to travel to distant places, he places a fragment of wasárowa in a glass of water and drinks it to prevent susto, which may occur while traversing unknown places. This practice is done in particular to prevent harm from the influence of water springs. If a person has been overcome by susto, a decoction from wasárowa is considered an effective remedy. The stick of wasárowa wood is scraped or shaved down with a knife and boiled in water. The whole body is then bathed one or more times from head to foot with this water. A bit of the liquid may also be drunk. The efficacy of this plant is not limited to human beings. It is widely used to protect crops from pests and weather damage.[1051]

The ovoid drupes of *Karwinskia humboldtiana* contain a toxic principle partially characterized as a quinone.[1052] The substance causes paralysis in man and nausea, progressive weakness, and death in animals. In spite of reported animal and human toxicity, the berries of *K. humboldtiana* are occasionally consumed as starvation food.[1053]

Wasía

Ligusticum porteri Coult. & Rose, *Conium maculatum* L.

TAR. *wasía, wa*sí, wa*siga*[1054]

SP./NAH. *chupate, chuchupaste, chuchupastle, chuchufate, chuchupate, hierba de cochino*[1055]

Named *chuchupaste* by the Mexicans, it is one of the several medicinal herbs that people from urban central Chihuahua buy from the wandering Tarahumara herb dealers. Pennington identified *wasía* as *Conium maculatum*, a perennial herbaceous flowering plant of the family Apiaceae, the poisonous hemlock native of Europe and the Mediterranean region. He claims that it is one of the plants used by the Tarahumaras as a piscicide.[1056] The lookalike *Ligusticum porteri* is native to the region. *Ligusticum* grows in Bakasórare near

Norogachi and is also called wasía. Both plants belong to the family Umbelliferae.[1057]

According to the eighteenth-century Franciscan *Relaciones*, chuchupaste or wasía was considered an important medicinal plant.[1058] Fr. Falcón Mariano reported in his 1777 "Relación de Wawachiki" that Tarahumaras extracted from *guasiaca* [sic] or chuchupate a liquid used to treat stomach pains and flatulence. Nowadays chuchupaste is used by Tarahumaras from Norogachi, Bakasórare, Kwechi, and Samachique as a favored remedy for chawiste (the common cold) and fever.

In another use, fragments of palo hediondo or the chuchupate root were wrapped in a piece of cloth and tied around a newborn child's neck; the strong smell was supposed to protect the infant from disease.[1059] Salmón claims that some Rarámuri still carry a piece of the root to ward off snakes and sorcerers.[1060]

The stem or roots can be chewed or smelled "to sneeze out the illness." Pneumonia and headaches are also in the list of uses.[1061] Anthelmintic and antibacterial effects have been mentioned.[1062] Pennington described how upland Tarahumaras crush and boil the roots of *Ligusticum porteri* to prepare a compound used as a lotion for rheumatic joints.[1063] The lotion is also used to wash wounds and in a poultice for bites of poisonous animals.[1064] The decoction may also be ingested for the same purpose.[1065] The roots of *C. maculatum* are used as a general medicine.[1066]

Finally, it is reported that wasía is applied to a ceremonial cross to protect animals from lightning.[1067]

The antipyretic and analgesic effects of the *Ligusticum* tea may depend on its content of coumarins, which lower body temperature. The possible presence of aromatic oils and resins may play a role as active components in the plant.[1068]

Watorí

Ficus petiolaris H.B.K.
    TAR. *watorí, chitori*
    SP. *tezcalama, tescalama*[1069]

The milk from the stem of *watorí* placed on a rag and applied as a poultice that supposedly alleviates body pain.[1070] The milky juice from its bark also is used by western canyon Tarahumaras as a wash for wounds, and the latex is applied on fractured bones and to persons suspected of suffering from majawá.[1071]

A substance present in the latex of *Ficus* sp. appears to physically bind to the flesh of infected wounds and thorn punctures, supposedly to protect them from infection and larvae.[1072]

Watosí

Salix bonplandiana H.B.K.

TAR. *watosí, batosí*

SP. *sauce, sáuz*

The inflorescence and cortex of this tree contain principles used to relieve spasm, fever, and joint aches. Most likely these effects are due to salicylates contained in it. A decoction of flowers and bark is used to relieve anxiety. The flower is used to ease uterine pain. *SEMARNAP* adds that a mixture of decoctions of white and red sauces is applied to ailing joints to ease the pain.[1073]

Wawachí

Unidentified

TAR. *wawachí, wawá*

SP. *jara blanca*,[1074] *vara blanca, hierba de San Pedro*

The name of *Wawachérare*, an important Tarahumara pueblo, derives from the name of this plant: *wawaché*="There is wawachí." Pennington reports that a wash for rheumatic pains is made by steeping leaves of *wawachí*.[1075]

Wetajúpachi

Krameria sp., *Krameria grayi* Rose and Painter

TAR. *wetajúpachi, chakate*[1076]

The skin of *Krameria* was pounded or ground and mixed with suet as an ointment for aching teeth.[1077] Its use for toothache seems to have survived in the western canyons.[1078] Salmón reports that *Krameria grayi* contains kramerine, krameriatannic acid, rhatanone, ratahine, and aminoacids.[1079]

Wichá, Garó or Garóko, and Bejoké

wichá: *Acacia farnesiana* Sprague & Riley, *Acacia cymbispina* (L.) Willd., *Ceanothus buxifolius* Willd.; **garó (garóko)**: *Mimosa* sp., *Mimosa dysocarpa*, Benth. *Pisonia capitata* (S. Wats.) Standl. *Celtis pallida* Torr.; **bejoké**: *Prosopis* sp., *Prosopis juliflora* DC., *Prosopis glandulosa* Torr.[1080]

TAR. **wichá**: *wichá, wicháka, wichayó, wichagó, wichabó*; **garó (garóko)**:

ga'ró,[1081] *galóko, agóko*,[1082] *garároa, go'rarowa, gorárowa, goróroa, karároa, galagá*,[1083] *galó chókame*,[1084] *kochílari*;[1085] **bejoké**: *bejoké, bejogé, beogé*[1086]

SP. **wichá**: *vinorama, espino, huizache vinolo, winolo, vainoro*; **garó** (*garóko*): *vainoro, gatuño, vinorama, bainoro* [sic] *prieto*;[1087] **bejoké**: *mezquite*

Wichá means "thorn, or spine." Therefore, the terms listed may be employed for a diversity of plants of the family Leguminosae from genera as diverse as *Mimosa, Acacia,* or *Prosopis* (*mezquite*). The term *garó* designates several plants characterized by their thorny stems and branches. *Garó chókame* designates the *vainoro prieto,* a plant of the family Nictaginaceae. The terms *garároa, go'rárowa, gorároa,* and *korároa* designate one or several *gatuños* or *vinoramas. Bejoké* is the common Tarahumara name given to mezquites.

The thorny forest formed primarily by *Acacias* or mountain mahogany and *Mimosas* on the slopes of the canyons is called by local people *monte mojino*. The natives distinguish *vinorama* from *vinolo, winolo,* or *vainoro* by its thorns. While the spines of the former are flat, the ones of the second are channeled, resembling a miniature *weja* (gourd) to the observer. Pennington identifies *vinola* as *Acacia cymbispina* and reserves *vinorama* for *A. farnesiana*.[1088] *Wichagó* has been described as a shrub with straight, long thorns up to seven centimeters long and *garó*, as the vinorama of the Urique Canyon.[1089] Some thorny plants grow also in the transitional areas in the uplands; *garároa*, a many-branched shrub, grows in thick stands along streams near Sitanápuchi, Narárachi and Norogachi with *Mimosa disocarpa*.[1090] Mezquite shrubs grow more abundantly in the eastern hill country on the borderland with the central Chihuahuan deserts.

In Norogachi, a decoction prepared from leaves of *wichagó* is appreciated as a tea, which compares to the ones made from laurel or *basigó*. The sweet taste of the leaves of *wichagó* makes them a favorite food for the goats. The large roots of *Mimosa dysocarpa* are crushed and placed, by means of a basket, in slow-moving streams as a piscicide;[1091] this use is confirmed by informants in Norogachi.

An alkaloid, mimosine, may be responsible for the piscicide action of *garároa*. We do not know whether its therapeutic effects are attributable to the same substance.[1092] Mezquite contains a mucilaginous substance that may replace arabic gum in all its medicinal purposes. Treating mezquite gum with alcohol, arabine or arabic acid is obtained. In addition, mezquite extracts contain abundant mineral salts.[1093]

The spines of *wichá* (*Acacia cymbispina* and *A. farnesiana*), which is distinguished by its handsome yellow flowers, are used to make a drink taken for kidney disorders.[1094] The flowers are pulverized, mixed with fat or oil, and rubbed on the head to relieve headache or on bruises to aid healing. The flowers are used to prepare a wash for eye ailments.[1095]

The bark and spines of wichá were cooked to make a decoction that relieves the pain of insect bites; they are particularly useful to treat the sting of the scorpion. If additional treatment was needed, peeled bark was fastened around the neck.[1096]

The leaves of a wichagó or *wichabó* are presently used by the Tarahumaras of Norogachi to treat pneumonia. Finally, Bye claims that the roots of another *wicháoko* or *wicháko* (*Ceanothus buxifoliusi*) are used to prepare a tea for diarrhea or as a general medicine.[1097]

The leaves of *garó chókame* (*Pisonia capitata*) are ground with water, and the juice that results is squeezed by means of a cloth into fresh water. This preparation is then slightly warmed and taken for fever.[1098] However, Bye claims that the bark, not the leaves, is used to prepare the remedy. In Norogachi, gorárowa or gatuño is prepared into a tea taken to treat intestinal worms.

The bark and leaves of mezquite are used to prepare a wash and a poultice for eye ailments.[1099]

Wichásuwa

Cheilanthes kaulfussii Kunze
 TAR. *wichásuwa, wichásua*[1100]
 This plant blooms with minute inflorescences; its leaves are boiled into a decoction for urinary disorders.[1101] It is also used for pulmonary troubles.[1102]

Wichurí

Mammillaria heyderi Muehl.,[1103] *Mammillaria craigii* Lindley, *Mammillaria grahamii* Engelm. var. *oliviae* (Orcutt) Benson, *Echinocereus salm-dyckianus* Scheer, *Echinocereus triglochidiatus* Engelm.,[1104] *Echinocereus dasyacanthus* Engelm., *Coryphanta compacta* (Engelm.) Britton & Rose
 TAR. *wichurí, uchurí, wichuchi*,[1105] *wichuwá, ri'wé*,[1106] *kapírame*
 SP. *biznaga*,[1107] *peyote de San Pedro*
 Wichurí is commonly considered one of the plants used by Tarahumaras

as substitutes for *jíkuri* (*Lophophora williamsii*); therefore, scholars and informants contend that it shares jíkuri's attributed mystical and medicinal properties. For example, Bye claims that *Mammillaria grahamii* is reported to be the actual jíkuri in the region of the canyon of Batopilas. The word *wichurí* has been thought to derive from the Tarahumara word *wichuwama*, "to go crazy"; however, it most likely just means "spiny" or "thorny," which both are characteristics of the plant.[1108] The fruits of *Mammillaria heyderi* and *Echinocereus dasyacanthus* (*kapírame*) are incorporated in the Tarahumara diet. The fruits are ground and added to atole, or gruel, for flavor. The plant was said to have a curious medicinal use: the spines are removed, the plant is cut in half and roasted in for a few minutes; then, the soft center is squeezed in the ear in to treat earache or deafness.[1109] More recently, a fresh raw slice of *wichuwá* has been used as a poultice for inflammations and burns. Also, wichurí in alcohol is rubbed on ailing joints.[1110] It is also used as a stimulant for the runners in kickball races. The cottonlike hairs of *wichurí*, *wichurí rosá*, or *algodón de biznaga* also have various general medicinal uses.

Special magic powers are attributed to wichurí. It is a powerful cactus, used mainly by the most experienced shamans; it is supposed to help them enter the other reality to retrieve souls and fight sorcerers.[1111] The shaman who eats it supposedly gains a special visual power that makes him able to locate wizards and witches.[1112] Some Rarámuri claim that eating the top part of the plant is more effective at promoting colorful visual visions and the feeling of traveling to distant places. However, they warn if the persons partaking of it are not prepared, it may drive them to madness.

Bye noted that Bruhn identified n-methyl-3–4-dimethoxyphenethylamine in an alkaloid fraction of *M. heyderi*. He also reports preliminary information, from McLaughlin, about the presence of tryptamine derivatives in *Echinocactus triglochidiatus* and, quoting Raffauf, macromerine in *Coryphanta*.[1113]

Wiígame
Solanum diversifolium Schlecht.
    TAR. *wiígame, rerowi sawara*[1114]

This spiny plant is boiled and the resulting liquid is rubbed upon the body for relief of rheumatic pains; the mixture is also taken as a tea to cure colds.[1115] The plant is also taken in a tea for aches and pains.[1116]

Wipá or Wipáka

Nicotiana tabacum L., *Nicotiana glauca* Graham, *Nicotiana rustica* L., *Nicotiana trigonophylla* Dun.

TAR. *wipá* or *wipáka*,[1117] *makuchi, bawará*

SP. *tabaco, papanti, cornetón*

Since the sixteenth century, tobacco was widely celebrated in Europe as a panacea for a variety of human ailments, but there is no reason to suppose that these alleged therapeutic virtues were learned from the Indians, who first cultivated the plant.[1118] The Tarahumaras appreciate the stimulant effects of smoking tobacco, consuming inexpensive Mexican brands or rolling in maize leaves the *wipáka* (*Nicotiana tabacum*) that they cultivate, or, occasionally, collecting and smoking the harsh wild tobacco (*makuchi*) that grows in the sierra. Pennington identified *cornetón* as *Nicotiana glauca*; he also identified *makuchi* or *tabaco de coyote* as *Nicotiana trigonophylla*.[1119] Zingg assigns to this last species the Tarahumara name *bawará*.[1120] In my experience, however, cultivated or commercial tobacco is always called *wipá* by the Tarahumaras, and *makuchi* is the name of any wild tobacco, regardless of the species. Tobacco is mystically associated by the Tarahumaras with snakes, which is an old Mesoamerican belief.[1121]

In spite of containing the very toxic non-oxygenated pyridine alkaloid nicotine, tobacco is frequently used by the Tarahumaras as a remedy.[1122] Tarahumaras hold tobacco in very high regard as a remedy for snakebites. They recommend that everyone always should carry wipá in their pockets. Some claim that rattlesnakes will cry, revealing their presence at the very smell of *wipá*. When a rattlesnake bites anyone, tobacco smoke is blown onto the bite. Bye also reports that the leaves of *Nicotiana rustica* are used in a poultice applied onto bites of other poisonous animals.[1123]

Wild tobacco (*Nicotiana glauca*) leaves are applied directly to the head to relieve headaches, the sticky surface of the leaf causing it to stick like a plaster.[1124] Pennington concurs in this use and application but points out that the leaves are heated before applying them.[1125] Finally, tobacco—which Zingg called the most important ceremonial plant for the Tarahumaras—is ritualistically employed in many curing ceremonies as a general preventative agent.[1126]

Wisaró

Populus tremuloides, Michx., *Populus angustifolia* James, *Populus alba* L., *Populus* sp.

TAR. *wisaró, wizaró, usaró*[1127]

SP. *alamillo*

The inner bark of *Populus* sp. (as well as that of *Salix* sp.) contains salicylin (2-(hydroxymethyl) phenyl-B-D glucopyranoside), the aspirin precursor.

The bark of the quaking aspen or other *álamos* is boiled into an infusion used to facilitate childbirth; the remedy is especially useful if there is a problem, such as delay in the delivery of the placenta (*kimara*) or retention of placental fragments.

Bye reports that the bark of *wisaró* is used in preparing a general medicinal tea.[1128] By the same token, Deimel posits that Tarahumaras use wisaró as a ritual general remedy.[1129] Cardenal reported some more specific indications: stomachache, fever, sore throat, and to stop the flow of milk when a woman has swollen breasts.[1130]

Witzora

Unidentified

TAR. *witzora, u'itzora*

The leaves of *witzora*, or *u'itzorai*, are prepared into a tea for coughs.

Zacate Limón

Andopogon citratus DC.,[1131] *Cymbopagon* sp.[1132]

TAR. *gasará júkame*[1133]

SP. *zacate limón, te limón*

Burgess and Mares report that Tarahumaras at Bakusínare plant *te limón*, or *gasará júkame*, which is used to prepare a tea.[1134] Bye identified the grass used to prepare this beverage as *Cymbopagon* sp.[1135] Mestizo housewives in Norogachi favor *zacate limón* for the relief of fever and some Tarahumaras have learned about the remedy and enthusiastically use it.

Cabrera reports an essential oil and mineral salts among the active pharmacological principles in zacate limón.[1136]

Zempoal

Tagetes erecta L.

   SP. *zempoal, zempoalxóchitl, cempasúchil, flor de muertos*

   NAH. *zempoalxóchitl*[1137]

The yellow flowers used throughout México on the Day of the Dead to honor dead relatives and friends are used as an antidiarrheic in Norogachi where they are appreciated also as an ornate plant and planted in the mestizos' gardens. A single flower or even half of a flower is boiled in a cup of water, making a rich tea that then is drunk and, often, after a single cup, diarrhea will stop. Some western medical settings, such as the San Carlos Clinic in Norogachi, have used this remedy extensively with good results. *Zempoal* is used also in herbal combinations; for example, with *cáscara de granada* to treat dysentery. Cardenal claims that a decoction of zempoal is sprinkled to scare away pests from the fields.[1138]

Morton reports that the flowers of *T. erecta* are rich in xanthophyll esters. Petals and roots contain α-terthienyl.[1139] Orellana adds that the plant contains also an essential oil, resin, and tannin.[1140]

Zempoalillo

Unidentified[1141]

   SP. *zempoalillo*

Less appreciated than the garden *zempoal, zempoalillo* ("little zempoal") grows wild near Norogachi. It is used in a tea for stomach ailments and diarrhea.

ELEVEN

The Tarahumaras
A Conventional Medical Perspective

In this section, health issues and epidemiology of conditions affecting the Tarahumaras are discussed from a conventional medical perspective. Most of this information derives from my experience of ten years practicing medicine in the area. In addition to working in facilities providing direct medical care to the Indian population as described in the book's introduction, I became familiar with the network of health services provided by the INI—the National Indigenist Institute—with clinics in Samachique, Guachochi, Turuachi, and Bakiríachi and fourteen health dispensaries. To craft this chapter, I also examined data from the health centers in Urique and Cerocahui, the regional hospital in Guachochi, and mission hospitals in Sisoguichi and Creel. Of course, the main source of information was Clínica San Carlos in Norogachi where I labored for five years.

The information presented has some limitations. The area is assumed to be of difficult access due to the mountainous characteristics of the land and poorly developed system of roads. Other factors include inadequate or unavailable laboratory facilities. For many years, Mexican official health programs did not fully reach Tarahumara Country or were limited to vaccination campaigns. Reliable survey data and reports or published medical literature also has been relatively scarce.

One more factor affecting the reliability of available information may be a bias that tends to focus on negative aspects, overlooking some of the outstanding physiological qualities of the Rarámuri, such as their characteristic athletic and graceful physique. These assets should include their capacity for energy expenditure, reaching the upper limits of voluntary human effort; the low rate of conditions such as hypertension; and a cardiovascular system that withstands extremes of endurance.[1] Some scholars have reported the absence of coronary risk factors such as maladaptive lipid and lipoprotein levels.[2] Lumholtz had noted long ago "the wonderful health these people enjoy is really their most attractive trait. They are healthy and look it. . . . In the highlands, where the people live longer than in the barrancas, it is not infrequent to meet persons who are at least a hundred years old."[3]

In spite of the limitations listed, the account may detect areas needing to be investigated, awaken public and scientific curiosity, and stimulate a commitment to carry out systematic investigations. Last but not least, the account may increase an appreciation of the problems faced by the Rarámuri in their struggle for survival and a better existence.

Children's Health

It is commonly stated that infant mortality is high in Tarahumara communities. Some observers report rates twice as high as that of other Indian groups or ten times larger than the average for Mexico.[4] Rates as high as 50 percent often have been reported.[5] The paucity of comparative information meeting methodological standards makes such assertions difficult to assess. It has even been suggested that the high infant mortality rate may contribute to maintaining the delicate balance between available food resources and the size of the population in the Tarahumara ecosystem. A significant decrease in the mortality rate would allow the population to grow beyond the size that can be supported by the available food supply. Resources would have to increase significantly; otherwise, large numbers of people would have to migrate.[6]

Measurement of the magnitude of infant mortality is complicated by several factors.[7] An operational and efficient official birth registry system is not available in most Tarahumara localities. This service, formally established in Mexico by President Juárez in 1859, has not effectively reached the Tarahumaras, whose births and deaths are seldom registered.[8] The main resource documenting births in the area is the baptism registry of Jesuit mission parishes. The Rarámuri, after three centuries of evangelization efforts, value highly their children's baptism and this ceremony is documented by the church. The church encourages the baptism of children shortly after birth; however, the Tarahumaras tend to delay considerably compliance with this advice and many infants die before having the opportunity to be taken to the church to receive their "water name," as they call the Catholic baptism.

Aware of these limitations, I tried to obtain an estimate of the infant mortality rate among Tarahumara and mestizo children in the town of Norogachi and its vicinity. Children born in 1975 were selected to survey these groups. The source of data was the "Books of Baptisms" of Norogachi that record baptisms at the town's church and also in locations away from the parish church but that receive pastoral visits from the priests to conduct baptisms. Information recorded in these books includes the name and date of birth of the child, the date when the baptism took place, the name of the parents, the place of birth, and the names of the godparents. From this information, cultural affiliation may be inferred.

As part of the survey, the living status of children who had been registered in the baptism books was inquired about within a time frame covering one year from the date of the birth. The living status of the children was learned through a network of informants familiar with the whereabouts of the children's parents and able to reach them and inquire about the living status of their child.

Data was obtained from three geographic areas—the Norogachi township, location of the main church; the Mariana Póbora, consisting of four smaller towns loosely affiliated with Norogachi; and other communities not so closely affiliated, including Rochéachi, Aboréachi, and Narárachi. The information about mestizo children was easier to obtain given that they usually live in towns. On the other hand, the Tarahumaras typically living in less accessible rancherías in the mountains, which made collecting information more difficult.

Using the approaches mentioned, the living status of 116 children (out of 278 recorded in the books) was obtained. Fifty-seven of them were Tarahumara; fifty-seven, mestizo; and two, of different ethnic and cultural background. From these groups, a total of five (9 percent) Tarahumaras and two (4 percent) mestizos were known to have died.[9] In spite of the serious methodological problems of this survey, one can infer from it that Tarahumara infant mortality at one year is at least twice as high as the mestizos' for the same age group, but the estimate of 50 percent for Tarahumara infant mortality at five years is most likely an exaggerated figure.[10]

Breast-Feeding Practices

Tarahumara mothers breast-feed their children until they are two years old, or even longer.[11] In some instances, this process is interrupted by the birth of a second child. It has been noted that the first born, who has been displaced and loses its source of life and growth, many times becomes irritable, aloof, and withdrawn and cries constantly. The infant may refuse to eat the conventional food consisting of parched ground corn (kobishi) and atole (corn gruel beverage). This food supplies caloric intake but limited protein. The infant may then develop either *kwashiorkor*, or marasmatic malnutrition, depending upon whether it refuses all food or accepts the kobishi and atole. The parents may attribute this condition to sickness, overlooking the fact that the child may be responding to being physically and emotionally displaced.[12] In cases when these children are hospitalized, they seem to respond favorably to the care from surrogate mothering given by nurses and Tarahumara auxiliaries.

Congenital Disease and Neglect of Sick Children

Severely mentally retarded children appear to be particularly vulnerable to neglect, not only because they are unable to feed themselves but also because the Tarahumaras appear to be afraid of the abnormal and tend to avoid them. The parents sometimes do not dare to approach the affected child, resulting in severe neglect. In one case, I observed a child whose face had been bitten by a pig. His face was carefully treated surgically and satisfactory aesthetic results were obtained. The mother, however, remained afraid of him, did not feed him properly, and the child eventually died.

A similar fate appears to be suffered by many children born with congenital malformations, such as those with cleft palate. Children who are severely mentally retarded seem to live for only a short time.[13]

Abortion and Infanticide

The Tarahumaras consider abortion objectionable and it is rarely attempted, although some women have tried to accomplish it by violently massaging their abdomen.[14] On the other hand, they use plants with the purpose of facilitating menstruation. Infanticide among the Tarahumaras has been reported since the times of Father Fonte.[15] Lumholtz reported that in rare instances a woman may sit on her child right after birth to end its life, but this practice does not appear to be an institutionalized one in their culture.[16] Bennett reported the case of a Tarahumara woman who killed her baby, the product of incest, by "letting her fall on the rocks."[17] Other cases have been mentioned, but most probably they were the result of passive neglect rather than planned infanticide. I exhaustively inquired about this issue and was unable to find an instance of voluntary infanticide among the Rarámuri of Norogachi. On the other hand, Mull and Mull claimed that 95 percent of the women they interviewed knew of someone who had committed infanticide; the methods used included letting the child fall from a cliff, burying it alive, abandoning it in the forest, and placing the baby in a hollow tree.[18]

Nutrition

The Rarámuri people's extraordinary physical fitness seems paradoxical when considered within the context of their often-reported malnutrition. When present, malnutrition may have been caused by infectious diseases rather than being the result of a lack of food. For instance, diarrhea is common among children who then become malnourished. A study by Cerqueira (1975) in Sisoguichi concluded that the Tarahumara diet is sufficient in quality and quantity,[19] which suggests that clinically observed malnutrition may be the result of disease rather than dietary deficiencies.[20]

Monárrez and Martínez estimated the prevalence of malnutrition among children less than five years of age, finding that malnutrition is higher in frequency during the second year of life and affects more boys than girls. The authors estimated that 53.1 percent of the children exhibited mild malnutrition, 23.8 percent had moderate malnutrition, and 1.3 percent were severely affected.[21]

Cerqueira noted that when corn is processed for pinole (parched ground corn), zein and other proteins are hydrolyzed by several methods into a mixture of amino acids different from that obtained when maize is prepared for tortillas using calcium oxide as the hydrolyzing agent. In this last instance, the denaturalized protein renders free niacin, which in turn permits better utilization of tryptophan, an important amino acid, scarce in the Rarámuri's maize-based diet.[22] So preparing and consuming maize in more than one way leads to a more balanced diet.

It is an old observation that the Tarahumaras consume several complementary sources of nutrients and proteins, such as ants, fish, lizards, squirrels, snakes, and deer, some of them considered aesthetically objectionable by other cultures.[23] Edible herbs, like *mekuásari* and *wasoriki*, also supply vegetable nutrients. The basic and often only food that runners consume during a race or women consume during labor is pinole, or kobishi, which is high in calories.

It should be noted that since colonial times the land of the Tarahumaras has suffered periodic famine. Some areas, such as canyon country and western Tarahumara Country, have been more regularly affected by periods of food scarcity. During the years 1992 to 1994, a severe drought, which continued until 2003, was responsible for extensive malnutrition.

Infectious Diseases

The occurrence of epidemics in the region has been chronicled since colonial times.[24] Dysentery outbreaks were common, as well as smallpox and measles.[25] These diseases were big killers, often targeting children and pregnant women. Fortunately, in modern times, diseases like smallpox have been eradicated, although as recently as the early 1950s many individuals still exhibited the facial scarring characteristic of previous smallpox infections.[26] However, common infectious diseases, particularly influenza, have been prevalent and have caused a large number of deaths.[27]

Practicing medicine among the Tarahumaras, one hears the words *rosówa*, meaning cough, and *witabúa*, meaning diarrhea, with such frequency that one might think practicing medicine there is the art of treating such symptoms. Depending on the season, respiratory tract infections and gastrointestinal problems take turns as leading causes of morbidity and mortality. While providing medical care in Cerocahui in 1972–1973, I observed a cyclic pattern of morbidity. From November to January, many children were affected by severe coughing spells, expectoration, and shortness of breath. Such illnesses, especially in the highlands and canyons, often were thought to be *tosferina*, or whooping cough, due to their severity. However, children vaccinated against pertussis, as well as those not immunized, were affected. Other bacterial or

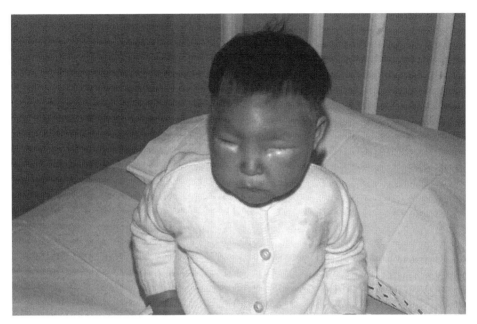
Tarahumara girl suffering from kwashiorkor-type malnutrition.

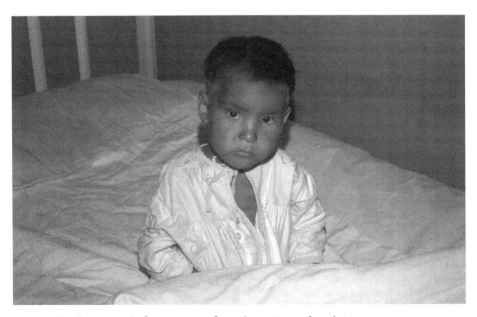
Same Tarahumara girl after treatment for malnutrition and nephritis.

viral agents may have caused many of the cases. Severe outbreaks of influenza, some associated with significant mortality, have been common, affecting broad segments of the population. The symptoms of these conditions appear to be particularly severe among the Tarahumaras. Mixed epidemics of measles and chicken pox coexisting with the diseases mentioned have been observed, followed by outbreaks of mumps and gastroenteritis.

The observations from Cerocahui could have been made in any other place in the land of the Tarahumaras. Mull and Mull, for example, reported that in 1985, nine persons died of measles at the mission hospital at Sisoguichi.[28]

In regard to other infectious diseases, some investigators have noted that in contrast with their counterparts in the United States, Sisoguichi adolescents present greater serologic evidence of exposure to mumps. Evidence of salmonella infections contributing to significant morbidity and mortality have been observed in a large proportion of the adolescent population.[29] Gajdusek reported an epidemic of typhoid fever at Sisoguichi in 1950 and I observed several cases of salmonellosis in Cerocahui.[30] Mull and Mull, while studying the use of health resources in a local facility—the Clínica Santa Teresita in Creel—reported fifty-three typhoid cases in just one month. The largest number of admissions for gastroenteritis occurred in the months of July through October, while the fewest were in the months of November to June. Amoebic dysentery is common, although systematic surveys have not been carried out to assess the magnitude of this problem. On small samples of subjects observed, Mull and Mull reported 9.2 percent of Tarahumara (n=88) and 1.7 percent of mestizo (n=90) children two years old and younger who visited the mission clinic Santa Teresita in 1981–1982 required hospitalization for treatment of gastroenteritis, pneumonia, or both. In their data, 36.4 percent of the Tarahumara children treated died in the hospital while the mortality rate was 21.1 percent in the mestizo children. This high mortality was probably due to the fact that only very severely ill children are brought to the hospital. The pattern for pneumonia was a mirror image of the one for enteric illnesses with most of the admissions occurring in the months of July through October.[31]

Gajdusek and Rogers reported that in the Tónachi region in 1950, a major outbreak occurred of what the Jesuit missionaries called "typhus," but bacteriologic support was not available to confirm this impression.[32] However, the authors found specific complement-fixing antibodies for typhus in the sera

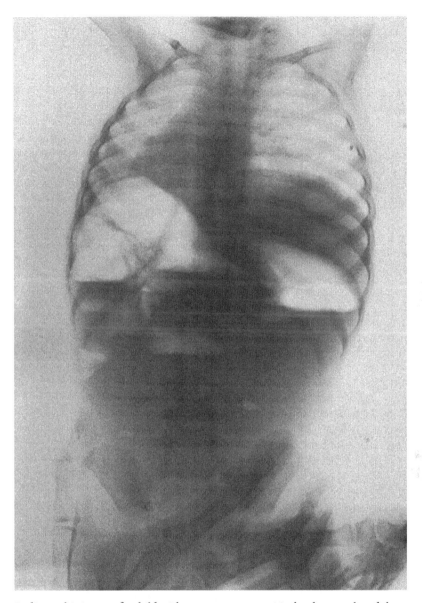

Radiographic image of a child with severe gastroenteritis that has paralyzed the intestinal function, a condition called ileus, demonstrated in the image by the presence of liquid and gas levels.

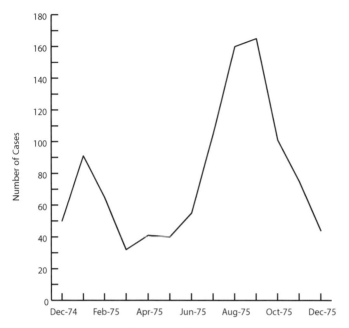

New patient clinic visits for acute infectious gastroenteritis, Clínica San Carlos, Norogachi, Chihuahua, 1975.

of Sisoguichi's children, confirming the presence of exposure to this disease among the Rarámuri. Seven out of fifty-five individuals were positive for *Brucella abortus*, suggesting that the illness is endemic in the area.

In Norogachi, influenza occurring epidemically, mainly in winter, was the most common illness, with 341 cases, followed by bronchitis, tonsillitis, gastroenteritis, wounds, pneumonia, dental caries, enterobiasis (pinworms), measles, amebiasis, mumps, abnormal vaginal bleeding, tuberculosis, lumbago, peptic ulcer syndrome and gastritis, and chronic rheumatism.

The following tables, summarizing data from the mission facilities in Creel, Sisoguichi, and Norogachi, as well as from the INI (Instituto Nacional Indigenista) hospitals, clinics, and health dispensaries in Samachique, Guachochi, Turuachi, and Bakiríachi, illustrate the frequency of the conditions observed in the region and particularly the high representation of infectious disorders.

Acute illnesses in rainy season and in winter

	% of visits	Seasonal Incidence	Contribution to Mortality
RAINY SEASON			
Acute Infectious Gastroenteritis	35	50	72
Acute Amoebic Dysentery	25	15	19
Influenza, Rhino-pharyngitis	16	13	0
Trauma	15	2	4
Miscellaneous Infections	5	5	1
Exacerbation of Chronic Illness	3	7	4
TOTAL	99	93	100
WINTER			
Influenza, Rhino-pharyngitis	60	75	2
Acute Bronchitis	20	15	5
Acute Bronchopneumonia	10	19	71
Acute Gastroenteritis	5	35	13
Trauma	2	5	1
Miscellaneous Infections	1	5	2
Exacerbation of Chronic Illness	1	7	6
TOTAL	99	161	100

Note: The table presents the percentage of visits due to a "broad group of illnesses" to clinics, hospitals or health posts. "Broad group of illnesses" is a term used by Mexican Public Health systems that roughly corresponds to that of International Classification of Diseases (ICD) systems and to that of "major illness categories" as used, for example, by Mull and Mull (1987). Seasonal incidence is an estimate of point incidences in September and January. Contribution to the general mortality of each group is estimated in the last column. The figure 161 for the winter total seasonal incidence represents the progression of a condition from bronchitis or pharyngitis to pneumonia.

Chronic Illnesses

Tuberculosis has, in spite of continued efforts to eradicate it, a high prevalence in the whole Tarahumara Country. It still accounts for 1 percent of all clinic visits and contributes importantly to general mortality. Mull and Mull report that at Clínica Santa Teresita in Creel there was a striking dissimilarity between Tarahumara and mestizo patients: while for the Tarahumara group (n=938) 10.7 percent of the initial clinic visits and 9.4 percent of the follow-ups were for pulmonary tuberculosis, for mestizos (n=5,200) the figures were 1.2 percent

Chronic illnesses

	% of Clinic Visits	Contribution to General Mortality
Pulmonary Tuberculosis	1	2.7
Metabolic Diseases	1	0.5
Intestinal Parasitosis	9	0.1
Pregnancy Complications	1	0.2
Urinary Infections	11	0
Rheumatic Illnesses	31	0
Gynecological Problems	7	0.01
Sensorial Organ Problems	6	0
Vascular Diseases	1	0.1
Neoplasms	0.5	0.1
Senility	6	1.6
Alcoholism	0.5	0.4
Scabies	15	0
Other	3	0.28
TOTAL	100	6

Note: Percentage of visits to clinics, hospitals, or health outposts throughout the year due to a broad group of illnesses. Diabetes is practically inexistent among the Tarahumaras, but diabetes affects acculturated individuals with an incidence similar to that observed in other American Indian groups. These figures exclude the enormous incidence of conjunctivitis. In Norogachi, Sunday after Sunday, the nuns dispense gallons of eyedrops that Tarahumaras request to treat *busichí okorá*. Malignancies are relatively rare among the Tarahumaras. In Norogachi, we had to amputate the leg of a young Tarahumara man to remove a osteosarcoma. In ten years I found only two basocellular carcinomas of the skin.

and 1.4 percent respectively. More Tarahumaras (37) required hospitalization than mestizos (15). Finally, of the admitted Rarámuri, three died but no mestizo deaths occurred.[33]

The climatic conditions of the canyon area favor the growth of mosquitoes and malaria, known locally as *bazo*, continues to affect the Tarahumaras in the most remote and isolated hamlets, in spite of the indefatigable efforts of CENEP, the National Commission for the Control of Malaria. As recently as 1976, official sources considered malaria to be the sixth leading cause of morbidity in the state of Chihuahua; however, the last death from malaria was reported in 1962.

Gajdusek and Rogers found that the Tarahumaras had serological evidence of past contact, usually during childhood or adolescence, with three types of poliomyelitis viruses. Poliomyelitis viruses probably contributed to the acute gastroenteritis episodes mentioned above. The authors mentioned also found cases of seropositivity to the herpes virus and psittacosis-LGV group.[34]

Sexually transmitted diseases are not common and when they occur are attributed by the Tarahumaras to sexual contact with mestizos. Lumholtz mentioned that syphilis was common, but Gajdusek and Rogers using a treponemal immobilization test found only one case positive for syphilis in a large survey of tests done on Tarahumaras and no cases of congenital syphillis.[35] Malignancies and diabetes are seldom observed.

A high prevalence of goiter has been noted in the canyons.[36]

Trauma

Trauma is an important source of morbidity and mortality among the mestizos and Tarahumaras of Cerocahui.[37] Murray studied the pattern of accidents occurring among Tarahumaras and mestizos in Norogachi, noting that most injuries were produced by axes, burns, or as the result of fights.[38] In *Cerocahui: Una Comunidad en la Tarahumara*, I commented on the growing number of accidents that occurred while Tarahumaras worked in sawmills or on roads under construction or repair.[39]

Hachazos, the wounds produced by axes, are injuries strongly related to the Tarahumara pattern of occupational activities; the ax is the usual tool employed for chopping wood and carving the rod beams characteristic of their ga'ri; it is not uncommon to see women cutting firewood with axes and occasionally experiencing injury. Tarahumara men place a special pride in their skill to use this tool. Using it as a weapon against others is considered a very serious crime. Typically, the lesions suffered are on the dorsum of the left foot, since that foot is the one used by right-handed individuals to secure the piece of wood. Not uncommonly, the accident occurs when the man is chopping the last piece of wood for the day, probably as a consequence of tiredness or natural distraction since the daily labor is almost over. An ax wound represents a favorable environment for infestation with fly larvae and thence infestation of the wound is frequently observed. Turpentine is commonly used to protect and soothe the lesion but it does not prevent the infestation.

Occasionally, the wooden ball used in the races traumatizes the head of a racer or spectator. Falls are very common in such mountainous and irregular ground. For the Tarahumaras and other canyon dwellers, slipping and rolling down into a gorge is frequently lethal.

Snakebites

Snakebites are relatively common in the area; the usual complaint is *sayawi ikíreke*, "A snake bit me." In the Clínica San Carlos in Norogachi, I treated approximately two hundred snakebites in a relatively short period of time, in spite of the claim that snakebites and scorpion bites are uncommon in the area.[40] The Tarahumaras, living closer to nature, seem to experience this injury more often than the mestizos, who tend to live in towns. Serious consequences however are rare; only one person among those observed by me became severely ill. This person suffered a massive edema of the lower limbs; required antivenin, anti-inflammatory medication; antihistaminic, anti-infectious agents; and debridement. These experiences contrast with my previous experience at the Hospital Central Universitario del Estado de Chihuahua in Chihuahua City where individuals bitten by snakes often developed massive necrosis requiring toe amputations. It appeared as if the venom of the Chihuahuan desert rattlesnake is more aggressive than that of the species from the Tarahumara region.

Mushroom Poisoning

Poisoning with mushrooms takes two or three lives every year in the land of the Tarahumaras. The victims are usually children but occasionally adults are the victims. The Tarahumaras have extensive knowledge of poisonous mushrooms; they usually collect only the edible mushrooms, called in Spanish *hongos de aguas*, or in Tarahumara, *wekogí*.[41]

Alcohol Abuse

Neumann, in a letter to a priest in Bohemia in 1691, depicted the Tarahumaras as "great drunkards (who) have several wives and are very superstitious." This conception, far from being overcome, periodically resurfaces. I have reported in the past that in some areas—for instance, the Urique Canyon—the severity of infectious diseases, nutritional deficits, and in particular the occurrence of trauma are compounded or even caused by the consumption of alcoholic beverages.[42] Deimel took issue with these views, pointing out that information derived from communities such as Cerocahui are not representative because in that area, there has been a considerable acculturation into drinking patterns that differ significantly from those of the Tarahumaras. He pointed out that common indicators of alcoholism, such as high frequency of diseases of the liver, including cirrhosis, high blood pressure, heart problems, and delirium tremens are not commonly observed among the Tarahumaras.[43] The absence of alcoholism—that is, the absence of cases that would meet the diagnostic criteria for alcohol dependence of the *Diagnostic and Statistical Manual of Mental Disorders* or the World Health Organization's *International Statistical Classification of Diseases*—among the Tarahumaras has been widely noted by scholars.[44] Furthermore, Zeiner, Paredes, and Cowden in a series of studies on the metabolism of alcohol by the Tarahumaras concluded that cultural factors might outweigh biological factors in the incidence of alcoholism and alcohol abuse in ethnic groups. Kennedy observed that the relative high cost and the low stability of batari, a beverage that takes several days to be prepared and fermented, that has to be ingested right at its point of maturity, and is not commercially available, are limiting factors.[45] Among the Tarahumaras, drinking takes place only at socially defined situations, such as the performance of collective work and the celebration of religious and ceremonial events. Furthermore, it is known that the Tarahumaras do not customarily consume distilled beverages with high alcohol content.

Mental Illness and Suicide

As mentioned earlier, mental illness has been infrequently observed among the Tarahumaras.[46] This probably is due to their high level of tolerance for unconventional behavior and thinking. This circumstance makes it possible for the mentally ill to find a niche in the community in which they may remain relatively undisturbed.

Lumholtz observed that suicide is never committed unless a person is intoxicated or angered by some slight or by jealousy. He noted though that there was an epidemic of suicides reported near Guachochi, with several men hanging themselves with their girdles; one of whom even suspended himself by the feet. Isolated suicides are known however, and the general belief is that they occur always in the same families.[47]

Acceptance of Conventional Medical Care

Reports such as Mull and Mull's and my own study indicate that although the Tarahumaras do not fully accept conventional medical facilities, many do utilize these services.[48] Such utilization is not only an alternative to their own medical practices but sometimes appears to be complementary to them. For instance, regarding childbirth, when the native resources seem to fail, the laboring woman is taken to the nearest clinic or the husband and neighbors try to bring the conventional physician or nurse to their hamlet. The same thing occurs with other types of medical attention, such as pediatric care: they don't fully accept it but may use it to complement their own practices or if their practices fail.[49]

As an example of the pattern of services provided and the type of conditions addressed at the San Carlos clinic from 1974–1978, almost 50 percent of the total outpatient consultations were given to treat infectious diseases: gastroenteritis, bronchitis, pneumonia, parasitosis, and skin infections. The rest was divided as follows: 20 percent, general disease, arthritis, trauma, problems related to chronic malnutrition, senility, and neuroses; 15 percent minor surgery; 15 percent dental care, primarily extractions; and 5 percent, administration of vaccines.

Of the cases requiring inpatient care, 30 percent of the cases were for acute childhood illnesses; 25 percent, chronic patients; 18 percent, acute adult illnesses; 15 percent, surgical patients; and 12 percent maternity patients.

Finally, in those places where conventional medical services are available, patients expect from the "white doctor" the type of attention they consider can be obtained from this source—things like shots, eye drops, pills, or surgery—while they look to "folk" practices and healers for what belongs to more traditional fields, such as lifting a sunken fontanel or performing a Jíkuri ceremony. Unfortunately, negative expectations about hospitals and clinics, such as "The

Volume of clinical work at Clínica San Carlos, Norogachi, Chihuahua

	Outpatients	Inpatients	Surgeries	Births
1974–75	10,260	305	38	32
1975–76	10,551	310	55	40
1976–77	8,741	313	49	41
Mean	9,851	315	47	38

Procedures necessitating the OR included D&Cs (34), complex sutures (31), tumor ablations (14), open fracture reductions (6), thoracocenteses (6), herniorrhaphies (6), C-sections (5), breast abscess drainages (5), minor ophthalmic procedures (4), scar removals (4), laparocenteses (4), circumcisions (4), bullet removals (4), amputations (3), and tendon repair, ovarian cyst ablation, exploratory laparotomy and Bartholin's cyst ablation (one of each).

clinic is the place one goes to die," are still prevalent in the region. In places like Norogachi, Guachochi, and Creel, this attitude is changing as a result of the continuous presence of conventionally trained physicians and nurses among the Rarámuri.[50]

The information just presented suggests that at the time of my observations, modern health care had only partially reached the land of the Tarahumaras. Since 1980, Tarahumara Country has witnessed the expansion of government health programs and also the appearance of new players in the health arena, such as a hospital in Samachiki sponsored by a church other than the Catholic Church. The mission clinics have grown in sophistication, installing equipment, including newer X-ray machines, and laboratories, which previously were very primitive or nonexistent. Permanent electricity has arrived in many towns, such as Norogachi, and roads have continued growing and improving. Better communications and better roads have facilitated not only the availability of medical and other supplies and enabled emergency transportation to regional and city hospitals but also have fostered the migratory movement of Tarahumaras and mestizos to the city and have increased tourism in the region. Unfortunately, alcohol, drugs, and criminal activity also have found more and better routes to the heart of the land of the Tarahumaras. These changes have had a profound effect on the health of the Rarámuri.

Notes

Introduction

1. Sejourné, *The Burning Water*, 499.
2. Gamio, "Papel de la antropología," 31.
3. Naranjo, "Influencia de la medicina aborigen," 255.
4. Bannerman, *Traditional Medicine*, 229–233, and Porras Carrillo, "La Sierra Tarahumara," 26.
5. Aguirre Beltrán, *Medicina y magia*, 36, 274; Vogel, *American Indian Medicine*; and Guerra, "La evaluación farmacológica," 43.
6. Although there is no evidence that wild yam was used to control fertility (Vogel, *American Indian Medicine*, 245).
7. Or diosgenins (Lewis and Elvin-Lewis, *Medical Botany*, 318; García Rivas, *Plantas medicinales*, 40).
8. Marshall, *Great Events*, 430.
9. Galeffi, "Recenti acquisizioni," 561–576.
10. González-Rodríguez, "Mitos étnicos," 32–35; Robles "La actividad misionera," 37; Brambila, "Psicología y educación," 199; and Irigoyen-Rascón, *Cerocahui*, 95.
11. Lookout, "Alcohol and the Native American," 31.
12. Díaz-Infante, personal communication, January 1, 2000; Ramos and Iturbide, "Migración," 49–55.
13. Rarámuri, from the Tarahumara words *rará* or *tará*=foot or leg and *mama*=run. The word "Tarahumara" also appears to have a legitimate etymology: *rara* and *tara* are forms of the same word, while *-muri* derives from *mama* ("to run," singular verb) and *-júmari* from *jumama* (plural verb="several people running"). The obsolete term "Tajúmare," used in some old documents, may also be explained by this etymology. See also Brambila, *Diccionario rarámuri-castellano*, 537; Ocampo, "*Historia de la misión*, 43; Márquez Terrazas, *Origen de la iglesia en Chihuahua*, 30; Valiñas, "Lengua, dialecto e identidad," 105; and Deimel, *Tarahumara*, 12.
14. Heisenberg's Uncertainty Principle states, in a simple version, that the combined

position and momentum of a subatomic particle are not definitely measurable since mere observation introduces quanta of energy that change both characteristics of the particle.
15. Author of *Pedro Páramo* and *El Llano en Llamas*.
16. Clínica San Carlos is managed by the Sisters of Charity of St. Charles Borromeo congregation of Catholic nuns based in Vienna, Austria. The sisters, trained in Europe, are the attending nurses. The thirty-bed hospital, founded in 1959, is supported by Mexican, American, and German charities. Periodic short visits of Mexican and American specialists support the work of the resident physician and nurses. The hospital had a potent diesel-powered generator, an old but operational X-ray machine, a basic clinical laboratory, and a well-equipped operating room. Clínica San Carlos provided the only medical service available within a 20–30 kilometer radius of Norogachi, an area with a population of 6–8,000. Murray has chronicled my initial medical work in Norogachi in his Ph.D. dissertation.
17. Zeiner, Paredes, and Cowden, "Physiological Responses, 1."
18. The original team was composed of Brothers Carlos Preciado, Manuel Villalobos, Juan Arciniega, Jorge Ibarra, and their chaplain, F. Mauricio Rivera, S.J.
19. Compare García Manzanedo, "Notas sobre la medicina tradicional," 61–70, and Mull and Mull, "Differential Use," 245–264.
20. The project resulted in the publication of Escobedo Chávez, *Bibliografía básica del estado de Chihuahua*.
21. The project resulted in the preparation of Ramos and Iturbide's B.A. dissertation, published later as "Migración tarahumara a la ciudad de Chihuahua" in *Los Rarámuri hoy: Memorias*, 49–55.
22. Published as Irigoyen-Rascón, ed., *Memoria del seminario de lenguas habladas en el estado de Chihuahua: Ra'ítzari, la palabra*.
23. The works of Christelow and González-Rodríguez on the Tarahumaras; Caraveo on the Sumas; Polzer and Sheridan and Naylor on the Tarahumaras; and Griffen and Spicer on the northern Mexican tribes contributed significantly to the formulation of the Tarahumara process history presented in this book.
24. Kroeber, *Anthropology*, 316–320.
25. Similar to the concepts proposed by meaning-centered medical anthropology, which postulates all illness realities as being fundamentally semantic, hermeneutic, or interpretative (Good and Good, "The Meaning of Symptoms," 167).
26. Committee on Form and Style, *CBE Style Manual*.
27. Lumholtz, *Unknown Mexico*, 1:356–379; Bennett and Zingg, *The Tarahumara*, 135–138; Anderson, *Peyote*, 90–93; Pennington, *The Tarahumar*, 171, 186; Bye, "Ethnoecology of the Tarahumara," 189–205; Artaud, *México y viaje*, 304–327; Thord-Gray, *Tarahumara-English, jíkuri*; Brambila, quoted in Irigoyen-Rascón, *Cerocahui*, 125–129; González Rodríguez, *Tarahumara*, 115–125; and finally, Deimel, "Die Peyoteheilung," 155–163, and *Tarahumara*, 80–83.

1. The Tarahumara Ecological Habitat

1. 94,831 square miles. Schmidt, *Geographical Survey*, 9; Martínez et al., *Mi Chihuahua*, 12. Compare Almada, *Geografía del estado*, 12.
2. INI, "Censo indígena," 101.
3. Total population of the area (INEGI, *XII censo general*). Several authors have advanced their own calculations about the total number of Tarahumaras: Lumholtz (*Unknown Mexico*, 1:119), 25,000; Artaud (*México y viaje*, 301), Martínez, J. Rosario C. ("El Tarahumara" in Sariego, *El indigenismo*, 96), 28,000; Bennett and Zingg (*The Tarahumara*, vii) and Gajdusek (*The Sierra*, 15) 40,000; Christelow (Introduction, 1) 40–50,000; Passin ("Place," pt. 1, 361) 40,000 to 50,000; Plancarte (*El problema*, 102) 44,000; Spicer (*Cycles*, 44) calculated 18,000 at the end of the colonial period, 35,000 in 1930, and 50,000 by 1950; Ocampo (*Historia*, 124), 46,500 by 1920; Kennedy ("Tesgüino Complex," 620), Gutiérrez and Gutiérrez ("50,000 Tarahumaras," 1), González Rodríguez (*Tarahumara*, 77), Murray ("The Tarahumara Project," 181), Mull and Mull ("Differential Use," 246, 253, and "Infanticide," 113), and Paredes and Irigoyen-Rascón ("Jíkuri," 122), 50,000; De Velasco (*Danzar o morir*, 29), 50,000 to 60,000; González Rodríguez (*El noroeste*, 9) and Koehler ("Crossing the Divide," 37), 60,000; Encyclopaedia Britannica's *Book of the Year* 1995 (780) 70,000; Rice ("Drought," 7A), 78,000; Monárrez Espino and Martínez Salgado ("Prevalencia de desnutrición," 9) 80,000.
4. INI, *Características de la población indígena*; INEGI, *XII censo general*.
5. Schmidt, *Geographical Survey of Chihuahua*, 12.
6. Pennington, *The Tarahumar*, 33.
7. Daugherty ("Railroad Log," 45) relates that the volcanic rocks of the Sierra Madre have been gently folded and tilted over large areas; the gross structure is best described as that of an anticlinorium. Later block faulting created northwest-trending ridges and valleys along the sierras. Pennington (*The Tarahumar*) localizes these processes stating that the western portion of Tarahumara Country is made up of several thousand feet of volcanic material that covered a surface of folded Cretaceous and older rocks in Tertiary times. On the other hand, the eastern portion is composed mainly of Cretaceous sediments with freshwater Quaternary sediments imposed here and there. Daugherty adds that these volcanic rocks can be generalized as a thick sequence of acidic pyroclastics (mostly rhyolites and latites) overlying more basic flow rocks (andesites and basalts).
8. García Gutiérrez and García Gutiérrez, "Geologic Setting," 131. These authors observe that in the western Sierra Madre, metallic deposits occur either in the andesite or in the intrusive portions that form the lower parts of the sierra. The famous mine in La Bufa exclusively produced copper. The old mining sites at Ocampo, Cusihuiríachi, Pinos Altos, Guazapares, Guadalupe y Calvo, Urique, Batopilas, and Moris gave gold and silver. Potrero de Bojórquez yielded copper, tungsten and molybdenum and La Reforma copper, lead and zinc.

9. Father Carlos Vallejo found marine fossils near Rituchi, a place well within the sierra. The significance of this finding has yet to be determined.
10. Brambila, *Bosquejos del alma tarahumara*, 12; Almada, *Geografía del estado*, 57.
11. Gajdusek, "The Sierra Tarahumara," 20; Wampler, *New Rails to Old Towns*, 143.
12. Some American writers have taken the side of the Mexican canyons. For example, Koehler ("Crossing the Divide," 35) pointed out that the system of interconnected gorges of the Copper Canyon has a volume nearly four times that of the Grand Canyon, and that the major barrancas are all — some almost one and a half times — deeper than the deepest point of their Arizona rival.
13. Dikes: Pennington, *The Tarahumar*, 31; benches: Pennington, *The Tepehuan of Chihuahua*, 28.
14. De la Peña, *Chihuahua económico*, 2:347–353.
15. Maderey, "Aspectos hidrológicos," 110.
16. Ibid. It did not cut deep canyons except when it is about to reach the Río Bravo. There, on the Mexican side of the border, the river has carved Cañon del Pegüis, an impressive counterpart to the American Big Bend National Park.
17. Ibid.
18. Bassols, *El noroeste*, 142.
19. Ibid.
20. Brondo Whitt, *Nuevo León*, 10. Lumholtz (*Unknown Mexico*, 1:130) reported that the waterfall height was only 298.7 meters tall, attributing the measurement to a mining expert at Pinos Altos. Recently, Comisión Nacional Geográfica claimed that the actual height of the fall is 311 meters (Instituto Chihuahuense de la Cultura, "Basaséachi," 7E; Ching Vega, "Cascada de Basaséachi," 14A). Fisher and Verplancken (*Chihuahua, Mexico*, 38) argue that the waterfall height has been re-estimated as 246 meters and that the nearby Piedra Volada waterfall is even taller: 453 meters. Cascada Piedra Volada was unknown, according to Lazcano Sahagún (*Barrancas del cobre*, 59) until 1995. See also Hancock, *Chihuahua: A Guide*, 95.
21. Bassols, *El noroeste*, 142.
22. Bassols, *El noroeste*, 135.
23. Neumann (Letters to an Unknown Priest, 70–72) described the damage that the cruel winter of 1681–82 caused. In modern times, De la Peña (*Chihuahua económico*, 1:62) reported on the memorable storm of 1945. In 1976 and 1979, there were severe snowstorms affecting a vast area of central Tarahumara Country; the latter storm particularly caused isolation, hunger, and the death of animals and people. As it may be expected because of its latitude, the northern sierra country usually receives more snow than the central uplands.
24. Since 2002, Instituto de Ecología AC/CEISS (Centro de Investigación sobre la Sequía) has a web page that displays the monthly Standard Precipitation Index (SPI). See also Rice, "Drought," 3A.
25. Maderey, "Características hidroclimáticas," 118.
26. De la Peña, *Chihuahua Económico*, 1:20; Bassols, *El noroeste*, 137.

27. Lartigue, *Indios y bosques*, 130; Weaver, "Changes in Forestry Policy," 9; and Dibildox et al., *Declaration of the Diocese*, 1–3.
28. Bassols, *El noroeste*, 138.
29. De la Peña; *Chihuahua económico*, 1:21.
30. Maderey, "Características hidroclimáticas," 124.
31. Pennington, *The Tarahumar*, 31.
32. De la Peña, *Chihuahua económico*, 2:21.
33. The kit fox or zorra norteña is *Vulpes macrotis* and the gray fox is *Urocyon cinereoargenteus* (Schmidt, *Geographical Survey of Chihuahua*, 37).
34. Compare Schmidt, *Geographical Survey of Chihuahua*, 37.
35. Schmidt (*Geographical Survey of Chihuahua*, 37) identifies the Chihuahuan *gato montés* or bobcat as *Lynx rufus*.
36. Pennington, *The Tarahumar*, 94.
37. Schmidt, *Geographical Survey of Chihuahua*, 36.
38. Pennington, *The Tarahumar*, 94.
39. Lumholtz, *Unknown Mexico*, 2:274.
40. "They are like wizards; when they sing something is about to happen" (Palma, Erasmo, *Donde cantan los pájaros*, 6).
41. Brambila, *Diccionario rarámuri-castellano*, 557.
42. Lumholtz (*Unknown Mexico*, 1:324) reported that a hummingbird stripped of its feathers, dried, and wrapped in wool (seed hair) from the *pochote* tree was a powerful malefic talisman in the hands of a sorcerer. He also said that *simuchí* is often mentioned in Tarahumara songs as "a good and mighty hero-god." While few traces, if any, of this belief survive among the modern Tarahumaras, its connection with Mesoamerican religion is important. Huitchilopóchtli, the bloodthirsty solar war god of the Aztecs, was identified by them with the hummingbird (in Náhuatl, *huitzilin*). Use of love charms made from mummified hummingbirds is still common in Oaxaca.
43. Brambila, *Diccionario castellano/rarámuri*, 110.
44. Lumholtz, *Unknown Mexico*, 1:54. Although hopeful conservationists still expect to find a live specimen, search expeditions have been fruitless and sighting rumors unconfirmed. The last confirmed sighting was in 1957; the species was declared "endangered" in 1970.
45. Schmidt, *Geographical Survey of Chihuahua*, 35, 36.
46. Brambila, *Diccionario castellano/rarámuri*, 275.
47. Brambila, *Diccionario castellano/rarámuri*, 516.
48. *Pichicuate, pichicoate,* or *pitzocoate* derives from the Náhuatl words *pitzáhuac,* thin object, and *cóatl,* snake (Cabrera, *Diccionario,* 109).
49. *Rana tarahumarae,* the Tarahumara frog, was originally described by Boulenger in 1917 at Yoquivo and Barranca del Cobre; its habitat extended to the American Southwest where it became extinct in the second part of the twentieth century (Kimberleigh et al., "A Proposal").

50. Lumholtz (*Unknown Mexico*, 1:120) reported the presence in the Rio Verde of three kinds of fish: suckers, catfish, and Gila trout. He noted that the Tarahumaras believed some fish would change into otters when they become old.
51. Brambila, *Diccionario rarámuri-castellano*, 520.
52. Tarahumaras apply this name also to their Tepehuan neighbors (Irigoyen-Rascón, *Cerocahui*, 21, and Brambila, *Diccionario rarámuri-castellano*, 512).

2. A Historical Review of the Tarahumara People

1. Waters and Stafford, "Redefining the Age of Clovis," 1122–1126.
2. Di Peso, *Casas Grandes*, 64. Compare Ascher and Clune, "Waterfall Cave," 270–274, and Gamboa, "Arqueología en la Sierra," 39.
3. Carlos Vallejo, S.J. (1976), related that during one of his surveys at the Arroyo de Basuchil site, a bone hunter from nearby Ciudad Guerrero told him he had found a human skull among the mammoth bones. The skull was exhibited for some time on the shelf of a store in Ciudad Guerrero. Mammoth and other megafauna remains have been also reported in the vicinities of Yepómera and Temósachi (Chihuahua) and Nácori (Sonora). (Lumholtz, *Unknown Mexico*, 1:23, 118; Brondo Whitt, *Nuevo León*, 341, 382; Di Peso, *Casas Grandes*, 66–70). The date of extinction of the megafauna is not known for northern Mexico, but we know that it survived, at the latest, until 1000 B.C. (Hester, "Late Pleistocene," 58–77; Di Peso, *Casas Grandes*, 69). Nonetheless, it is important to remember that in South America, some species of mastodon (such as *Cuvieramus postremus*) survived until the fourth century of the Christian era (Pieris, "Proboscidea," 15:1–3).
4. Irwin-Williams, "Post-pleistocene Archaeology," 9:37.
5. Ascher and Clune, "Waterfall Cave," 270–274; Gamboa, "Arqueología en la Sierra Tarahumara," 39.
6. Di Peso, *Casas Grandes*, 66.
7. Dick, *Bat Cave*, 37; Irwin-Williams, "Post-pleistocene Archaeology," 9:40; Gummerman and Haury, "Prehistory: Hohokam," 77.
8. Some authors believe squash to be autochthonous from the Río Conchos basin. Di Peso (*Casas Grandes*, 78n1) argues that squash might have been known by the northern Chichimecans only since about 3000 B.C., while some Mesoamerican groups were using squash as food at least since 6000 B.C.
9. Woodbury and Zubrow ("Agricultural Beginnings," 43–60) render a detailed account of what is known about the spreading of these domesticated species to the American Southwest. Bye ("Ethnoecology," 77–78) reports that teosinte, the ancestor of maize, still grows in Tarahumara Country. He describes also the process of hybridization that led to the development of the modern varieties of corn. Bye studied the permanence among Tarahumara crops of other relatively primitive strains of maize.
10. Bennett and Zingg, *The Tarahumara*, 356–359, and also in Zingg's *Report on the Archaeology of Southern Chihuahua*, 17.

11. Di Peso, *Casas Grandes*, 83–85.
12. Ascher and Clune, "Waterfall Cave," 270–274; Di Peso, *Casas Grandes*, 83–85.
13. As Pennington (*The Tarahumar*, 12) points out, if the Basketmaker association is correct, Zingg's Transitional Culture could be dated back as much as two thousand years. Pennington correlates Ascher, Clune, and Cutler's findings of assemblages dating from 1000 to 1600 A.D. with the later part of Zingg's transitional phase. This could mean that Di Peso's association of the Waterfall Cave artifacts with Zingg's Transitional Phase is imprecise since Zingg's construct requires the absence of pottery as a characteristic trait of the Basketmaker horizon.

    Foster, Michael ("Loma San Gabriel," 38), and Lazalde (*Durango indígena*, 7, 38) date the beginning of the Loma San Gabriel culture between 50 B.C. and 100 A.D. (Compare Di Peso, *Casas Grandes*, 83–85.) The cultural proximity of this culture with the proto-Tarahumara is undeniable.
14. See Ratkay, *An Account of the Tarahumara Missions*, 26; Lumholtz, *Unknown Mexico*, 206, 397; Bennett and Zingg, *The Tarahumara*, 54–55; and Di Peso, *Casas Grandes*, 110, 264.
15. *Kokoyome* is a cryptic term; it refers not only to these structures, but also to the "ancient aliens" whose bones can be now seen in the caves. Cocoyome, of course, was also the name of an Indian tribe that inhabited the desert at Bolsón de Mapimí and was annihilated by the Spaniards. The Tarahumaras remember the kokoyome as a group of relentless enemies. See Lumholtz, *Unknown Mexico*, 192; Almada, *Diccionario de historia*, 103. On the other hand, *sonogori* is a storage house for the cornstalks and husks (Brambila, *Diccionario rarámuri-castellano*, 533), which may serve as emergency fodder or as food for the work oxen (Bennett and Zingg, *The Tarahumara*, 30). Mestizos and some Tarahumaras call the old ruins *sonogori* for their resemblance to these modern round constructions.
16. Di Peso, *Casas Grandes*, 293. Or 1150 A.D., according to Le Blanc, "Dating of Casas Grandes," 801; Lazalde, *Durango indígena*, 40, and Kelley, "Reconnaissance," 174–175.
17. First described by Hrdlička (*Physiological and Medical Observations*) in 1908, this cultural trait has been studied by Romano ("Deformación cefálica intencional," 201) who reports that eleven crania with intentional deformation have been found in the state of Chihuahua. I studied a skull with a classic three-lobulated deformation; the specimen apparently was the head of a mummy found in a cave in the Uruachi region in an excellent state of conservation.
18. Lumholtz, *Unknown Mexico*, 327–329.
19. Hayden, "La arquelogía de la Sierra del Pinacate," 281.
20. Sauer, *Distribution of Aboriginal Tribes*, 82.
21. Miller, "Nota sobre los lenguajes extintos," 50.
22. Lionnet, *Los elementos*, 11.
23. Kroeber, *Uto-Aztecan Languages*, 1–28; Steele, "Uto-Aztecan," 453; Miller, "Nota sobre los lenguajes extintos," 48.
24. From colonial Opata materials that Dr. Pennington generously let me examine.

25. Pennington (*The Tarahumar*, 13), however, recorded another tradition that makes the Tarahumaras come from the northeastern desert country. Lumholtz (*Unknown Mexico*, 1:297) also reported that the "Rarámuri were led by Tata Dios into the mountains having come from the northeast or east."
26. Pennington, *The Tarahumar*, 1–13.
27. Chihuahua City, originally San Francisco de Cuéllar, then San Felipe el Real, was founded on October 12, 1709, by the governor of Nueva Vizcaya, Don Antonio Deza y Ulloa, on lands ceded by Don Ildefonso de Irigoyen. We know little of the Indians that lived in the Chihuahua valley before the arrival of the Spaniards. The very preliminary work of Guevara Sánchez (*Un sitio arqueológico*, 1) permits us at least to establish the presence in the area of nomadic groups with southern and Casagrandian influences. See Almada (*Diccionario de historia*, 160) and Lumholtz (*Unknown Mexico*, 2:119).
28. North from the Chínipas, Decorme (*La obra de los Jesuitas*, 2:220) places the Hío, a group that, he states, spoke a Tarahumara dialect. Pécoro and Prado ("Historia," in González Rodríguez, *Crónicas*, 81–107) reported that the Varohíos found in the area of Chínipas in the late 1670s were actually Tarahumaras and Híos and that the Guailopos had replaced the true Chínipas after they abandoned the area following destruction of the original mission. See also Christelow, Introduction, 10; Ratkay, *An Account of the Tarahumara Missions*, 15 and Merrill, "La identidad ralámuli," 81.
29. *Annua, 1608*, in González-Rodríguez, *Crónicas*, 165. Compare Spicer, *Cycles of Conquest*, 28, and Merrill, *Rarámuri Souls*, 31.
30. Hackett ("Historical documents," 159) in Pennington, *The Tarahumar*, 19.
31. González Rodríguez, *Tarahumara*, 60.
32. González Rodríguez, *Tarahumara*, 150, and *Crónicas*, 23–30.
33. Fonte's report actually described the first Rarámuri migration registered by history (in González Rodríguez, *Crónicas*, 162).
34. Sheridan and Naylor (*Rarámuri*, 8) contend that the Indians "requested baptism and submitted themselves to mission life" because they thought that they could obtain certain benefits from the missionaries. Opportunism is, however, a weak explanation for the strategy when one attends to the events that would be unfolding later. As a matter of fact, the early missionaries believed in the good faith of the Indian invitations (see for example Fonte's *Annua, 1608* and *Annua, 1611* in González Rodríguez, *Cronicas*, 178–181, 186–187); they would later pay a high price for such naïveté.
35. The Laws of the Indies required that each *reducción* would have a minister (law 3), a church with door and key (law 4 — Philip III, Madrid, October 10, 1618), two or three singers and a *sacristán* (law 6), a *fiscal* who gathers the Indians for the catechism (law 7) and Indian alcalde (major) and several *regidores* (law 15). Law 10 is particularly interesting since it recommends the foundation of towns close to the mining centers. Laws 21 (Philip II, Madrid, May 2, 1563) and 22 prohibit Spaniards, blacks, mestizos, or mulattos from living in the pueblos or among the Indians. *Recopila-*

ción, vol. 3, laws VI-II-3–VI-II-29; Palacios, *Notas a la recopilación*, 372–377; Porras Muñoz, *Iglesia y estado*, 260–261).

36. Neumann, *Historia seditionum*, C1.
37. Sánchez-Téllez and Guerra, *Pestes y remedios*, 53.
38. Riva Palacio et al., *México a través*, 5:207–208. Decorme (*La obra*, 1:20), following Alegre, dates the beginning of the epidemic in 1575 and gives an estimated death toll of more than 800,000; Riva Palacio et al., following Cavo, calculate more than 2,000,000.
39. We do not know whether the epidemic of European influenza that decimated the Indian population throughout the Caribbean Islands actually reached North America (Sánchez-Téllez and Guerra, *Pestes y remedios*, 53). The spread of smallpox, on the other hand, has been better tracked. According to some historians, smallpox (*viruela* or *viruela negra*) was carried into New Spain in 1520 by an "anonymous Black man whose name it is not known, but who anyway became exceedingly famous" (Trabulse, "Los hospitales," 1422; Mörner, *Race Mixture*, 32). According to others, it was brought specifically by Juan Guía, a slave of Pánfilo de Narváez (Riva Palacio, *México a través*, 3:272). This disease, known by the Indians as *hueizáhuatl* ("grand leprosy") or *teozáhuatl* ("God's pox"), counted among its victims the Aztec emperor Cuitláhuac and the king of Michoacán, Zuangua (Riva Palacio, *México a través*, 5:302). The first epidemic was followed immediately thereafter by another that the Indians called *tepitonzáhuatl* ("little leprosy"). This second epidemic has been identified as measles.

    Exanthematic typhus (1526) and measles (1531) also inflicted great damage on the Indian population. The *matlazáhuatl* (typhus) epidemics of 1545, 1576, and 1578 were no less catastrophic, as were those of 1588 and 1589. Riva Palacio et al. emphasize that these epidemics afflicted only the Indians and not the Spaniards and mestizos. Compare Martínez Caraza, *El norte bárbaro*, 28.

    Closer to Tarahumara territory, Velasco (González Rodríguez, *Crónicas*, 35) reported that by 1593 smallpox and measles—*cocoliztli*—inflicted countless deaths among the Sinaloan Cáhitas. The same epidemics struck the mission of Parras, Coahuila, in 1612, 1622, 1652, and 1664 (Decorme, *La obra*, 2:33). Father Castini (*Carta annua, 1623*, in González Rodríguez, *Crónicas*, 51–52) reported that in October of 1623 the pestilence, which "killed more than 8,500 of their people" in Sinaloa and part of the region of Chínipas, acquired full strength, reaching disaster proportions. Rodrigo del Castillo in 1662 described how the population of San Miguel de las Bocas, Durango, had been decimated by disease in "recent times," and how *cocoliztle* had pounded the Salineros, a tribe of the desert (González Rodríguez, *Crónicas*, 249–260). In 1731, Father Echeverría reported that smallpox was decimating the Tubar Indians (González Rodríguez, *Crónicas*, 382), who were also afflicted by smallpox or cholera in 1617 and 1851.
40. González Rodríguez, *Tarahumara*, 161.

41. Nonetheless, Neumann (*Historia seditionum*, D2) would claim that the Tarahumaras blamed baptism for the death of their children. For example, the Indians justified the martyrdom of Father Básile at Villa de Aguilar in 1652 by attributing to the priest the death of a child that he had just baptized (Decorme, *La obra*, 2:270; Neumann in González Rodríguez, *Révoltes des indiens*, 68). Father Salvatierra (1680) wrote, "It is to be known that in past years, when Father Nicolás de Prado entered to baptize infants, the devil placed in their [the western Tarahumaras'] hearts an apprehension that children would die if baptized, and that the priest by the act of blowing on them, killed them" (González Rodríguez, *Crónicas*, 68).
42. Decorme, *La obra de los Jesuitas*, 2:271.
43. Dunne, *Early Jesuit Missions in Tarahumara*, 94.
44. Neumann, *Historia seditionum*, D.
45. Decorme, *La obra de los Jesuitas*, 2:303. Possibly the same "great epidemic" reported by Father Abbé (1744, in Sheridan and Naylor, *Rarámuri*, 83–84). Deeds ("Resistencia indígena," 58–50) places the epidemic in 1725 and claims that most experts identify the disease as typhoid.
46. Father Pedro Méndez accompanied the Martínez de Hurdaide expedition in 1601; to him are credited the first baptisms of Tarahumaras. Father Wadding, also known as Godínez, visited the Chínipa Indians by 1618–1619 (González Rodríguez, *Crónicas*, 44).
47. Almada, *Geografía del estado*, 106 and *Resumen de historia*, 56; González Rodríguez, *Crónicas*, 46–47.
48. Decorme, *La obra de los Jesuitas*, 2:224–225; Riva Palacio et al, *México a través*, 131.
49. Pécoro and Prado (in González Rodríguez, *Crónicas*, 81) claim that loyal Chínipas "did not desert the faith" but resolved to leave and resettle in the Christian towns of San Andrés Conicari, Concepción de Vaca, and San José del Toro, where they survived intermingled with the Sinaloa Indians. Varijíos and Guazaparis occupied the territory left by the fleeing Chínipas. The valley of Chínipas was then populated with Guailopos or Chínipas Nuevos Indians among whom the missionaries would work later.
50. Griffen (*"Procesos de extinción,* 716) states that the jurisdictional dispute that arose and lasted long years between Franciscans and Jesuits over the limits of the area comprised by the Conchos and Tarahumara missions had a profound impact upon the Indians and was very important in the definition of the geopolitical geography of the area. See also Spicer, *Cycles of Conquest*, 232.
51. Sheridan and Naylor, *Rarámuri*, 70n64.
52. Hackett (*Historical Documents*, 2:159, in Pennington, *The Tarahumar*, 19). Compare Christelow, Introduction, 7, which affirms that Tarahumaras were working at the Santa Bárbara mines as early as the end of the sixteenth century.
53. Sheridan and Naylor, *Rarámuri*, 8.
54. Sheridan and Naylor, *Rarámuri*, 11–12.

55. San Miguel de las Bocas played an important role in the development of the Tarahumara missions, both as a base of operations for the initial *entradas* and as the site where the Jesuits destined for the Tarahumara missions would spend time studying the language. González Rodríguez (*Crónicas*, 250) claims that the first contacts with the Tarahumaras of Bocas del Río Florido can be dated to 1612–1615, but Rodrigo del Castillo (1662, in González Rodríguez, *Crónicas*, 250) reported that Bocas was "formed" in the year of 1630 with gentile Tarahumaras "brought from inland." Márquez (*Satevó*, 37) dates this "second foundation" on 1630 or 1631. See also Christelow, Introduction, 11; Decorme, *La obra*, 2:254; Sheridan and Naylor, *Rarámuri*, 14; and Dunne, *Early Jesuit*, 37. Pennington (*The Tarahumar*, 4n5) quotes from a document in Parral Archives written much later (1685) that suggests again that this San Miguel Bocas was founded more than once.
56. Griffen, "Procesos de extinción," 708, and "Some Problems in the Analysis," 44–51.
57. The term *repartimiento* was first applied to the action of distributing among the Spaniards the goods and lands taken from the Moors when they were expelled from Spain (Quillet, *Diccionario enciclopédico*, 7:502). In the New World, it meant dividing up the Indian labor force. Since 1509, King Ferdinand V (in Martínez Marín, "El reparto," 1105) had recommended that the *adelantado*, governor or pacifier, would divide up the Indians among the Spanish settlers so they "have them and enjoy their tributes." See also Palacios, *Notas*, 391–398; Herring and Herring, *A History of Latin America*, 191–192; and Mörner, *Race Mixture*, 91–93.
58. Pennington, *The Tarahumar*, 18–23. For example, as late as 1744, a letter from the governor of Nueva Vizcaya to the viceroy shows that repartimientos were alive and well. The governor intended to increase the ratio of the repartimiento from 4 percent to a whole third of the inhabitants of each Tarahumara pueblo. Furthermore, the document proposed to educate children only until they would be of laboring age and to periodically rotate the working crews so the two thirds of the people remaining in the *reducción* could work their own fields and those of the mission (Sheridan and Naylor, *Rarámuri*, 89–101; see also Thomas, *Teodoro*, 123).
59. Conquistadors were authorized to enslave rebellious Indians only if the Indian had previously protested allegiance to the Spanish crown. In spite of this, Governor Guajardo Fajardo was accused in his *juicio de residencia* (trial of residence) of having illegally sold (as slaves) four hundred Indians (Escribanía de cámara, in Porras Muñoz, *Iglesia y estado*, 564).
60. Pennington, *The Tarahumar*, 19.
61. Pascual, José, "An Account of the Missions," in Sheridan and Naylor, *Rarámuri*, 19.
62. Dunlay, "Indian Allies," 238–259; Martínez Caraza, *El norte bárbaro*, 20.
63. Estrada, Letter, in Sheridan and Naylor, *Rarámuri*, 74. Original in Archivo Histórico de Hacienda, Mexico City (278, 7, DRSW 2889).
64. Griffen, "Procesos de extinción," 717.
65. The mission of Nombre de Dios, presently within the metropolitan area of Chihua-

hua City, was founded in 1697 with Concho Indians, but Norteños and Tarahumaras also resided there (Almada, *Resumen de historia*, 87). This mission had as visitas (hamlets with small chapels) San Antonio de Chuvíscar (a Tarahumara hamlet) and San Juan Bautista de Norteños. Fr. Mariano de Mora (1778) reported the residence in the cabecera (the main pueblo) and its two visitas of 108 Indian families. Nombre de Dios accounted for 109 souls (Tarahumaras, Conchos, and Norteños), Chuvíscar for 123 (Tarahumaras), and San Juan Bautista for 180 (Norteños).

66. Neumann, Letters to an Unknown Priest, Feburary 20, 1682, p. 61.
67. The threat of an alliance of the rebellious Tarahumaras with the fiery Tobosos persisted until the final extinction of the latter. Governor Guajardo Fajardo decisively defeated the Tobosos, fortified at the top of a cliff (*peñol*) at San Miguel de Nonolat. Only then did he resume his campaign against the Rarámuri rebels. When Guajardo returned to Tarahumara Country, he had to battle the Tarahumaras in another peñol, this time at Pichachi. The Rarámuri were initially able to repulse the Spaniards, but soon thereafter were routed by Guajardo's forces.
68. In Sheridan and Naylor, *Rarámuri*, 17–18. According to several sources, this cacique was killed by Governor Luis de Valdés. See Decorme, *La obra de los Jesuitas*, 2:263–265. The word *tlatol* comes from the Náhuatl *tlatolli*, according to León Portilla ("Pensamiento," 841) and means "word, discourse, narration, story or exhortation." See also Molina, *Vocabulario*, 14; González Rodríguez, *El noroeste*, 289.
69. In Sheridan and Naylor, *Rarámuri*, 65.
70. In 1649, Governor Diego Guajardo Fajardo began his campaign against the Tarahumaras burning 4,000 *fanegas* (Spanish bushels) of corn and 300 rancherías forcing the Indians to deliver their chieftains Supegiori, Tepox, Ochavarrí, and Bartolomé (Zepeda "Alzamiento," 244–245; Decorme, *La obra*, 2:267; Dunne, *Early Jesuit*, 55; and Spicer, *Cycles*, 32).
71. Presenting battle from the top of a peñol was not an exclusively Tarahumara strategy; apparently all Chichimecan groups relied on it. For example, in 1519 the Indians of Calco, Tlamanalco, and Oaxtepec used it in defending against Hernán Cortés's campaign (*Cartas*, 153–155). The conquest of the Gran Chichimeca witnessed many other examples, such as the campaign of El Mixtón and the Acaxée revolts (Santarén, 1604, quoted in González Rodríguez, *El noroeste*, 135–150). In battles with a definite importance for our area of study, such as the ones of Nonolat and Pichachi in 1652, Peñol del Diablo in 1684 (Di Peso, *Casas Grandes*, 869), and Tres Castillos in 1890, the Apaches used peñoles as defensive fortresses. In Tarahumara Country, in the battle of Tomochi (1649), Guajardo Fajardo could not take the peñol and was defeated. According to González Rodríguez (in Neumann and González Rodríguez, *Historia de las rebeliones*, 65n51), Tarahumara rebels used Peñol de Sopechí as a fortress at least three times (1650, 1652, and 1697) (see Neumann, Letters to an Unknown Priest, April 23, 1698, p. 127) and the battles of Corodéchi (September, 1687) and Wébachi (Neumann, *Historia seditionum*, H6) are also examples of this strategy's efficacy.

72. Decorme (*La obra*, 2:271) reports that the poison in their arrows was lethal even if one had got only a scratch. He also remarks that an Indian revealed the antidote to the Spaniards. See also Neumann, Letters to an Unknown Priest, February 20, 1682, p. 17; Letter to Father Stowasser, July 29, 1686, pp. 126, 132, and *Historia seditionum*, A6.
73. Apparently the governor wanted to put some healthy distance between him and his archenemy, the bishop of Durango, Diego de Evía y Valdés. Although the bishop declared in one of his letters (Letters to the King, April 2, 1652) that he left Parral to prevent "fires and ruin" and in another (Letters to the King, April 6, 1652), that he was attempting to avoid feuding with the governor. The fact is that since the time of Luis de Valdés, the bishop had been feuding with the civil government and the religious orders (see Porras Muñoz, *Iglesia y estado*, for details of these conflicts). Among his accusations against the religious orders were that they invented rumors of revolts and exaggerated the pressing needs for supplies, protection, etc., of missionaries and soldiers to benefit themselves with royal favors.
74. Evía y Valdés, Letters to the King, April 6, 1652.
75. Decorme, *La obra de los Jesuitas*, 2:277–278.
76. Originally, Fathers Gamboa and Barrionuevo were in charge of this "accelerated missionization," but their poor health prevented them from performing the advances so impressively accomplished by Guadalajara and Tardá. Márquez (*Satevó*, 38–39; *Origen*, 157–161), Sheridan and Naylor (*Rarámuri*, 32), and Spicer (*Cycles*, 33) give great importance to the intervention of Don Pablo, a Tarahumara governor who participated in a meeting in 1673 with other Indian chiefs, Spanish authorities, and missionaries at Parral for the initiation of this definite entrada in Tarahumara Country.
77. Decorme, *La obra de los Jesuitas*, 2:292.
78. Father Pécoro founded Santa Teresa de Guazapares, Magdalena de Témoris, and Valle Umbrosa, exploring, in addition, to the east, the rancherías of Cuiteco and Cerocahui. The missions of Chínipas and Tarahumara thus came into contact with each other (Decorme, *La obra*, 2:229).
79. Caraveo, "Cambio demográfico," 25–31. Di Peso (*Casas Grandes*, 866) contends that the refugees from the Manso-Suma conflict in 1680 helped to spread the Great Southwestern Revolt of New Mexico deep into Nueva Vizcaya.
80. Neumann, *Historia seditionum*, C2 et seq.
81. Christelow, Introduction, 63–66.
82. Neumann, *Historia seditionum*, C3.
83. Neumann, *Historia seditionum*, D4. Sheridan and Naylor's careful analysis has found such omens did take place, but were only loosely related in time to the revolts. The earthquake has not been verified, the comet occurred in 1765, and the eclipse a year *after* the rebellion (*Rarámuri*, 68).
84. Compare Cramaussel and Álvarez Suarez, "La creación."
85. Caraveo, "Cambio demográfico," 30–31.

86. Spicer, *Cycles of Conquest*, 34.
87. Spicer, *Cycles of Conquest*, 35.
88. See for example the declaration of Gov. Gabriel del Castillo in 1697 in Sheridan and Naylor, *Rarámuri*, 64–65.
89. Spicer, *Cycles of Conquest*, 36.
90. Neumann, *Historia seditionum*, I-5. See also Neumann, Letters to an Unknown Priest, 1698, p. 137, and *Historia seditionum*, I; Abbé, 1744, in Sheridan and Naylor, *Rarámuri*, 21; Esteyneffer, *Florilegio medicinal*, 1–973; Foster, George, "El legado," 17; and González Rodríguez, *El noroeste*, 536.
91. Márquez, "Notas sobre historia de la Tarahumara," 41. Of course, there were remarkable exceptions, such as that of Father Glandorff, who was a humble man of God endowed with extraordinary faculties and regarded as one of the most saintly Jesuits of all times. Another exception was Brother Esteyneffer (Steineffer) whom Neumann (*Historia seditionum*, I-5) describes as a truly religious man endowed with great charity.
92. Sheridan and Naylor, *Rarámuri*, 102.
93. Bancroft, *History of the North Mexican States and Texas*, 372–478; Spicer, *Cycles of Conquest*, 37–40.
94. Vogel, *American Indian Medicine*, 59.
95. Pennington, *The Tarahumar*, 23.
96. See Giner Rey, *Apuntes*, 41–66; Irigoyen-Rascón, *Cerocahui*, 63–64; and Peña Moyrón, "Una Atlántida."
97. Lumholtz, *Unknown Mexico*, 1:119–120.
98. Ocampo, *Historia de la misión*, 137, 152, 153, 186, and 209. Other Jesuits, among them Brothers Velásquez, González Ochoa, and Vega, excelled as nurses or dentists.
99. Brambila, *Bosquejos del alma tarahumara*, 11; Díaz Infante, *100,000 kilómetros*, 22–24.
100. Spicer, *Cycles of Conquest*, 40.
101. See Gajdusek, "The Sierra," 15, and Fontana and Schaefer, *Tarahumara*, 6.
102. Irigoyen-Rascón, *Cerocahui*, 137–163.
103. Champion, "Acculturation among the Tarahumaras," 561–563.
104. Miguel Rivas, personal. communication, 1979.
105. Ramos and Iturbide, "Migración," 25–50, and De Vries, "El caso," 54.
106. See Dibildox et al., *Declaration of the Diocese*, and Weaver, "Changes in Forestry Policy," 1.

3. Rarámuri, the People and Their Culture

1. Irigoyen-Rascón, *Cerocahui*, 86.
2. Vivó, *Razas y lenguas indígenas*, 11.
3. Ratkay, *An Account of the Tarahumara Missions*, 26.

4. Neumann, Letters to an Unknown Priest, February 20, 1862.
5. Lumholtz, *Unknown Mexico*, 1:235.
6. González Rodríguez, *Tarahumara*, 87.
7. Kennedy (*Tarahumara*, 75) highlights that the Tarahumaras of Inápuchi wear a distinctive dress, hand-sewn or woven by the females in the family, which serves as a symbol of the cultural separateness of the gentile Tarahumara. Father Uranga's quest (*Uirichiki*, 9), on the other hand, was rewarded by a couple of ephemeral sightings of gentile Tarahumara women from Wirichiki wearing the woolen skirts the women wore in bygone days. More recently, Lévi ("La flecha," 127–153) has rediscovered the woolen skirts in the remote canyons of El Cuervo.
8. Pennington, *The Tarahumar*, 224; Fontana and Schaefer, *Tarahumara*, 36–45, and Díaz Infante, "La cultura tarahumara reflejada en su lengua," 12–13.
9. Pennington, *The Tarahumar*, 225.
10. See Urbina, Antonio de, "Relación de Serocahue" (in Del Paso y Troncoso, *Relaciones del siglo XVIII*); Lumholtz, *Unknown Mexico*, 1:245, and Bennett and Zingg, *The Tarahumara*, 57.
11. Sheridan and Naylor (*Rarámuri*, 72) consider that the Tarahumaras' transhumance in colonial times was not necessarily a refusal to accept the discipline of Christian village life as the missionaries interpreted it; rather it was an adaptive response to an environment where arable land was scarce. Neumann (Letters to an Unknown Priest) wrote in 1681, "They change their place of residence three or four times in the course of a year. In particular if death occurs the house is destroyed, and the site is not inhabited thenceforth." Lumholtz (*Unknown Mexico*, 2:162) reported, "While most of the Tarahumares live permanently on the highlands, a great many of them move for the winter down into the barranca." In modern times, Murray ("The Tarahumara Project," 180) has preferred the concept "semi-sedentary transhumance."
12. Brambila, "Psicologia y educación del Tarahumar," 201.
13. Gajdusek, "The Sierra Tarahumara," 33.
14. Ruíz, in Uribe and León, *Estudio comparativo*, 218; Díaz Infante, "La cultura Tarahumara," 14.
15. *Rimuma* means not only "to dream," but also, according to Brambila (*Diccionario rarámuri-castellano*, 476), "to think deeply, to meditate or reflect."
16. Merrill, *Rarámuri Souls*, 14.
17. Brambila, *Supersticiones y costumbres de los Rarámuri*, 13.
18. Neumann, Letters to an Unknown Priest, February 20, 1682; Ratkay, *An Account of the Tarahumara Missions*, 26.
19. Kennedy, "Tesguino Complex," 623; West, Paredes, and Snow, *Sanity in the Sierra Madre*, 164–166; Greenblatt, "LJ West's Place," 9–10; Paredes and Irigoyen-Rascón, "Jíkuri, the Tarahumara Peyote Cult," 121.
20. Kennedy, "Tesguino Complex," 627.
21. Ratkay (*An Account*, 40) affirmed that the Tarahumaras of Carichí did worship idols.

One of them was a large snake, but also they kept sacred stones. The missionary enumerated a series of powerful demons with noxious and healing attributes, as well as fauns that lived in the forest. The name of these demons or fauns was *testsani* and their wives', *uribi*. The Lord of the Underworld was called *terégori*.

22. Neumann, Letter to Father Stowasser, July 29, 1686.
23. Merrill, *Rarámuri Souls*, 113–114. This report actually resembles very much Mesoamerican ways of thinking about an afterlife. Merrill ("The Concept of Soul," 35–36), on the other hand, minimizes the influence of Catholic catechesis on the Rarámuri belief system, a contention I believe unfortunate, since the missionaries and their agents (*rezanderos*, lay catechists, etc.) have been teaching about the existence of levels of afterlife—i.e., purgatory, limbo, heaven.
24. Murray, "The Tarahumara Project," 180.
25. Irigoyen-Rascón and Paredes, "Biosocial Adaptation," 11–12.
26. Especially comparing with the massive damage caused by timber companies and illegal exploitation of the forest. Cf. Weaver, "Changes in Forestry Policy," 3–7, and Dibildox et al., *Declaration of the Diocese*, 1–3.
27. Brouzés, "La nourriture partagée," 12.
28. Preciado et al., *Informe Kwechi*, 109.
29. Brouzés, "La nourriture partagée," 45.
30. Lumholtz, *Unknown Mexico*, 1:183.
31. Irigoyen-Rascón, *Cerocahui*, 95.
32. Or Swadesh's (*Indian Linguistic Groups*, 12) Macro-Nawa or Steele's ("Uto-Aztecan," 449) Taracahitan.
33. INEGI, *XI censo*; *XII censo*; INI, 1995 (*Atlas de las lenguas indígenas*). The early missionaries produced a number of "artes" and "vocabularies." The works of Steffel (1809), Tellechea (1826), Gassó (1926), and Ferrero (ca. 1930) have historical importance. Of particular relevance is the work of Father David Brambila, S.J. (*Gramática rarámuri*, *Diccionario rarámuri-castellano*, and *Diccionario castellano-rarámuri*) with grammatical and lexical analyses comparable only to those available for modern languages. Thord-Gray published a Tarahumara-English dictionary. The Summer Linguistics Institute linguists, particularly Simon Hilton and Paul Carlson, in addition to their biblical translations, prepared glossaries and phraseologies of great value. Burgess has published a number of Tarahumara tales and legends in Tarahumara language and, with Mares Trías, a book about the foodstuffs of the Tarahumara. Lionnet (*Los elementos*) has been the pioneer of modern linguistic analysis and his works have been used by multiple authors. In *Memoria*, I studied the verbs *simea* or *simama* (to go). There are two available methods to study Tarahumara, Father Llaguno's *Assimil*, and the audiolingual (unknown author). Along with Erasmo Palma, I published a method of medical Tarahumara for Spanish-speaking physicians (*Chá okó*). Burgess has published collected materials in the dialect of western Tarahumara Country and Díaz Infante a study of the Chinatú dialect.

34. Miller, "Nota sobre los lenguajes extintos," 50.
35. Valiñas, "Lengua, dialectos e identidad," 116–117.
36. Preciado et al., *Informe Kwechi*, 102–107; Passin, "Place of Kinship," pt. 2, 471; Paredes, West, and Snow, "Biosocial Adaptation and Correlates," 165; Merrill, *Rarámuri Souls*, 22.
37. Bennett and Zingg (*The Tarahumara*, 220–223), Passin ("Place," pt. 1, p. 235; pt. 2, p. 370), and Brambila (*Gramática*, 20–21) have studied this terminology. Passin (pt. 1, p. 370) places the Rarámuri system within Spier's ("Distribution," 69–88) Yuman type, as it uses differential terms to name younger and older brothers and sisters, and as an old Uto-Aztecan language as it preserves an eight-term nepotic-avuncular reciprocal system. Brambila is an authoritative reference for specific kinship terms. Merrill ("Tarahumara Social," 290) places Tarahumaras' kinship terminology within the Neo-Hawaiian.
38. Neumann, Letters to an Unknown Priest, February 20, 1682.
39. Bennett and Zingg, *The Tarahumara*, 201.
40. The pueblo's see, also called *cabecera*, is usually composed of a number of buildings: the church (*teyópachi* or *riobá*) with its atrium, the *komerachi* or *casa de comunidad* (a communal room for tribal meetings, trials, etc.), the cemetery (*kamposántochi*) and a more or less well-delimited plaza. The *komerachi* is many times used as jail too (as in Norogachi) The term *convent*, used commonly to designate buildings next to the church, is sometimes appropriate, but other times it simply designates the sacristy and the priest's habitation. In modern times a school and one or several general stores—which take advantage of the gatherings and church functions to sell their goods to the Indians—are also part of the pueblo.
41. Bennett and Zingg, *The Tarahumara*, 203.
42. Preciado et al., *Informe Kwechi*, 49.
43. In Norogachi, the siríame grande has permanent headquarters in the town itself and his auxiliaries exert their authority, one to the east, the other on the western portion of the pueblo territory. The minor siríames will always consult the *wa'rura* before making important decisions. Brambila (*Diccionario castellano-rarámuri*, 579) calls these auxiliary siríames *wákame*.
44. Preciado et al., *Informe Kwechi*, 50.
45. Lumholtz, *Unknown Mexico*, 1:140.
46. Preciado et al, *Informe Kwechi*, 84.
47. Fried, "Relation of Ideal Norms," 289.
48. Tarahumaras consider it a serious crime to use an ax to hit someone or cause wounds with a knife. Bennett and Zingg (*The Tarahumara*, 203) noted that witchcraft used to be a serious offense too. As a matter of fact, even as late as the 1940s in Norogachi several trials of *sukurúame* and individuals possessing witchcraft paraphernalia, such as *rusíwari*, were brought to the attention of the native or even the municipal authorities. Ching Vega ("La Tarahumara," 1) relates the case of a *sukurúame*

(wizard) from Tewerichi who was imprisoned at Carichí after being tried for having threatened to destroy the whole community using witchcraft. Apparently in that case Mexican authorities respected the Rarámuri verdict.
49. Preciado et al., *Informe Kwechi*, 85. Compare Kennedy, "Tesguino Complex," 631.
50. Compare Bennett and Zingg, *The Tarahumara*, 210.
51. Preciado et al., *Informe Kwechi*, 90.
52. Merrill, *Rarámuri Souls*, 101–102.
53. Fried, "Relation of Ideal Norms," 288; Merrill, "Tarahumara Social Organization," 290.
54. Merrill, *Rarámuri Souls*, 25.
55. Bennett and Zingg, *The Tarahumara*, 206.
56. Merrill, "The Concept of Soul," 84.
57. The Tarahumara word *ropiri* seems to derive from the Náhuatl *topil*. Cabrera (*Diccionario*, 140) defines *topile=alguacil*, the one who carries the cane, from *topille*=cane, justice stick.
58. Merrill, "The Concept of Soul," 85.
59. Diaz Infante, "La cultura tarahumara," 60. See Brambila, *Gramática*, 155. Also see Van Gennep (*Rites of Passage*, 169) on the use of special or sacred language. Merrill (*Rarámuri Souls*, 83) reports that Tarahumara orators employ prominently two kinds of figures of speech: metonomy and synecdoche.
60. Nevertheless this is the Tarahumaras' "national sport"; known as *neotzamia*, or in Spanish, *chirinola*, gossip and prevarication were acknowledged by Lumholtz (*Unknown Mexico*, 1:245) and continue in modern times, being constantly indulged in. Passin ("Tarahumara Prevarication," 235–247) meticulously analyzed prevarication as a serious problem when obtaining information from Tarahumara informants.
61. González Rodríguez, "Jesús Hielo Vega," 487–488; Robles, *Mujé narí nurema sinéame*, 6–3.
62. Ocampo, *Historia de la misión*, 368–370.
63. In the 1970s, Father Carlos Díaz Infante revived the Misa-Yúmari in Norogachi, and since then other Jesuit liturgists have incorporated some of its features into the mass.
64. Brouzés, "La nourriture partagée," 26–27.
65. Ovalle, "Bases programáticas," 10–21; Presidencia de la República, *Convenio IMSS-COPLAMAR*, 1–15.
66. Irigoyen-Rascón, *Cerocahui*, 66–67.
67. Almada, *Geografía del estado*, 17.
68. *Presidentes de sección municipal* are usually mestizos. In Norogachi in 1989 the community elected a Tarahumara, Juan Gardea, as *presidente seccional*. This historical development was followed by the election of his brother Marciano Gardea three years later and, again three years later, another brother, Cruz Gardea, became president. Unfortunately, on June 22, 1996, Cruz Gardea was murdered while trying to stop a brawl among mestizos (Pietrich, "De cinco balazos," 1).

4. Affiliative Social Activities of the Tarahumara People

1. Kennedy, "Tesguino Complex," 620–640.
2. Palma, Erasmo, *Donde cantan los pájaros*, 10–15.
3. Kennedy, "Tesguino Complex," 630–637.
4. Almada, *Diccionario de historia*, 238.
5. Balke and Snow, "Anthropological and Physiological Observations," 293.
6. Groom, "Cardiovascular Observations," 304–314.
7. Pennington, *The Tarahumar*, 168n37.
8. Bennett and Zingg, *The Tarahumara*, 386.
9. Pennington, "La carrera de bola," 18. Ratkay (*An Account*, 33–34) described the game of *ulama*, played by Tarahumaras at Carichí. Two teams tried to take a ball of the size of a large quince—made of the hardened sticky gum of a tree—to the goal line of their rivals. For this purpose they could catch the ball only with their thighs and shoulders. They could play all day long. Such a description fits well with games presently played among other groups, such as *pelota mixteca* played in Oaxaca and the Sinaloan ulama (Cuéllar Zazueta, "Ulama," 60–63), and in pre-Columbian times throughout Mesoamerica. The modern *palillo*, *tákuri*, or *ra'chuera*—resembling lacrosse—played by the western and canyon Rarámuri may derive from ulama, but is played with a wooden ball that is tossed into the air with a spoon-shaped club. The teams try to hit the ball with their clubs and drive it to the goal zone. The game is strenuous and players suffer frequently from serious injuries and fractures. See Burgess McGuire, *Rarámuri riěcuara*, 27–38.
10. Di Peso, *Casas Grandes*, 591–592.
11. Campos and Gaeta, *El rutuburi*, 13.
12. In Irigoyen-Rascón, *Cerocahui*, 115–116.
13. Deimel, "Narárachi, Zwischen Traditionalismus und Integration," 1–9.

5. Great Life Occasions and Ceremonies

1. Bennett and Zingg, *The Tarahumara*, 233.
2. Irigoyen-Rascón, *Cerocahui*, 93.
3. A belief proceeding from Mesoamerica (Ortiz de Montellano, *Aztec Medicine*, 143).
4. Bennett and Zingg, *The Tarahumara*, 234–235.
5. Brambila, *Diccionario rarámuri-castellano*, 527.
6. Irigoyen-Rascón, *Cerocahui*, 86.
7. The physiological advantages of the "vertical" delivery are well known. See Engelmann, *Labor among Primitive Peoples*, 130–151.
8. Lumholtz, *Unknown Mexico*, 1:272.
9. In Lumholtz's day (1902), the umbilical cord was cut "with a sharp reed or a sharp-edged piece of obsidian, but never with a knife, for in that case the child would be-

come a murderer and could never be a shaman" (Lumholtz, *Unknown Mexico*, 1:272, and Van Gennep, *Rites of Passage*, 51).
10. Bennett and Zingg, *The Tarahumara*, 234.
11. Ocampo, *Historia de la misión*, 53.
12. Deimel, "Narárachi, Zwischen Traditionalismus und Integration," 1–9.

6. Major Festivities of the Tarahumaras

1. Rubén Amador, personal communication, June 1, 1979.
2. In Samachique, the first procession must be composed mostly of women, and the image of Our Lady of Guadalupe is taken out but not the one of Jesus. For the first midnight procession, the woman carrying the image of the Virgin of Guadalupe has a white handkerchief set on her head secured by a white hoop; she is accompanied by the rezandero and two women who cry when they pray; another maroma carries a matraca and another one incense; they are also accompanied by two children with candles and ocote. Marching and dancing by the sides of the procession are the evil ones—the pariseo and one carrying a flag (Rubén Amador, personal communication, June 1, 1979).
3. In Samachique, the pariseos return to the temple after destroying the arches and there they are confronted by the moros' bascoreros; they mock-fight for a while and finally capture the Judas. The pariseos are divided to destroy the arches into two groups that actually compete to see which one will throw down more arches. A winning team is proclaimed.

7. Loss-of-Health Conceptual Schemes of the Tarahumaras

1. Foster, George, "El legado," 6. Compare Villa Rojas, *Los elegidos de Dios*, 378–382, and Castillo and González, *La salud de las comunidades*, 5.
2. The teachings of that school, known as the humoral school, can be summarized as follows: The universe is composed of the four elements: air, water, fire and earth. Each of the elements has a particular quality: coldness, wetness, heat, and dryness respectively. The elements have a counterpart in the composition of the four humors of the body: phlegm, black bile ("atrabilious" from Latin *atra*=black or "melancholia" from Greek *melanos*=black), yellow bile, and blood. Excessive amounts or deficiencies of the humors will produce different diseases or characterize the various psychological temperaments. Or in the words of Hippocrates, the father of medicine: "The body is composed by blood, phlegm, yellow bile and black bile; this constitutes its nature and creates disease and health. Man is essentially healthy when these elements are in a just relation of krasis (κρασίς=*krasis*, or temperament), strength and quantity . . . then the balance is perfect" (*Nature of Man*, 174).
3. Foster, George, "El legado," 6.

4. Aguirre Beltrán, "Medicina y salubridad," 13.
5. Villa Rojas and Redfield, *Chan Kom*, 65; López Austin, "Cosmovisión y medicina náhuatl," 22–24, and "La dualidad 'frío-caliente,'" 18–20; Sejourné, *The Burning Water*, 87.
6. For the Tarahumaras, cold and heat, as properties of an herb, seem to exist at the same level as bitterness, "sneeze" production, chile-likeness, or acidity. Also, some plants that are regarded as "cold" by some are thought to be "hot" by others; however, certain herbs do have properties that should always be evaluated by considering their dialectic polarization with regards to the heat/cold complex.
7. Which is by itself pathogenic. For example, in cases of fright, or majawá, exposure to death or human bones and in some cases of bewitchment—sipabuma—the impact may not be sufficient to cause an actual loss of the soul. Therefore these patients may be treated with remedies considering that their iwigára remains within them and has not wandered off.
8. From *iwima*=to breathe; *eká, iká*=wind. Similar to Greek ψυχή, ής or Latin *anima* (Mateos, *Etimologías griegas*, 95; *Etimologías latinas*, 73).
9. Merrill, "The Concept of Soul," 131, and *Rarámuri Souls*, 93, 113.
10. Although rare, suicide does exist among the Tarahumaras, and it is conceived as a form of familial madness.
11. A remarkable exception to this was that of a young patient from Rochéachi who was brought to the clinic with the fully developed syndrome. This was the second time the patient had suffered an "iwigara loss." The first time she had been rescued by a doctor who, using a strong suggestive technique, told the patient that he was going to cure her by administering a "special IV serum," which the patient's husband described to me as a reddish solution. Following these leads, I dissolved an ampoule of vitamin B-12 in normal saline and administered it to the patient through an IV line. The treatment, again, was successful. Obviously, the expectation of receiving again the mysterious "serum" was embedded in the psychological format of the syndrome for this patient.
12. Brambila, *Diccionario rarámuri-castellano*, 536–537.
13. Bennett and Zingg, *The Tarahumara*, 338.
14. Snakes, particularly rattlesnakes, are not only pathogenic agents; they also can convey healing. Rattlesnakes are actively hunted by Tarahumaras because their dried and ground meat is easily sold to Mexicans who resell it, sometimes encapsulated, as a remedy for cancer and rheumatism.
15. Lumholtz, *Unknown Mexico*, 1:310. Not only are serpents denizens of the river ponds; in Charco Barbero, about a mile downstream from Urique, a strange being with pig feet is said to live at the bottom of the river (Irigoyen-Rascón, *Cerocahui*, 81). See also Bennett and Zingg, *The Tarahumara*, 121–124.
16. Brambila (*Diccionario rarámuri-castellano*, 388) argues that *no'pi*—or in Spanish, *salamanquesa*—is an animal that, superficially resembling a small serpent, is actu-

ally four-legged and, therefore, a lizard with, as he reports, "a greenish beautiful back and blue tail; Tarahumaras say it is *rayénari bukura* (belonging to the sun). Some Tarahumaras in Cerocahui call centipedes *no'pi*."
17. Bennett and Zingg, *The Tarahumara*, 124.
18. Palma, Erasmo, *Donde cantan los pájaros*, 12.
19. García Manzanedo, "Notas sobre la medicina tradicional," 69.
20. Brambila, *Diccionario rarámuri-castellano*, 262.
21. García Manzanedo, "Notas sobre la medicina tradicional," 69.
22. Burgess McGuire, "Leyendas tarahumaras," 103.
23. Bennett and Zingg, *The Tarahumara*, 137.
24. See also Irigoyen-Rascón and Jesús Manuel Palma, *Rarajípari: La Carrera*, 24, 35.
25. Merrill, *Rarámuri Souls*, 122.
26. Brambila (*Diccionario rarámuri-castellano*, 500) claims that the bird *ru'síwari* sings like *tochapi*, the small bird that announces snowstorms. García Manzanedo ("Notas sobre la medicina tradicional," 68) reports that "the illness known as *ru'síwari* is the result of contact with a being whose form and size resemble that of a small frog. Its color is whitish and, according to some informants, spotted."
27. Brambila, *Diccionario rarámuri-castellano*, 500.
28. Burgess McGuire, *¿Podrías vivir como un Tarahumara?*, 13.
29. Burgess, "Leyendas tarahumaras," 72.
30. Lumholtz, *Unknown Mexico*, 1:284.
31. Brambila, *Diccionario rarámuri-castellano*, 478. Brambila (*Diccionario rarámuri-castellano*, 210, 478) also lists *itibiri* and *ripibíwari*, defined as rheumatic pain caused by exposure to whirlwinds. In Norogachi, however, rheumatic pains are called, independently from their putative cause, *reuma* or *riuma*, and only rarely is the word *ripibíwari* used.
32. Bye, "Ethnoecology of the Tarahumara," 133.
33. García Manzanedo, "Notas sobre la medicina tradicional," 66.
34. Burgess McGuire, "Leyendas tarahumaras," 95.
35. García Manzanedo, "Notas sobre la medicina tradicional," 69.
36. Mull and Mull, "Differential Use of a Clinic," 113. Specifically, these authors explain the Tarahumaras' fear of lightning as a valuable defense in the context of the real possibility of lightning strikes.
37. The favorite is the Roman crucifix. This is characterized by the ends of the poles of the cross ending in a trilobulated fashion (a crossed cross); the crossbones rest under the feet (indicating that Jesus is dead) and the INRI sign is nailed on the upper beam above Jesus's image. On the back of the cross there may be a heart transfixed by a dagger. Brambila (*Supersticiones y costumbres*, 6) relates the use of crucifixes to appease or scare away the wandering *anayáwari* (ancestors) who bother the living at night. He remarks the belief that larger crucifixes have more power than smaller ones. Paradoxically, the crucifix is also the symbol of the sukurúame.
38. García Manzanedo, "Notas sobre la medicina tradicional," 70.

39. Brambila, *Diccionario rarámuri-castellano*, 139, 293, 539.
40. Irigoyen-Rascón, *Cerocahui*, 91.
41. The Tarahumara term *namuti* has the double meaning of animal and thing, which contributes sometimes to make patients' descriptions confusing; i.e., *ta namuti patza enaro* ("a small *animal* [or *thing*] walks [moves about] within me").
42. Latido is a common complaint throughout Mexico and beyond. At times it corresponds just to the aortic pulse that the patient perceives through a thin abdomen, but in others it may correspond to multiple intra-abdominal pathologies.
43. Brambila, *Diccionario rarámuri-castellano*, 268, 520.
44. Bye, "Ethnoecology of the Tarahumara," 133.
45. Bennett and Zingg, *The Tarahumara*, 133.
46. Lumholtz (*Unknown Mexico*, 1:242) reported the apparent absence of tapeworms among the Rarámuri and speculated that, since their sheep did have tapeworms, surely drinking tesgüino prevented human infestation. See also Gajdusek, "The Sierra," 37.
47. García Manzanedo ("Notas," 68) describes also *rurusí*, a condition caused by the introduction through the skin of a worm that inhabits swamps, ponds, and waterfalls. The condition courses with generalized edema, particularly severe on the face.
48. García Manzanedo, "Notas sobre la medicina tradicional," 63–64.
49. Bennett and Zingg, *The Tarahumara*, 265.
50. Several attempts to make a conventional disease out of the traditional empacho have been made at several points in the history of Mexican medicine. Peón y Contreras ("Empacho," in Domínguez et al., *El medico práctico*, 211), for example, wrote in 1889: "The term *empacho* is given to the permanence for a variable time of undigested foodstuffs in any part of the digestive tract. Common symptoms are abdominal discomfort which may become intermittently acute and severe pain, diarrhea, anorexia, flaccidity and exhaustion.... In small children who suck milk in excess and frequently regurgitate, digestive processes become lousy and masses of cheese move into the intestine where they cannot be digested. These masses act as true foreign bodies and, in emaciated babies, one can palpate them in the abdomen." Anyhow, the concept goes far beyond the Mexican borders: Pérez de Zárate (*Medicina folklórica panameña*, 60), for example, lists the remedies used to treat empacho in rural Panamá. These include common laxatives such as senna, jalap, malva, but also boy's urine. Empacho also is known in Mexican American communities in the United States (Kay, "Health and Illness," 133).
51. Interestingly, bezoars have been considered powerful amulets by many Indian tribes. The *piedra bezal* (bezoar stone), during colonial times, was searched for by the Indians in Sonora in the stomach of deer as a precious object, both an amulet and base to prepare "remedies." Spanish conquistadors also held in high esteem the medicinal properties of these stones—some of them of a weight of ½ pounds—that were swiftly commercialized and used as presents for the royal or ecclesiastical authorities (González Rodríguez, *Etnología y misión*, 52, 53, 119, 140). See Wintrobe et al. (*Har-

rison medicina interna, 1644–1655) for the modern pathological significance of bezoars.
52. Groom, "Cardiovascular Observations," 304–314; Irigoyen-Rascón and Erasmo Palma, *Chá okó*, 16–17.
53. Also called *chiote* from the Náhuatl *xixiotqui*=leper.
54. De Ramón (*Diccionario popular*) in 1896 defined *alferecía* as a childhood illness characterized by convulsions and loss of consciousness. In a previous work, I defined *alferesía* as susto, asustamiento, and even as a state of hysteria.
55. Paredes, West, and Snow, "Biosocial Adaptation," 170.
56. Burgess, "Leyendas tarahumaras," 107.
57. Palma, Erasmo, *Donde cantan los pájaros*, 92–93.
58. Brambila, *Diccionario rarámuri-castellano*, 599. Merrill (*Rarámuri Souls*, 206n8) lists the terms *ke richoti* and *uchuwátiri* as subtle alternations for *lowíame*.
59. Possibly this is the same as *wichuwáka* or *hierba loca*. Some Tarahumaras claim that it is used by the sukurúame or just evil people to poison people. For this purpose, it is added to batari so the victim will drink it without detecting its bitter flavor.

8. Rarámuri Healers

1. Deimel, "Narárachi, Zwischen Traditionalismus und Integration," 1–2, and Deimel, "Pflanzen zwischen den Kulturen," 41–64.
2. Mull and Mull, "Differential Use of a Clinic," 245.
3. Christelow, Introduction, 56; Lumholtz, *Unknown Mexico*, 2:311.
4. Bennett and Zingg, *The Tarahumara*, 252.
5. Healers who use *bakánowa* also perform a ceremony that requires rasping; therefore, they are also called *si'páame*.
6. Eventually one may run into a Tarahumara *curandera*. However, this will surely be a product of acculturation since among the Mexicans, healers are predominantly female.
7. Merrill, *Rarámuri Souls*, 119.
8. Paredes, West, and Snow, "Biosocial Adaptation and Correlates," 12.
9. Bye, "Ethnoecology of the Tarahumara," 192.
10. The sucking tubes of the *waníame* are made from *Phragmites communis* (Pennington, *The Tarahumar*, 179).
11. From the verb *wanimea*, meaning "to expel, or to take out" (Brambila, *Diccionario rarámuri-castellano*, 581).
12. Or as in the case of *ru'síwari*, the stone passed to another shaman. "The *waníame* takes it [the *sukí*] out and gives it to the *si'páame*, to whom it belongs" (Brambila, *Diccionario rarámuri-castellano*, 525).
13. Aguirre Beltrán (*Medicina y magia*, 50) contends that the shaman, the patient and the observers don't see in the extracted objects a material form: "Their mentality, trained on emotional grounds, helps them perceive in such objects a spiritual, super-

natural being from which stones, paper and worms are but disposable containers. The foreign body is just the vehicle used by sorcerers or gods to manifest their anger."
14. Plancarte, *El problema indígena tarahumara*, 72.
15. Merrill, *Rarámuri Souls*, 76.
16. Bye, "Ethnoecology of the Tarahumara," 208.
17. Wagner (*Plantas medicinales*, 84–85) chronicled the story of Marcial Vega, the original Indio de los Capomos, stating that he, resisting economic offers of considerable importance, tenaciously kept the secret of his remedy to cure rabies. More recently the secret has passed to his descendants.
18. Irigoyen-Rascón, *Cerocahui*, 87.

9. The Jíkuri Ceremonial Complex

1. Although in this area, the si'páame are called *peyoteros* and usually come from Tecorichi.
2. Irigoyen-Rascón, *Cerocahui*, 120–121; Burgess McGuire, *¿Podrías vivir?*, 12.
3. Lumholtz, *Unknown Mexico*, 1:366.
4. Prado, "Peyote," in Brambila, *Supersticiones y costumbres*, 14; Irigoyen-Rascón, *Cerocahui*, 32.
5. Bennett and Zingg, *The Tarahumara*, 291–292.
6. During one of the rituals, a terrible problem arose when *mensia*, the local *owirúame*, probably moved by professional jealousy, challenged *juaniserio*, the visiting *si'páame*. He invited juaniserio to fistfight or to compete in dancing with him. First, the community pretended to ignore such an outrageous attitude, but as he persisted, the siríame had to intervene and put the envious healer in jail.
7. Lumholtz (*Unknown Mexico*, 2:364) claims that this fire is so important that the ceremony itself was known also as "dancing around the fire." See also Deimel, "Die Peyoteheilung," 155.
8. Curiously, the block is called *sonorá* (lungs). One would expect that *surá* (the heart) would be emphasized since according to Rarámuri mystical anatomy this is the seat of the soul (*iwigá*). However since the term *iwigá* means also "to breathe," the lungs may also be thought of as the soul's residence. During one of the ceremonies, someone forgot to hang the sonorá and the ceremonies were ordered to be stopped until this, called a terrible omission, was corrected.
9. In the last ceremony the author participated in, a very dramatic event occurred. A young Tarahumara man, who was quite intoxicated with batari, was making fun of the ceremonies, particularly the movements that the participants have to do to go in and out of the sacred circle. So he jumped, once and again, from outside the circle to inside and vice versa, bragging loudly, "See, nothing happens to me!" The si'páame, who so far had tolerated such behavior, suddenly raised his head, staring at the man. At that precise moment, the young man fell to the floor—as if struck by lightning, suffered a convulsion, and stopped breathing. It took half an hour of vigorous CPR to bring an otherwise healthy

man back to life. Needless to say, one cannot imagine a more convincing demonstration for those present about the power of Jíkuri and the si'páame, and the consequences of defying the ritual prescriptions: the desgracia del Jíkuri.
10. Artaud, *México y viaje*, 317–318.
11. Each communicant drinks about three ounces of the infusion; the total amount of peyote used is of more or less 100 grams. Lumholtz (*Unknown Mexico*, 1:364) wrote that "at an ordinary gathering, a dozen or two of the plants suffice" for a ceremony. If this observation was accurate, there has been a dramatic decrease over the last one hundred years in the dosage of peyote used by the Tarahumaras. Deimel ("Narárachi," 3–4, and "Die Peyoteheilung," 155) states that jíkuri is taken in small doses while alcohol is ingested in huge ones so the latter attenuates the effect of the former.
12. Brambila, *Diccionario rarámuri-castellano*, 525.
13. Gaeta (in Campos and Gaeta, *El rutuburi*) observed a similar ritual in a tutuguri dance. Both *ba'wí ówima* and *wejazo* may be remnants of old Mesoamerican purification rituals.
14. Lumholtz, *Unknown Mexico*, 2:477.
15. Lumholtz (*Unknown Mexico*, 1:331) reported in 1902 that the shaman invokes the aid of all the animals, mentioning each by name, especially the deer and the rabbit, asking them to multiply so that the people may have plenty to eat.
16. Burgess McGuire, *¿Podrías vivir?*, 16.

10. Compendium of Tarahumara Herbal Remedies and Healing Practices

1. The terms listed under *aborí* represent different forms of the same word (√*aorí*) that varies according to the Tarahumara language alternation and rules for transitional sounds as proposed by Brambila (*Gramática*, 3–11). This diversity has no semantic importance in modern Tarahumara, but it might have had it in the past, as it still does in Northern Tepehuan (*gáyi*=juniper, *gágayi*=junipers). So forms like *kawarí* or *wakarí*, and especially *gayorí*, could represent former plural or frequentative forms (intensives) of the primitive word √*aorí*.
2. Synonyms: *Juniperus mexicana* Schltdl. et Cham. (*Juniperus deppeana* var. *deppeana* Steud., or *Juniperus tetragona* Schltdl.).
3. Pennington, *The Tarahumar*, 36–38; Bye, "Ethnoecology of the Tarahumara," 161.
4. Brambila, *Diccionario rarámuri-castellano*, 3; SEMARNAP, *Diagnóstico de productos*, 1–3; Díaz Infante, "Dialecto tarahumara de Chinatú," 100; Bye, "Ethnoecology of the Tarahumara," 161; Bennett and Zingg, *The Tarahumara*, 73.
5. Brambila, *Diccionario rarámuri-castellano*, 576.
6. From the Náhuatl *tlazcan* (cypress), transformed first into *tlázcate* and then into *tázcate* or *táscate* (Cabrera, *Diccionario*, 123). The term, however, could derive also from *tácatl*, which according to Molina, in 1571, means "plant or tree" (*Vocabulario*, 90).
7. Not to be confused with the *ahuehuete* or *sabino* (*Taxodium mucrognatum* Ten.) that

is a common tree in southern Mexico so remarkably represented by the Santa María del Tule tree in Oaxaca: the thickest tree in the world!
8. Hrdlička, *Physiological and Medical Observations*, 250.
9. Easily explained by the abundance of ornamental *thujas* in the Chihuahuan cities and also because the *arbor vitae*, or white cedar (*Thuja occidentalis* L.), superficially resembles a *tázcate*. By the same token, some Tarahumaras touring Chihuahua City refer to the ornamental Fletcher cypresses and *Cupressus funebris* Endl. with the names *aborí* or *waá*. Vogel (*American Indian*, 273–275) has exhaustively examined the use by North American Indian groups of *Thuja occidentalis* as a remedy, which has parallels with the Tarahumaras' use of *aborí*. Also it is important to remember that *Thuja occidentalis* is—like *Artemisia absinthium* L. (*Ro'sábari*)—an ingredient used to flavor absinthe and a source of α-thujone a substance that blocks the γ-aminobutyric acid type A (GABAa) receptor chloride channels in the brain and may cause convulsions. (Hold et al., "Alfa-thujone," in Harris, "Starry Night," 979)
10. Irigoyen-Rascón, *Cerocahui*, 92.
11. Salmón (*Sharing Breath*, 365) also attributes anthelmintic and antiaterogenic effects to this concoction.
12. Pennington, *The Tarahumar*, 179; Bye, "Ethnoecology of the Tarahumara," 161.
13. Hrdlička, *Physiological and Medical Observations*, 250.
14. Pennington, *The Tarahumar*, 171; Bennett and Zingg, *The Tarahumara*, 73, 337.
15. Burgess and Mares, *Ralámuli nu'tugala go'ame*, 470.
16. Bye, "Ethnoecology of the Tarahumara," 134–135.
17. Many of the belief systems associated in Mesoamerica with the *temazcalli*—Aztec sweat house—complex are shared by the humbler Tarahumara *wikubema*.
18. Lara in Brambila, *Supersticiones y costumbres*, 5–6.
19. Bennett and Zingg, *The Tarahumara*, 73–74.
20. Bennett and Zingg, *The Tarahumara*, 73.
21. Brambila, *Supersticiones y costumbres*, 14.
22. *Enciclopedia Monitor*, 5352.
23. Debove, Pouchet, and Sallard, *Aide-mémoire de thérapeutique*, 738; Arnozan (*Précis de thérapeutique*, 294) argued, however, that in addition to pelvic congestion, abortion could be caused by direct action of savine oil on the medullary vesico-uterine centers.
24. Lewis and Elvin-Lewis, *Medical Botany*, 300.
25. Urbina, Antonio de, "Relación de Serocahue (Cerocahui)," in Del Paso y Troncoso, *Relaciones del siglo XVIII*; Pennington, *The Tarahumar*, 144; "Relación de Norogachi," in Pennington, *The Tarahumar*, 144.
26. Bennett and Zingg, *The Tarahumara*, 74.
27. Pennington, *The Tarahumar*, 13.
28. Pennington, *The Tarahumar*, 116; Bye, "Ethnoecology of the Tarahumara," 176.
29. Bye, "Ethnoecology of the Tarahumara," 139.

30. Pennington lists this plant among those of the family Loganiacea. Bye classifies the genus *Mascagnia* in the family Malpighiaceae.
31. Bye, "Ethnoecology of the Tarahumara," 176.
32. Pennington, *The Tarahumar*, 187; Bye, "Ethnoecology of the Tarahumara," 176.
33. Bennett and Zingg, *The Tarahumara*, 140; Pennington, *The Tarahumar*, 184; Bye, "Ethnoecology of the Tarahumara," 296.
34. Bennett and Zingg, *The Tarahumara*, 140. Bye ("Ethnoecology," 179) reports that the bark is used boiled into a tea as a general remedy and Salmón (*Sharing Breath*, 374) considers it a blood purifier and aid for stomach problems. Vogel (*American Indian*, 318) and Lewis and Elvin-Lewis (*Medical Botany*, 376) report on the use of the root bark of this tree by the Menominees and the Meskwakis as a sacred medicine and panacea.
35. Standley, *Trees and Shrubs*, 531; Pennington, *The Tarahumar*, 184n113.
36. Pennington, *The Tarahumar*, 213.
37. Bye, "Ethnoecology of the Tarahumara," 207.
38. Robalo, *Diccionario de aztequismos*, 456.
39. *Enciclopedia de México* (3:62–63) lists another seven *Erythrinas* and at least eleven other unrelated plants that are also called *colorín* in different parts of Mexico.
40. Ratkay in 1683 (in González Rodríguez, *Tarahumara*, 113, and Reynolds, *Some Letters*) was possibly the first European to hear about the giant. Lumholtz (*Unknown Mexico*, 2:299) recorded an early version of the tale. Irigoyen-Rascón (*Cerocahui*, 113–115) published Brambila's recording of Erasmo Palma's version of the tale. Burgess McGuire ("Leyendas," 160) also recorded an interesting version of the tale.
41. Irigoyen-Rascón, *Cerocahui*, 116–118.
42. Bennett and Zingg, *The Tarahumara*, 170.
43. Zingg, *Report*, 1–95, in Bye, "Ethnoecology of the Tarahumara," 208.
44. Watson, "List of Plants Collected by Palmer," 425, in Pennington, *The Tarahumar*, 182.
45. Bye, "Ethnoecology of the Tarahumara," 174.
46. Bye, "Ethnoecology of the Tarahumara," 207–208; Lewis and Elvin-Lewis, *Medical Botany*, 43, 164.
47. Bennett and Zingg, *The Tarahumara*, 170; Pennington, *The Tarahumar*, 210; Bye, "Ethnoecology of the Tarahumara," 174.
48. Hargreaves et al., "Alkaloids," 570–572, in Bye, "Ethnoecology of the Tarahumara," 207–208; Lewis and Elvin-Lewis, *Medical Botany*, 43, 164.
49. Brambila, *Diccionario rarámuri-castellano*, 30; Burgess McGuire and Mares Trías, *Ralámuli nu'tugala go'ame*, 357.
50. The use of $a^*rí$ by the Tarahumaras has been reported since 1777. Fr. Falcón Mariano in his "Relación de Guaguachicque (Wawachiki)" (in Sheridan and Naylor, *Rarámuri*, 111) described a beverage, highly prized by the Tarahumaras, consisting of saltpeter, a *gomilla* or *hari*, and a touch of lye.

51. La Llave, "Materia Medica," 147–152; Lumholtz, *Unknown Mexico*, 1:229.
52. Urbina, Manuel, "*Aje, Axin o Ajin*," 363–365.
53. Quillet, *Diccionario enciclopédico*, 2:581.
54. Urbina, "*Aje, Axin o Ajin*," 363–365; *Enciclopedia de México*, 1:364.
55. Bennett and Zingg, *The Tarahumara*, 150; Bye, "Ethnoecology of the Tarahumara," 174, 281; Burgess McGuire and Mares Trías, *Ralámuli nu'tugala go'ame*, 357–358.
56. Irigoyen-Rascón, *Cerocahui*, 92.
57. Burgess McGuire and Mares Trías, *Ralámuli nu'tugala go'ame*, 357, 498; Brambila, *Diccionario rarámuri-castellano*, 30.
58. Lumholtz, *Unknown Mexico*, 1:229.
59. Bennett and Zingg, *The Tarahumara*, 131.
60. McGregor and Meza, "Los insectos," 29–31.
61. Sahagún, *Historia general*, 10:155; Hernández, *Historia natural*, in Ortiz de Montellano, *Aztec Medicine*, 154–155; Pennington, *The Tarahumar*, 183.
62. Brambila, *Diccionario rarámuri-castellano*, 31.
63. Besides *Cucurbita foetidissima*, other plants, such as *Cucurbita radicans* Neud. and *Cucurbita digitata* Gray, are known in Mexico as *calabacilla* or *calabacita*.
64. *Enciclopedia de México*, 2:451; Bye, "Ethnoecology of the Tarahumara," 243; Watson, "List of Plants Collected by Palmer," in Bye, "Ethnoecology of the Tarahumara," 243; Brambila, *Diccionario rarámuri-castellano*, 31.
65. Pennington, *The Tarahumar*, 212; Brambila, *Diccionario rarámuri-castellano*, 31; Bye, "Ethnoecology of the Tarahumara," 243.
66. Brambila, *Diccionario rarámuri-castellano*, 31.
67. Pennington, *The Tarahumar*, 169; Irigoyen-Rascón and Jesús Manuel Palma, "Rarajípari," 54.
68. Wall et al., "Steroidal Sapogenins: VII," 1–7, and "Steroidal Sapogenins: XII, 503–505," in Bye, "Ethnoecology of the Tarahumara," 244.
69. Bye, "Ethnoecology of the Tarahumara," 145, 170, 244.
70. Cabrera, *Plantas curativas*, 300.
71. Bennett and Zingg, *The Tarahumara*, 171.
72. Pennington, *The Tarahumar*, 44.
73. Wall et al., "Steroidal Sapogenins: VII," 1–7, and "Steroidal Sapogenins: XII, 503–505," in Bye, "Ethnoecology of the Tarahumara," 244.
74. Cardenal, *Remedios y prácticas*, 61.
75. Bye, "Ethnoecology of the Tarahumara," 179.
76. Pennington, *The Tarahumar*, 181.
77. Salmón, *Sharing Breath with Our Relatives*, 366.
78. Brambila, *Diccionario rarámuri-castellano*, 46; Bye, *Plantas psicotrópicas*, 53, "Ethnoecology of the Tarahumara," 162–204; Bennett and Zingg, *The Tarahumara*, 295.
79. Bennett and Zingg, *The Tarahumara*, 136; Deimel, "Narárachi, Zwischen Traditionalismus und Integration," 5.

80. Bye, *Plantas psicotrópicas*, 62; Bye, "Ethnoecology of the Tarahumara," 204; Salmón, *Sharing Breath with Our Relatives*, 366.
81. Bye, "Ethnoecology of the Tarahumara," 205.
82. Ibid.
83. Deimel, "Narárachi, Zwischen Traditionalismus und Integration," 3.
84. Deimel, "Pflanzen zwischen den Kulturen," 61; Bye, "Ethnoecology of the Tarahumara," 204.
85. Burgess McGuire, *Leyendas tarahumaras*, 105.
86. Ibid.
87. Salmón, *Sharing Breath with Our Relatives*, 366.
88. Bennett and Zingg, *The Tarahumara*, 295.
89. Pennington, *The Tarahumar*, 128.
90. Tarahumaras also call *baká* (*Sorghum vulgare* Pers.) "sorghum" (Pennington, *The Tarahumar*, 43).
91. Brambila, *Diccionario rarámuri-castellano*, 49.
92. Bye, "Ethnoecology of the Tarahumara," 163.
93. Bennett and Zingg, *The Tarahumara*, 115.
94. Pennington, *The Tarahumar*, 95, 110.
95. Brambila, *Diccionario rarámuri-castellano*, 49.
96. Pennington, *The Tarahumar*, 128.
97. Brambila, *Diccionario rarámuri-castellano*, 39.
98. Bye, "Ethnoecology of the Tarahumara," 163.
99. Bennett and Zingg, *The Tarahumara*, 115, 145; Lévi, "La flecha," 139n6.
100. Cardenal, *Remedios y prácticas*, 104.
101. Pennington, *The Tarahumar*, 179.
102. Salmón, *Sharing Breath with Our Relatives*, 371.
103. Cardenal, *Remedios y prácticas*, 47.
104. Bye, "Ethnoecology of the Tarahumara," 163; Salmón, *Sharing Breath with Our Relatives*, 371.
105. Bye, "Ethnoecology of the Tarahumara," 61, 160, 222.
106. *Cola de caballo* (horsetail) is the Spanish name of *E. arvense* L., a plant with extensive medicinal uses—particularly in nephrology—and as a hemostatic and disinfectant. See Wagner, *Plantas medicinales*, 21.
107. Pennington, *The Tarahumar*, 178.
108. Bye, "Ethnoecology of the Tarahumara," 221–222.
109. Der Marderosian and Liberti, *Natural Product Medicine*, 312, in Salmón ("Cures of the Copper Canyon," 48). Salmón adds that Duke and Ayensu (*Medicinal Plants of China*, 295) claim that *E. hyemale* contains polyphenolic flavonoids with bactericidal activity.
110. Brambila, *Diccionario rarámuri-castellano*, 51.
111. Synonym: *Eriogonum jamesii* Bentham var. *undulatum*.

112. Pennington, *The Tarahumar*, 131, 180.
113. Bye, "Ethnoecology of the Tarahumara," 177; Pennington, *The Tarahumar*, 180 Brambila, *Diccionario rarámuri-castellano*, 54.
114. Watson, "List of Plants Collected by Palmer," 438.
115. Deimel, "Pflanzen zwischen den Kulturen," 61.
116. Brambila, *Diccionario rarámuri-castellano*, 54.
117. According to Salmón (*Sharing Breath*, 377), Rarámuri also call *Rumex crispus* L. "*basigó*." He states that *basigó* or *Rumex* is used boiled into a tea for diarrhea, and as a wash for skin irritations. The species contains oxalates and may be therefore toxic for animals and men (Blanco Madrid, Enríquez Anchondo, and Siqueiros Delgado, *Manual de plantas tóxicas*, 124).
118. Brambila, *Diccionario rarámuri-castellano*, 56.
119. Bye, "Ethnoecology of the Tarahumara," 242.
120. It is important not to confuse this plant with the Old World anises, *Pimpinella anisum* L. (family Umbelliferae) or *Illicum verum* Hooker (family Magnoliaceae). In Chihuahua City, herbalists call the Old World anise *anís estrella* while American *Tagetes* sp. (family Asteraceae) is referred to as *yerbanís*.
121. Bye, "Ethnoecology of the Tarahumara," 242.
122. Bennett and Zingg, *The Tarahumara*, 143.
123. From the Náhuatl: *iyautli*="offering flower," *iyaua*=to offer (Cabrera, *Diccionario*, 156).
124. Alfonso Reyes Ochoa, a refined poet and one of the "Seven Mexican Wise Men," considered repositories of great wisdom, wrote: "Today they bring only herbs in their bundle / healthy herbs that they trade for pennies / yerbanís, limoncillos, simonillo, / to alleviate troubled entrails, / as well as orejuela de ratón / for the ailment people call "bilis"; / yerba del venado, chuchupaste / and yerba del indio that restore the blood; / pasto de ocotillo for bruises; / contrayerba for swamp fever, / yerba de la víbora that cures colds; / necklaces of seeds of ojo de venado; / so effective for bewitchment! / and sangre de grado that tightens the gums / and grips, by their roots, loose teeth." ("Yerbas del Tarahumara" in *Obras completas* 10:121)
125. Ortiz de Montellano, *Aztec Medicine*, 194.
126. Urbina, Antonio de "Relación de Serocahue."
127. Bye, "Ethnoecology of the Tarahumara," 242.
128. Lewis and Elvin-Lewis, *Medical Botany*, 84.
129. Martínez, Maximo, *Las plantas medicinales*, in Bye, "Ethnoecology of the Tarahumara," 143.
130. Ortiz de Montellano, *Aztec Medicine*, 197, from Hegnauer, *Chemotaxonomie der Pflanzen*, 3:526, 528, and Steinegger and Hänsel, *Lehrbuch der Allgemeine Pharmakognosie*, 169, 174.
131. Pennington, *The Tarahumar*, 44, 213.
132. Lumholtz, *Unknown Mexico*, 1:251.

133. Cabrera, *Plantas curativas*, 230. Synonym: *Alternanthera echinata* Smith.
134. Brambila, *Diccionario rarámuri-castellano*, 72; Bye, "Ethnoecology of the Tarahumara," 183.
135. Deimel, "Pflanzen zwischen den Kulturen," 61.
136. *Tianguis-pepetla*, from the Náhuatl words *tianquiztli*, meaning "market, or plaza," and *pepetla*, which is a collective and plural form of *petatl* (carpet, rug). The name reflects the fact that the plant spontaneously grows on the floor of the market plazas. According to Cabrera (*Diccionario*, 137), the long form *tianguis-pepetla* is applied to *Alternanthera repens*, but Robalo (*Diccionario*, 260) contends that this name is to be given only to *Alternanthera achyrantha*. Both authors agree that the shorter *tianguis* should be applied only to *Plumbago pulchella*, a plant also known as *hierba del alacrán* (the scorpion's herb).
137. Oberti, "*Tianguispepetla*," 1.
138. Interestingly, the term *be'techókuri* can also be applied, in a sarcastic sense, to a lazy person who stays always at home, refusing to go out and work.
139. Bye, "Ethnoecology of the Tarahumara," 183; Pennington, *The Tarahumar*, 181; Oberti, *Tianguispepetla*, 1–2.
140. Hernández, *Historia natural*, 144–145.
141. Oberti, "*Tianguispepetla*," 1–2.
142. "Relacion de Guazapares," in Del Paso y Troncoso, *Relaciones del siglo XVIII*.
143. Pennington, *The Tarahumar*, 181; Bye, "Ethnoecology of the Tarahumara," 163.
144. Cardenal, *Remedios y prácticas*, 128.
145. Lasso de la Vega in Cabrera, *Plantas curativas*, 230, and Oberti, "*Tianguispepetla*," 2.
146. Oberti, "*Tianguispepetla*," 3.
147. Pennington, *The Tarahumar*, 193.
148. Analysis of the plant disclosed the presence of several alkaloids, among them boldine, isocoridine, sinoacutine, and pronuciferine (Torres Gaona, "Recientes investigaciones," 349).
149. Bye, "Ethnoecology of the Tarahumara," 165, 275.
150. *Enciclopedia Monitor* (928) and others, however, cluster the Forget-me-nots in the genus *Myosotis* (i.e., *M. scorpioides* L.=true forget-me-not; *M. stricta* Link ex Roemer & J. A. Schultes=strict forget-me-not).
151. Bye, "Ethnoecology of the Tarahumara," 165.
152. Quillet, *Diccionario enciclopédico*, 2:210.
153. *Sitagapi* is properly *Haematoxilon brasiletto*, but *brasilillo* is also called *sitagapi*.
154. Pennington, *The Tarahumar*, 183.
155. Bye, "Ethnoecology of the Tarahumara," 183.
156. Bennett and Zingg, *The Tarahumara*, 170; Pennington, *The Tarahumar*, 188; Bye, "Ethnoecology of the Tarahumara," 183.
157. Orellana, *Indian Medicine*, 229.
158. Lewis and Elvin-Lewis, *Medical Botany*, 80.

159. Lewis and Elvin-Lewis, *Medical Botany*, 363.
160. Bye, "Ethnoecology of the Tarahumara, 140; Orellana, *Indian Medicine*, 229.
161. Robalo (*Diccionario*, 348) derives the voice *calahuala* from the Náhuatl *calan=tlalan* ("under earth"). The term *calaguala* is used even in Panama (Pérez de Zárate, *Medicina folklórica*, 90) and in Chile (Gracia Alcover, *Vitaminas y medicina*, 263) where it is applied to *Polypodium trilobum*. *Enciclopedia de México* (454) and Wagner (*Plantas medicinales*, 92) identify the *calahuala* from central Mexico as *Polypodium aureum* L. (*Phlebodium aureum* [L.] Smith) reputed as pectoral and diaphoretic.
162. Bennett and Zingg, *The Tarahumara*, 177.
163. Ibid.
164. Bye, "Ethnoecology of the Tarahumara," 160.
165. Synonyms: *Laurus cinamomum* L. or *Cannella cinnamomum*.
166. Robalo (*Diccionario*, 44) claims that *cascalote* is a corruption of the Náhuatl *nacazcólotl* ("twisted ear"). Such an analogy would apply to the fruits of the cascalote described by *Enciclopedia de Mexico* (2:812, 5:1093) as a leguminous or caesalpinaceous tree whose pods or fruits are twisted like ears. These fruits are very rich in tannin (30 percent of tannic acid or 15 kilograms per tree); Bye, "Ethnoecology of the Tarahumara," 140.
167. Synonym: *B. saliciflora* (Ruíz Pavón) Pers.
168. *Baccharis pteronoides* DC. is commonly called in the literature *hierba del pasmo* (*B. neglecta* Britt. *jarilla de río*). Also, *B. bigelowii* A. Gray is said to have similar properties. *B. pteronoides* is known to be toxic to sheep and cattle (Blanco Madrid, Enríquez Anchondo, and Siqueiros Delgado, *Manual de plantas tóxicas*, 86).
169. Synonym: *Ambrosia cordifolia* Payne.
170. Brambila, *Diccionario rarámuri-castellano*, 102.
171. Bye, "Ethnoecology of the Tarahumara," 166.
172. Brambila, *Diccionario rarámuri-castellano*, 102.
173. Pennington, *The Tarahumar*, 191, 107.
174. Pennington, "Tarahumar Fish Stupefaction Plants," 98, 100, and *The Tarahumar*, 107, 109.
175. Pennington, *The Tarahumar*, 191; Brambila, *Diccionario rarámuri-castellano*, 102.
176. Pennington, *The Tarahumar*, 182.
177. Bennett and Zingg, *The Tarahumara*, 176.
178. Bye, "Ethnoecology of the Tarahumara," 166.
179. Lewis and Elvin-Lewis, *Medical Botany*, 56.
180. Brambila, *Diccionario rarámuri-castellano*, 109.
181. Bye, "Ethnoecology of the Tarahumara," 201.
182. Ibid.
183. Lionnet, *Los elementos*, 101.
184. Burgess McGuire and Mares Trías, *Ralámuli nu'tugala go'ame*, 353.
185. Brambila, *Diccionario rarámuri-castellano*, 358.

186. Pennington, *The Tarahumar*, 118; Bye, "Ethnoecology of the Tarahumara," 202.
187. Brambila, *Diccionario rarámuri-castellano*, 109.
188. Bennett and Zingg, *The Tarahumara*, 149–161.
189. Dawson, *How to Know the Cacti*, 91.
190. Pennington, *The Tarahumar*, 76.
191. Burgess McGuire and Mares Trías, *Ralámuli nu'tugala go'ame*, 354.
192. Pennington, *The Tarahumar*, 118, 130, 155. Burgess's *napisóla* is evidently the same *na'písora* recorded by Pennington and Brambila. My informants always call the small, ash-colored pitahaya (*Echinocactus*) "*na'písora*," while they call the larger, bearded Spanish pitahaya (*Cephalocereus*) "*barbona*." The photograph of a *napisóla* presented by Burgess McGuire and Mares Trías (*Ralámuli nu'tugala go'ame*, 354) is possibly *Echinocactus scheeri* or *E. enneacanthus* Engelmann.
193. Pennington, *The Tarahumar*, 118.
194. Lumholtz, *Unknown Mexico*, 1:188.
195. Pennington, *The Tarahumar*, 76.
196. Burgess McGuire and Mares Trías, *Ralámuli nu'tugala go'ame*, 348.
197. Irigoyen-Rascón, *Cerocahui*, 91.
198. Bye, "Ethnoecology of the Tarahumara," 166.
199. Cardenal, *Remedios y prácticas*, 108.
200. Pennington, *The Tarahumar*, 167.
201. Castetter and Bell, *Aboriginal Utilization*, 39, in Pennington, *The Tarahumar*, 167.
202. Bye, "Ethnoecology of the Tarahumara," 202; Späth, "Über das Carnegin," 1021–1024; Agurell, "Cactacea Alkaloids, 1," 213–214, in Anderson, *Peyote*, 122.
203. Bruhn and Lindgren, "Cactacea Alkaloids" 175–177, in Anderson, *Peyote*, 122.
204. Anderson, *Peyote*, 109.
205. Agurell, "Cactacea Alkaloids, 1," 213–214.
206. From the Náhuatl *chichic*=bitter and *quílitl*=pot green, or vegetable (Cabrera, *Diccionario*, 68). The Náhuatl-speaking peoples of central Mexico used the term *quilitl* in opposition to *tzácatl*, food for animals (Spanish *zacate*), and *xíhuitl*=poisonous or not edible herb (Spanish *jehuite*) Although the Tarahumara word *kiribá* (or *kiribáka*) or *giribá* designates *quelites*, vegetables or edible greens (Pennington, *The Tarahumar*, 124; Brambila, *Diccionario rarámuri-castellano*, 178, and Bye, "Ethnoecology of the Tarahumara," 110), the Rarámuri usually prefer to call each variety or species with its particular name; i.e., *wasorí*=quelite de aguas (*Amaranthus* sp.).
207. Bennett and Zingg, *The Tarahumara*, 168–169; Bye, "Ethnoecology of the Tarahumara," 190.
208. Pennington, *The Tarahumar*, 189; Bennett and Zingg, *The Tarahumara*, 168–169.
209. Pennington, *The Tarahumar*, 193.
210. Cardenal, *Remedios y prácticas*, 60.
211. Bennett and Zingg, *The Tarahumara*, 169.
212. Pennington, *The Tarahumar*, 189.
213. Pennington, *The Tarahumar*, 190.

214. Bennett and Zingg, *The Tarahumara*, 139.
215. Bye, "Ethnoecology of the Tarahumara," 181.
216. Tyler, Brady, and Robbers, *Pharmacognosy*, 457, in Orellana, *Indian Medicine*, 242.
217. Blanco Madrid, Enríquez Anchondo, and Siqueiros Delgado, *Manual de plantas*, 190.
218. Bye, "Ethnoecology of the Tarahumara," 131.
219. The term *chicura* has possibly derived from *chicoria* or *achicoria* (chicory) that is applied to the plants of the genus *Chicorum* of the family Compositae. *Chicorium intybus* L. (*achicoria amarga*) is a well-known Old World wild chicory. Achicorias are widely used in Europe as edible greens. An infusion from chicory is sometimes used in place of coffee and traditionally has been considered as an appetite stimulant.
220. Bye, "Ethnoecology of the Tarahumara," 166.
221. Lewis and Elvin-Lewis, *Medical Botany*, 84, 85. Caffeolyquinic acids, cinnamic acid, coumarin, germacranolides, glycosides, kaurenic acids, sesquiterpene lactones, stigmasterol, resins, and tannins have been listed among the chemical constituents of *Ambrosia* sp. See also *Encyclopaedia Britannica* (*Micropedia*), 1984, 7:387.
222. Bye, "Ethnoecology of the Tarahumara," 167. Bye ("Ethnoecology," 167) lists another two *Ambrosias* used as remedies by Tarahumaras: *A. acanthicarpa* Hook. (*u'rí* or *u'ríta* [TAR. *u'rí*=grape]) — its leaves are applied in a poultice for sores and wounds — and *A. psilostachya* DC. (*chipúnua*) — its leaves boiled into a tea for gastrointestinal ailments.
223. Pennington, *The Tarahumar*, 190.
224. Pennington, *The Tarahumar*, 95.
225. Pennington (*The Tarahumar*, 141). The presence of a cyanogenic glycoside is also reported by Lewis and Elvin-Lewis (*Medical Botany*, 18) who remind us that many plants of the family Rosaceae, even those as common as apples (*Malus*) and peaches (*Prunus*) also contain cyanogenic glycosides.
226. Pennington, *The Tarahumar*, 182.
227. Salmón, *Sharing Breath with Our Relatives*, 378.
228. Brambila, *Diccionario rarámuri-castellano*, 116.
229. Ibid.
230. Deimel ("Pflanzen zwischen den Kulturen," 62) identified another *rawichí* — *Macromeria viridiflora* DC. — and claims it is used as a remedy for burns.
231. Bennett and Zingg, *The Tarahumara*, 142.
232. Bye, "Ethnoecology of the Tarahumara," 167.
233. Pennington, *The Tarahumar*, 188.
234. Brambila, *Diccionario rarámuri-castellano*, 106, 116.
235. Bennett and Zingg, *The Tarahumara*, 161.
236. Pennington, *The Tarahumar*, 128, 192.
237. Pennington, *The Tarahumar*, 77; Brambila, *Diccionario rarámuri-castellano*, 116.
238. Bennett and Zingg, *The Tarahumara*, 142; Brambila, *Diccionario rarámuri-castellano*, 458.
239. Bye, "Ethnoecology of the Tarahumara," 169.

240. Bennett and Zingg, *The Tarahumara*, 161.
241. Cardenal, *Remedios y prácticas*, 103.
242. Pennington, *The Tarahumar*, 192.
243. Bennett and Zingg, *The Tarahumara*, 142.
244. Bye, "Ethnoecology of the Tarahumara," 167.
245. Pennington, *The Tarahumar*, 188.
246. Brambila, *Diccionario rarámuri-castellano*, 133.
247. Pennington, *The Tarahumar*, 189.
248. Pennington (*The Tarahumar*, 189) and Bye ("Ethnoecology," 172) quote Bennett and Zingg (*The Tarahumara*, 144) about the use of another wild mint, *Agastache pallida* (Lindl.) Cory. (family Labiatae), whose pungent leaves are placed in the nose to relieve congestion. They also claim that a tea made from it is drunk for colds and coughs. Lumholtz (*Unknown Mexico*, 2:314) wrote that headache was cured by a green herb called *pachoco*, "which they smell until they begin to sneeze." The use of plants that promote sneezing for the treatment of headache is a Mesoamerican trait. Ortiz de Montellano (*Aztec Medicine*, 150) explains that violent sneezing was intended to make the nose bleed. Whenever this method failed, Aztecs would proceed to pierce the head with a piece of flint. That author sees an Aztec etiological hypothesis confirmed by the treatment: headaches are caused by an excess of blood in the head. Other Indian groups also procured sneezing remedies, for example see for the Comanches' use of *Helenium microcephalum* (Vogel, *American Indian*, 236).
249. Brambila, *Diccionario rarámuri-castellano*, 133–134.
250. Cardenal, *Remedios y practices*, 64.
251. Bye, "Ethnoecology of the Tarahumara," 172.
252. Pennington, *The Tarahumar*, 167.
253. Brambila, *Diccionario rarámuri-castellano*, 237.
254. Burgess McGuire and Mares Trías, *Ralámuli nu'tugala go'ame*, 337–340.
255. Bye, "Ethnoecology of the Tarahumara," 249.
256. Brambila, *Diccionario rarámuri-castellano*, 134.
257. Bye, "Ethnoecology of the Tarahumara," 250.
258. Cardenal, *Remedios y prácticas*, 89.
259. Pennington, *The Tarahumar*, 127.
260. Brambila, *Diccionario rarámuri-castellano*, 134.
261. Bye, "Ethnoecology of the Tarahumara," 274; Burgess McGuire and Mares Trías, *Ralámuli nu'tugala go'ame*, 501.
262. Pennington, *The Tarahumar*, 127, 188.
263. Brambila, *Diccionario rarámuri-castellano*, 134.
264. Bye, "Ethnoecology of the Tarahumara," 249.
265. SEMARNAP, *Diagnóstico de productos*, 4.
266. The term *chúchaka* is also employed by the Tarahumaras to name the wild raspberry or blackberry, in Spanish *zarzamora* (*Rubus idaeus* L.var. *strigosus* Michx.)

(Pennington, *The Tarahumar*, 116). The author of the Franciscan "Relación de Guazapares" (1777) claims that the Indians called peyote *echuchaka*. This plant may be also a *chuchá*, reported by Falcón Mariano in 1777 (Sheridan and Naylor, *Rarámuri*, 112) as a poisonous plant, from which Tarahumaras made a balm to coat their arrows. The same authors state that Pennington speculates that the colonial *chuchá* may be *Ligusticum porteri* (Wasía). Finally, *chúchaka* might also be an old reduplicated form of *chu'ká*, meaning that these are one and the same plant.

267. Brambila, *Diccionario rarámuri-castellano*, 138.
268. Pennington, *The Tarahumar*, 193; Brambila, *Diccionario rarámuri-castellano*, 138.
269. Brambila, *Diccionario rarámuri-castellano*, 117.
270. Bennett and Zingg, *The Tarahumara*, 142.
271. Bye, "Ethnoecology of the Tarahumara," 241.
272. Salmón, "Cures of the Copper Canyon," 50.
273. These names are given also to *Chenopodium album* Moq., and also to *Ch. mexicanum* Moq., which is also known as *remébari* or *chináka* and is an edible green. *Chual* (Náhuatl *tzohualli*) is the name of several *bledos* or chenopods from which the Aztecs made a sacrificial bread (Cabrera, *Diccionario*, 78). Fr. Falcón Mariano wrote in 1777 that the Tarahumaras liked to eat the leaves of *bledos*, raw or cooked, and that a beverage was made from them. He noted that "drinking the leaves alone is not good because those who do so often swell up and get pustules on their faces and legs" (in Sheridan and Naylor, *Rarámuri*, 113, and Pennington, *The Tarahumar*, 116).
274. Bye, "Ethnoecology of the Tarahumara," 241.
275. Ibid.
276. An antibiotic similar to the senecionine or senecionine-n-oxide discussed by Lewis and Elvin-Lewis (*Medical Botany*, 135) for *Senecio triangularis* (groundsel).
277. Bennett and Zingg, *The Tarahumara*, 142.
278. Bye, "Ethnoecology of the Tarahumara," 241.
279. Bye, "Ethnoecology of the Tarahumara," 168, 241.
280. Deimel, "Pflanzen zwischen den Kulturen," 61.
281. Almada, *Geografía del estado*, 37.
282. Vogel, *American Indian Medicine*, 397.
283. Ibid.
284. Lewis and Elvin-Lewis, *Medical Botany*, 258.
285. *Enciclopedia de las hierbas*, 13.
286. Pennington, *The Tarahumar*, 116.
287. The plants of the genus *Smilax* are known in many Latin American countries as *zarzaparrilla* or *sarsaparilla*.
288. *Enciclopedia de México*, 2:1112.
289. Cabrera, *Diccionario de aztequismos*, 49.
290. Bennett and Zingg, *The Tarahumara*, 174; Bye, "Ethnoecology of the Tarahumara," 177.

291. Bennett and Zingg, *The Tarahumara*, 174.
292. Bye, "Ethnoecology of the Tarahumara," 177.
293. From the Náhuatl *cocóltic*=twisted, *mécatl*=rope, and *xíhuitl*=herb "twisted vine" (Cabrera, *Diccionario*, 49, and Ortiz de Montellano, *Aztec Medicine*, 181).
294. Bennett and Zingg, *The Tarahumara*, 174, 263; Bye, "Ethnoecology of the Tarahumara," 177.
295. Many of the references to *contrahierba* in the literature do not refer to the plants known with this name in Tarahumara Country, but to *Dorstenia contrajerva* W. However Bye ("Ethnoecology," 176) reports that the Tarahumaras use the roots of *Dorstenia drakeana* L. (Tarahunara *o'chala*) boiled into a tea in the treatment of fever.
296. Cabrera, *Plantas curativas*, 55.
297. Wagner, *Plantas medicinales*, 96.
298. "Relacion de Batopilillas, 1777" in Franciscan *Relaciones*, in Pennington, *The Tarahumar*, 194.
299. Bye, "Ethnoecology of the Tarahumara," 175.
300. Bennett and Zingg, *The Tarahumara*, 169.
301. Bye, "Ethnoecology of the Tarahumara," 252.
302. Pennington, *The Tarahumar*, 150.
303. Pennington, *The Tarahumar*, 160; Brambila, *Diccionario rarámuri-castellano*, 36.
304. Pennington, *The Tarahumar*, 151.
305. Bennett and Zingg, *The Tarahumara*, 167; Bye, "Ethnoecology of the Tarahumara," 170.
306. From Náhuatl *copalli*=incense and *chichic*=bitter (Robalo, *Diccionario*, 364) or *copalli* and *tzin*=diminutive (Cabrera, *Diccionario*, 51).
307. Robalo, *Diccionario*, 364; Bennett and Zingg, *The Tarahumara*, 161.
308. Bennett and Zingg, *The Tarahumara*, 150, 151, 161, 169, 191; Pennington, *The Tarahumar*, 1, 160, 190; *Enciclopedia de México*, 3:268.
309. Bye, "Ethnoecology of the Tarahumara," 252.
310. Pennington, *The Tarahumar*, 150, 190.
311. Bye, "Ethnoecology of the Tarahumara," 170.
312. Cabrera, *Plantas curativas*, 57.
313. Pennington, *The Tarahumar*, 160.
314. Bennett and Zingg, *The Tarahumara*, 161.
315. Needless to say, commercial quinine and quinidine are obtained from *Cinchona* sp.
316. Watson, "List of Plants Collected by Palmer," in Gray, "Contributions," 379, in Pennington, *The Tarahumar*, 190.
317. "Relación de Guazapares" and "Relación de Chínipas," 1777, in Del Paso y Troncoso, *Relaciones del siglo XVIII*; also in Pennington, The *Tarahumar*, 190.
318. Hrdlička, *Physiological and Medical Observations*, 250.
319. Bennett and Zingg, *The Tarahumara*, 169.
320. Bye, "Ethnoecology of the Tarahumara," 252.

321. Pennington, *The Tarahumar*, 191.
322. Bennett and Zingg, *The Tarahumara*, 168; Pennington, *The Tarahumar*, 184.
323. Bye, "Ethnoecology of the Tarahumara," 170.
324. Bennett and Zingg, *The Tarahumara*, 168; Pennington, *The Tarahumar*, 184.
325. Turner, Nancy, and Szczawinski, *Common Poisonous Plants*, 227, in Salmón, "Cures of the Copper Canyon," 48.
326. Bye, "Ethnoecology of the Tarahumara," 140.
327. Willaman and Schubert, *Alkaloid-bearing Plants*, in Bye, "Ethnoecology of the Tarahumara," 146.
328. Bye, "Ethnoecology of the Tarahumara," 235; Pennington, *The Tarahumar*, 192.
329. Bye, "Ethnoecology of the Tarahumara," 23.
330. Deimel, "Pflanzen zwischen den Kulturen," 61.
331. Bye, "Ethnoecology of the Tarahumara," 123.
332. *Epasote* or *epazote* comes from the Náhuat *épatl*=skunk and *zotl* or *tzotl*=debris, or filth, meaning then "skunk's feces" (Cabrera, *Diccionario*, 79).
333. Burgess McGuire and Mares Trías, *Ralámuli nu'tugala go'ame*, 171.
334. Bye, "Ethnoecology of the Tarahumara," 166.
335. Brambila, *Diccionario rarámuri-castellano*, 424.
336. Bye, "Ethnoecology of the Tarahumara," 123.
337. Watson, "List of Plants Collected by Palmer," 437; Pennington, *The Tarahumar*, 180.
338. Bye, "Ethnoecology of the Tarahumara," 141.
339. Bye, "Ethnoecology of the Tarahumara," 166.
340. Preciado et al., *Informe Kwechi*, appendix 1.
341. Bye, "Ethnoecology of the Tarahumara," 140.
342. Deimel, "Pflanzen zwischen den Kulturen," 62.
343. Bye, "Ethnoecology of the Tarahumara," 141.
344. Velázquez Moctezuma, "Epazote," 6.
345. Bye, "Ethnoecology of the Tarahumara," 141.
346. Lewis and Elvin-Lewis, *Medical Botany*, 291–292.
347. Velásquez Moctezuma, "Epazote," 6.
348. Lewis and Elvin-Lewis, *Medical Botany*, 291.
349. *Synonim Opuntia tomentosa* Salm-Dyck var. hernandezii (DC.) Bravo.
350. Bennett and Zingg, *The Tarahumara*, 165.
351. Pennington, *The Tarahumar*, 118.
352. Burgess McGuire and Mares Trías, *Ralámuli nu'tugala go'ame*, 363–368.
353. Brambila, *Diccionario rarámuri-castellano*, 426.
354. Bye, "Ethnoecology of the Tarahumara," 275.
355. Bye, quoted in Burgess McGuire and Mares Trías, *Ralámuli nu'tugala go'ame*, 498.
356. Burgess McGuire and Mares Trías, *Ralámuli nu'tugala go'ame*, 363–368.
357. Bennett and Zingg, *The Tarahumara*, 166.
358. Ibid.

359. Bye, "Ethnoecology of the Tarahumara," 165–166. The Badianus Manuscript (1554) describes similar uses of *nopal* by the Aztecs (Vogel, *American Indian Medicine*, 204). Cabrera (*Plantas curativas*, 163) reports that in other parts of Mexico the stems, sliced longitudinally, are heated and applied on an abscess to favor its maturation.
360. Bye, "Ethnoecology of the Tarahumara," 166.
361. Cabrera, *Plantas curativas*, 163.
362. Salmón, *Sharing Breath with Our Relatives*, 364.
363. Lewis and Elvin-Lewis (*Medical Botany*, 346) report that the bark of *Ipomoea arborescens* (Humb. & Bonpl. ex Willd.) G. Don., an *escorcionera* from Sinaloa, is used as an antidote for rattlesnake bites. They discuss also a more illustrious *Ipomoea* known as "morning glory," or *ololiuhqui*. The seeds of *Ipomoea violacea* L. are ritually ingested by individuals of several tribes of central Mexico pursuing the hallucinogenic effect of an alkaloid contained in them. Zingg described a morning glory in Tarahumara Country — *Ipomoea purpurea* (L.) Roth. — but he did not mention any ceremonial or medicinal uses of this species, stating only that it is eaten as a pot green. Salmón (*Sharing Breath*, 364) claims that the roots of *I. purpurea* "are considered very powerful by the Rarámuri; if mishandled, they can make a person go crazy" (Bennett and Zingg, *The Tarahumara*, 140).
364. Spanish *escorzonera*, a plant "useful as a medicine and as a foodstuff," is often identified as *Scorzonera vulgaris* (Quillet, *Diccionario enciclopédico*, 8:512) but also as *Scorzonera hispanica* L., while *Picridium vulgare* Desf. is known as "French scorzonera." Whether the modern term *escorcionera* corresponds to the archaic *escuerconcera* (as it appears in the title of a classic work by the doctor Nicolás Monardes in 1545) or not is a dilemma that is very difficult to settle.
365. Pennington, *The Tarahumar*, 211.
366. Bye, "Ethnoecology of the Tarahumara," 167, 239.
367. Irigoyen-Rascón, *Cerocahui*, 90; Pompa, *Medicamentos indígenas*, 106; Cardenal, *Remedios y prácticas*, 69.
368. Bye, "Ethnoecology of the Tarahumara," 167, 239.
369. Bye, "Ethnoecology of the Tarahumara," 177.
370. Pennington, *The Tarahumar*, 187.
371. Bye, "Ethnoecology of the Tarahumara," 177.
372. Lewis and Elvin-Lewis, *Medical Botany*, 84, 300.
373. Cabrera, *Plantas curativas*, 91.
374. Kock-Weser, J., "Venenos más comunes," in Wintrobe et al., *Harrison medicina interna*, 723.
375. Synonym: *Covillea tridentata* (Moc. & Sessé ex DC.) Vail.
376. Throughout northern Mexico, *gobernadora* is used in baths, against rheumatic conditions, and, boiled into a decoction, as a diuretic (*Enciclopedia de México*, 5:792). Kay ("Health and Illness") reports on its use by Mexican Americans in a barrio of Tucson, Arizona, to treat kidney ailments, colic pains, and *pasmo*. Recently, in Chihuahua and

other parts of Mexico, claims about the efficacy of a *Larrea* sp. infusion in the treatment of cancer have appeared, provoking considerable interest among lay persons and the scientific community as well. Other uses reported are the dissolving of gallbladder and renal calculi, as an antiseptic and as an anti-ulcerous agent, and against dysuria and venereal disease (SEMARNAP, *Diagnóstico*, 4).

377. Bassols, *El noroeste*, 440; Schmidt, *Geographical Survey of Chihuahua*, 23.
378. Pennington, *The Tarahumar*, 37.
379. Vogel, *American Indian Medicine*, 296.
380. Brambila, *Diccionario rarámuri-castellano*, 189, 596.
381. Bennett and Zingg, *The Tarahumara*, 149; Pennington, *The Tarahumar*, 154.
382. Brambila, *Diccionario rarámuri-castellano*, 189.
383. Pennington, *The Tarahumar*, 183; Brambila, *Diccionario rarámuri-castellano*, 189.
384. Burgess McGuire and Mares Trías, *Ralámuli nu'tugala go'ame*, 75.
385. Irigoyen-Rascón, *Cerocahui*, 90.
386. Cabrera, *Plantas curativas*, 99.
387. Brambila, *Diccionario rarámuri-castellano*, 319.
388. Pennington, *The Tarahumar*, 117.
389. Bennett and Zingg, *The Tarahumara*, 167.
390. Pennington, *The Tarahumar*, 193.
391. Synonym: *Mentha arvensis*. More precisely: *Mentha arvensis* var. *canadensis* L. or *Mentha arvensis* var. *villosa* (Benth.) Stew. (Vogel, *American Indian*, 340).
392. Bennett and Zingg, *The Tarahumara*, 144.
393. Urbina, Antonio de, "Relación de Serocahue (Cerocahui)," in Del Paso y Troncoso, *Relaciones del siglo XVIII*.
394. Bennett and Zingg, *The Tarahumara*, 144.
395. Vogel, *American Indian Medicine*, 340.
396. Deimel, "Pflanzen zwischen den Kulturen," 61.
397. Pennington, *The Tarahumar*, 184.
398. Bye, "Ethnoecology of the Tarahumara," 171, 279.
399. For example, F. Joseph Pascual, S.J. (1652, in González-Rodríguez, *Tarahumara*, 166), related that the arrow poison used by Tarahumaras was so effective that "it would just take for the arrow to cause any bleeding to kill one." He relates that the lethality of the poison was offset only when a traitor revealed to the Spaniards the antidote. Velarde reported in 1716 that the Pimas had discovered an antidote for the arrow poison, a *jicamilla*, "as effective as that so-praised from Julimes." González Rodríguez speculates that the jicamilla from Julimes, Chihuahua, was *Jathropa* sp. (González-Rodríguez, *Etnología y misión*, 51).
400. Pennington, *The Tarahumar*, 108; Lumholtz, *Unknown Mexico*, 2:404.
401. Pennington, *The Tarahumar*, 108, from Bradley, "Yerba de la flecha," 365.
402. Watson, "List of Plants Collected by Palmer," 440.
403. Bye, "Ethnoecology of the Tarahumara," 171.

404. Cardenal, *Remedios y prácticas*, 76.
405. Pennington, *The Tarahumar*, 184.
406. Bye, "Ethnoecology of the Tarahumara," 164.
407. "Relación de Guazapares," "Relación de Batopilillas," and "Relación de Chínipas." All found in *Relaciones del siglo XVIII relativas a Chihuahua*.
408. Pennington, *The Tarahumar*, 184.
409. Cardenal, *Remedios y prácticas*, 76.
410. Bennett and Zingg, *The Tarahumara*, 167.
411. Bye, "Ethnoecology of the Tarahumara," 170.
412. Bye, "Ethnoecology of the Tarahumara," 282.
413. "Relación de Nabogame," in Pennington, *The Tarahumar*, 183, and "Relación de Guazapares" (in Del Paso y Troncoso, *Relaciones*), which describes its use to treat snakebites.
414. Irigoyen-Rascón, *Cerocahui*, 90; Bye, "Ethnoecology of the Tarahumara," 175.
415. Cardenal, *Remedios y prácticas*, 104.
416. Pennington, *The Tepehuan of Chihuahua*, 188.
417. Cardenal, *Remedios y prácticas*, 132.
418. Ibid.
419. Brambila, *Diccionario rarámuri-castellano*, 51.
420. (*Cuphea aequipetala* Cav.) is the *hierba del cáncer* of central Mexico, but *Acalypha phlioides* is also called *hierba del cáncer*. *Cuphea* is used there to cure wounds (Wagner, *Plantas medicinales*, 102) and, in a tea, for stomach and duodenal ulcers (*Enciclopedia de México*, 7:49). Additionally, the herb is said to enhance fertility and to be useful to treat dysentery, internal and external infections, mumps, cancer, insomnia, and many other disorders (SEMARNAP, *Diagnóstico*, 5).
421. Brambila, *Diccionario rarámuri-castellano*, 51.
422. Pennington, *The Tarahumar*, 76.
423. Lewis and Elvin-Lewis, *Medical Botany*, 301.
424. Irigoyen-Rascón, *Cerocahui*, 90.
425. Preciado et al., *Informe Kwechi*, appendix 1.
426. "Relación de Nabogame" (in Pennington, *The Tarahumar*, 194). "Relación de Guazapares" (in Del Paso y Troncoso, *Relaciones*) reports that *hierba del pastor* was used to treat buboes and *purgación* (gonorrhea). The former *relación* required one to add to the decoction *agua de las llagas* (water from the sores); the latter relación indicated that the herb had to be boiled and *serenada* (exposed to the dew) and then taken while fasting.
427. Bye, "Ethnoecology of the Tarahumara," 177.
428. Hewitt in Bye, "Ethnoecology of the Tarahumara," 174.
429. Pennington, *The Tarahumar*, 183.
430. Bye, "Ethnoecology of the Tarahumara," 174.
431. Ibid.
432. Bye, "Ethnoecology of the Tarahumara," 230.

433. Bennett and Zingg, *The Tarahumara*, 141.
434. Bye, "Ethnoecology of the Tarahumara," 161, 179; Cardenal, *Remedios y prácticas*, 112, 122.
435. Pennington, *The Tarahumar*, 180.
436. Bennett and Zingg, *The Tarahumara*, 176.
437. Pennington, *The Tarahumar*, 180.
438. Götzl and Schimmer, "Mutagenicity," 17–22.
439. Bennett and Zingg, The Tarahumara, 167.
440. Bye, "Ethnoecology of the Tarahumara," 279.
441. Lewis and Elvin-Lewis, *Medical Botany*, 38.
442. Bye, "Ethnoecology of the Tarahumara," 88.
443. Bennett and Zingg, *The Tarahumara*, 167.
444. Pennington, *The Tarahumar*, 184–185.
445. Lewis and Elvin-Lewis, *Medical Botany*, 20, 97. Salmón ("Cures," 50), quoting Marderosian, Liberti, *Natural Product*, 298, and Pusztai (*Plant Lectins*, 19), singles out toxalbumin ricin and ricin-agglutinin as the most toxic substances carried by *Ricinus communis*.
446. Cardenal, *Remedios y prácticas*, 81.
447. Burgess McGuire and Mares Trías, *Ralámuli nu'tugala go'ame*, 322–323.
448. Bennett and Zingg, *The Tarahumara*, 168.
449. Bye, "Ethnoecology of the Tarahumara," 182.
450. Pennington, *The Tarahumar*, 120, 189; Bye, "Ethnoecology of the Tarahumara," 288. Robalo (*Diccionario*, 109) identifies *Vitex molis* as the Aztecs' *coiotomatl*, which produces a fruit—sweet but that "somewhat locks a bit the throat." He claims that the root of this plant prepared into a tea and drunk in moderation cleans the bowels and attributes Sahagún for having discovered that it also purifies the milk of nursing mothers.
451. Bennett and Zingg, *The Tarahumara*, 168; Pennington, *The Tarahumar*, 189.
452. Bye, "Ethnoecology of the Tarahumara," 182.
453. Known in North American sources as milkweed, butterfly weed, or pleurisy root (Vogel, *American Indian*, 287), *Asclepias tuberosa* was a very important medicinal plant among many Indian groups with a wide range of uses in conditions from syphilis to dysentery. Tarahumaras use at least four other plants of the family Asclepiadaceae as remedies and two more—*A. glaucescens* and *A. brachystephana*—as food (Pennington, *The Tarahumara*, 127).
454. "Relación de Nabogame," "Relación de Chínipas," and "Relación de Batopilillas," quoted in Pennington (*The Tarahumara*, 188, 194). The "Relación de Serocahue (Cerocahui)" (Urbina, Antonio de, quoted in Del Paso y Troncoso, *Relaciones*, 19) also lists *inmortal* among the important Rarámuri medicinal plants. The "Relación of Guazapares" (quoted in Del Paso y Troncoso) calls this plant *pajiguecholi* and reports on its use for tumors.
455. Pennington, *The Tarahumar*, 188.

456. Cardenal, *Remedios y prácticas*, 81.
457. Lewis and Elvin-Lewis, *Medical Botany*, 18, 30, 52, 193.
458. Bye, "Ethnoecology of the Tarahumara," 245.
459. Bye, "Ethnoecology of the Tarahumara," 246.
460. Lumholtz, *Unknown Mexico*, 1:183.
461. Bennett and Zingg, *The Tarahumara*, 72.
462. Bye, "Ethnoecology of the Tarahumara," 246.
463. Burgess McGuire and Mares Trías, *Tase suhuíboa huijtabúnite*, 17.
464. Bennett and Zingg, *The Tarahumara*, 72; Bye, "Ethnoecology of the Tarahumara," 247.
465. Cardenal, *Remedios y prácticas*, 141.
466. Bye, "Ethnoecology of the Tarahumara," 247.
467. Possibly a "code name" used by Tarahumaras to prevent the casual listener from knowing what they are talking about. Compare Deimel, "Die Peyoteheilung" 156.
468. Sahagún in 1560 called those groups Teochichimecans and reported that "they first discovered and used the root called peyotl and ate and drunk it in place of wine." By the time of the European contact, peyote was known and used by Indians in central Mexico. Sahagún also reported that the disease known as *atonahuiztli* (aquatic fever) was treated by ingesting peyote and other hallucinogens (Sahagún, *Historia general*, 10:90; Laurencich Minelli, "Iconographie del peyotl," 256; and Ortiz de Montellano, *Aztec Medicine*, 158).
469. Pennington, *The Tarahumar*, 166. *E. micromeris* is called *jíkuri mulato* by some scholars (Pennington, *The Tarahumar*, 166). A more advanced vegetative stage identified by Robinson of the Gray Herbarium is called *jíkuri rosápari* (Bye, "Ethnoecology," 198). Lumholtz (*Unknown Mexico*, 2:373) was told that *jíkuri rosápari* is a good Christian who, when he sees an evildoer, gets very angry and either drives the offender crazy or throws him down precipices. Bye ("Ethnoecology," 199, 209) adds *Mammillaria heyderi*, *M. craigii*, and *M. grahamii* to the list of possible alternatives for *L. williamsii*.
470. Most scholars affirm that peyote is difficult to cultivate and consequently this forces the Indians to gather it from its natural habitat; however, at least one of the shamans from Narárachi cultivates the plant using a sandbox. Anderson (*Peyote*, 153) clarifies that although getting peyote to grow from seed takes a long time, cultivation from caespitose plants is relatively easy to accomplish.
471. "Relación de Guazapares," in Del Paso y Troncoso, *Relaciones del siglo XVIII*. Pennington (*The Tarahumar*, 186) adds that the "Relación de Chínipas" extols the use of jíkuri for "*la costumbre detenida*" — unspecific amenorrhea — which may indicate that it was believed to be an abortive remedy.
472. Alegre, *Historia de la compañía*, 465, in Anderson, *Peyote*, 15.
473. Hrdlička, *Physiological and Medical Observations*, 250.
474. Bennett and Zingg, *The Tarahumara*, 136.

475. Pennington, *The Tarahumar*, 186.
476. Deimel, "Die Peyoteheilung," 155.
477. Lumholtz, *Unknown Mexico*, 2:286–287; Bennett and Zingg, *The Tarahumara*, 137; Irigoyen-Rascón and Jesús Manuel Palma, *Rarajípari: La Carrera*, 120.
478. 3,4,5-trimethoxyphenethylamine (Mandell and Geyer, "The Euphorohallucinogens," 588).
479. Cabrera, *Plantas curativas*, 185; Monache, "Aspetti," 501–510; Oliverio and Castellano, "Psicofarmacología," 520–522; Anderson, *Peyote*, 105.
480. Peyote alkaloids are also classified as mono-, di- and trioxygenated phenethylamines, and tetrahydroisoquinolines.
481. Evans, "Botanical Sources," 104–119; Anderson, *Peyote*, 105; Pennington, *The Tarahumar*, 171n54.
482. Cabrera, *Plantas curativas*, 185.
483. Lumholtz, *Unknown Mexico*, 2:359.
484. Ludwig, "Altered States of Consciousness," 10–13, in Anderson, *Peyote*, 67. Many other psychological and psychedelic effects have been attributed to *Lophophora*, including feelings of dual existence; distortion of time and space; synesthetic mixing and enhancing the senses; afterimages; changes in brightness and saturation of perceived images; dimensionality; and many others (Anderson, *Peyote*, 70; Lisansky et al., "Drug-induced Psychoses," 81).
485. Vogel, *American Indian Medicine*, 190; Fisher, "Some Comments," 149–163; Unger, "Bibliography," 245–259; Hoffman, "Investigaciones," 28.
486. Bye and Constance, "A New Species," 44–47.
487. Bye, "Ethnoecology of the Tarahumara," 265, 288.
488. Bye, "Ethnoecology of the Tarahumara," 288; Bye and Constance, "A New Species," 44–47.
489. Viesca Treviño, *Estudios*, 55.
490. Bennett and Zingg, *The Tarahumara*, 167; Pennington, *The Tarahumar*, 184.
491. Lewis and Elvin-Lewis, *Medical Botany*, 38.
492. Orellana, *Indian Medicine*, 213.
493. Duke and Ayensu, *Medicinal Plants of China*, 309.
494. Synonym: *Teloxys graveolens* (Willd.) Weber.
495. Cabrera, *Plantas curativas*, 85.
496. Villacís, *Plantas medicinales*, 42–43.
497. Bye, "Ethnoecology of the Tarahumara," 166.
498. Bye, "Ethnoecology of the Tarahumara," 178, 284.
499. Deimel, "Pflanzen zwischen den Kulturen," 62.
500. Salmón, *Sharing Breath with Our Relatives*, 378.
501. Bye, "Ethnoecology of the Tarahumara," 139, 168.
502. Pennington, *The Tarahumar*, 181.
503. Bye, "Ethnoecology of the Tarahumara," 178.

504. Deimel, "Pflanzen zwischen den Kulturen," 61.
505. Bennett and Zingg, *The Tarahumara*; Pennington, *The Tarahumar*, 140.
506. Brambila, *Diccionario rarámuri-castellano*, 247.
507. Pennington, *The Tarahumar*, 191.
508. Brambila, *Diccionario rarámuri-castellano*, 247.
509. Brambila, *Diccionario castellano-rarámuri*, 435.
510. Brambila, *Diccionario rarámuri-castellano*, 552.
511. Bye, "Ethnoecology of the Tarahumara," 169, 277.
512. Cabrera (*Plantas curativas*, 179) described a *peonía del país* (*Cyperus esculento* L.) as a plant of mucilaginous flavor and aromatic smell. Its medicinal principles are contained mainly in the dark reddish or black oblong rhizomes. The herb, especially in an alcoholic infusion, is considered an effective diuretic and diaphoretic. European *peonía* or *peonia verdadera* is, of course, *Peonia officinalis* L.
513. Bennett and Zingg, *The Tarahumara*, 168; Bye, "Ethnoecology of the Tarahumara," 258; Salmón, "Cures of the Copper Canyon," 49.
514. Pennington, *The Tarahumar*, 191.
515. Salmón, *Sharing Breath with Our Relatives*, 384.
516. Bennett and Zingg, *The Tarahumara*, 189.
517. Salmón, "Cures of the Copper Canyon," 49; Claeys et al., *Umuhengerin*, 966.
518. Brambila, *Diccionario rarámuri-castellano*, 251.
519. Ibid.
520. Bye, "Ethnoecology of the Tarahumara," 183, 283.
521. *Enciclopedia de México*, 3:295.
522. Pennington, *The Tarahumar*, 190.
523. Bye, "Ethnoecology of the Tarahumara," 163.
524. About Hewitt's specimens: Bye, "Ethnoecology of the Tarahumara," 163.
525. Bye, "Ethnoecology of the Tarahumara," 139.
526. Bye, "Ethnoecology of the Tarahumara," 134, 286.
527. Burgess McGuire and Mares Trías, *Ralámuli nu'tugala go'ame*, 500.
528. From the Náhuatl *chil-tecpin* (chilli+flea), referring both to the size of the fruit and to the pungency of its bite (Robalo, *Diccionario*, 132).
529. Pennington, *The Tarahumar*, 67; Bye, "Ethnoecology of the Tarahumara," 88.
530. Bye, "Ethnoecology of the Tarahumara," 88, 180.
531. Salmón, *Sharing Breath with Our Relatives*, 380; Bennett and Zingg, *The Tarahumara*, 147.
532. Cardenal, *Remedios y prácticas*, 61.
533. Pennington, *The Tarahumar*, 193.
534. Bennett and Zingg, *The Tarahumara*, 169, 238, 246.
535. Bye, "Ethnoecology of the Tarahumara," 88, 134.
536. Bennett and Zingg, *The Tarahumara*, 265.
537. Lewis and Elvin-Lewis, *Medical Botany*, 85, 121.

538. Brambila, *Diccionario rarámuri-castellano*, 263.
539. Lionnet, *Los elementos*, 62.
540. Brambila, *Diccionario rarámuri-castellano*, 263.
541. Pennington, *The Tarahumar*, 133.
542. Bye ("Ethnoecology," 262) reports on other *Prionosciadium*s with ethnobotanical importance for the Tarahumaras: *P. madrense* S. Watson, a plant whose old rhizomes if eaten may cause one to go crazy; *P. serratum* Coulter & Rose, one of several *nawábori* whose rhizomes are dug out and eaten; and *P. townsendii*.
543. Bye, "Ethnoecology of the Tarahumara," 161, 263, 288.
544. Bennett and Zingg, *The Tarahumara*, 144; Urbina, Antonio de, "Relación de Serocahue," in Del Paso y Troncoso, *Relaciones de siglo XVIII*, 19.
545. Bennett and Zingg, *The Tarahumara*, 144; Bye, "Ethnoecology of the Tarahumara," 253.
546. Synonyms: *Bigognia stans* L., *Tecoma mollis* H.B.K., *Stenolobium stans* See., *Stenolobium tronadora* Loesener, and others.
547. Cardenal, *Remedios y prácticas*, 87.
548. *Tecoma* derivatives, including fluid extracts, tinctures, and purified active principles are extensively used in Mexico. See Lozoya, "Tronadora," 1–4; Wagner, *Plantas medicinales*, 115; and Villacís, *Plantas medicinales*, 126. Interestingly, Rodríguez and Machado (*Algunas plantas*, 414) place *Tecoma stans* among the plants presently being studied for their medicinal potential in Venezuela.
549. Pennington, *The Tarahumar*, 190.
550. "Relación de Guazapares," in Del Paso y Troncoso, *Relaciones del siglo XVIII*.
551. Lozoya, "Tronadora," 2.
552. Hammouda and Motawi and Jones, Fales, and Wildman in Lozoya, "Tronadora," 2; Lewis and Elvin-Lewis, *Medical Botany*, 218.
553. Formerly *Arctostaphylos uva ursi* L. The *Enciclopedia de México* (10:630) records the species *Arctostaphylos polifolia* H.B.K. and *A. arguta* (Zucc.) DC. as sharing economic and medicinal properties with *A. pungens*.
554. Brambila, *Diccionario rarámuri-castellano*, 279.
555. Pennington, *The Tarahumar*, 34–35.
556. Bennett and Zingg, *The Tarahumara*, 5, 139.
557. Bye, "Ethnoecology of the Tarahumara," 231.
558. Bye, "Ethnoecology of the Tarahumara," 149.
559. Irigoyen-Rascón, *Cerocahui*, 90.
560. Pennington, *The Tarahumar*, 187.
561. Bye, "Ethnoecology of the Tarahumara," 70.
562. Pennington, *The Tarahumar*, 163.
563. Irigoyen-Rascón, *Cerocahui*, 90.
564. Pennington, *The Tarahumar*, 122.
565. Villacís, *Plantas medicinales*, 97.

566. *Enciclopedia de México*, 10:630; Villacís, *Plantas medicinales*, 97.
567. Pennington, *The Tarahumar*, 106; Fernald, *Gray's Manual of Botany*, 428.
568. Pennington, "Tarahumar Fish," 97. One should not eat the leaves. As early as 1777, Urbina ("Relación de Serocahue") and Aréchiga ("Relación de Thomochic") in Del Paso y Troncoso (*Relaciones*, 20) reported that leaves of certain *nogales* growing near La Guáchara (the municipio of Urique) could kill cattle that eat them.
569. Irigoyen-Rascón, *Cerocahui*, 91.
570. Cabrera, *Plantas curativas*, 161.
571. The horned toad is not a true chameleon but, since it shares with it an impressive ability to change colors, it is called *camaleón* throughout northern Mexico. The Tarahumara word *láchimi* comes from *lá*, or blood, referring to the perception that the animal's eyes shed blood when it is caught. *Lá'chimi* is also the name of a squirrel (Brambila, *Diccionario rarámuri-castellano*, 281).
572. Bennett and Zingg, *The Tarahumara*, 127; Pennington, *The Tarahumar*, 135.
573. Bennett and Zingg, *The Tarahumara*, 127.
574. Bennett and Zingg (*The Tarahumara*, 127) reported other lizard remedies used by the Tarahumaras: Yarrow's scaly lizard (*Sceloporus jarrovii*) was a remedy for toothache: the live lizard was dropped in a hot *olla* (pot) over the fire and the ashes rubbed on the gums. Small greenish specimens of this lizard were caught by sterile women, held by the neck, and passed down the abdomen and between the legs to enable them to have children. Hrdlička (*Physiological and Medical*, 25) reported that Tarahumaras used sliced open bodies of lizards as poultices for broken bones. Cardenal (*Remedios y prácticas*, 142) claims that Tarahumaras tie lizards to their swollen arms or legs to combat inflammation. He also reports their blood being anointed to cure warts.
575. Bye, "Ethnoecology of the Tarahumara," 177, 283.
576. Bye, "Ethnoecology of the Tarahumara," 177.
577. The roots of *Phytolacca* sp. are decocted into a tea drunk to "purify the blood" and the flowers are used in a poultice for poisonous animal bites. The roots contain a glycoside; a glycoprotein; a resin; phytolaccatoxin; a water-soluble triterpene, phytolaccigenin; and a toxic alkaloid, phytolaccine (Salmón, "Cures of the Copper Canyon," 50, in Schmutz and Hamilton, *Plants that Poison*, 175; Fuller and McClintock, *Poisonous Plants*, 132; and Blackwell, *Poisonous and Medicinal Plants*, 243).
578. Bye, "Ethnoecology of the Tarahumara," 290.
579. Urbina, Antonio de, "Relación de Serocahue, (Cerocahui)" in Del Paso y Troncoso, *Relaciones del siglo XVIII*, 19.
580. Ferrari, "Ricerche sulle Laureacee del genere *Ocotea* dell'America latina," 374.
581. Bye, "Ethnoecology of the Tarahumara," 277.
582. Pennington, *The Tarahumar*, 192.
583. Bye, "Ethnoecology of the Tarahumara," 138, 168.
584. Bye, "Ethnoecology of the Tarahumara," 175.
585. Bye, "Ethnoecology of the Tarahumara," 160, 270.

586. Bye, "Ethnoecology of the Tarahumara," 160, 270.
587. Bye ("Ethnoecology," 270) also lists *Polypodium erythrolepis* Weath. and *P. thyssanolepis* A. Br. as ferns used medicinally by the Tarahumaras. Bye calls the latter *chawirate*, a name that Salmón (*Sharing Breath*, 377) applies to *Polypodium polypodioides* L., which he considers an antimicrobial and claims is added to batari to increase fermentation and enhance the medicinal qualities of the beer.
588. Bye, "Ethnoecology of the Tarahumara," 160.
589. Pennington, *The Tarahumar*, 178.
590. Bye, "Ethnoecology of the Tarahumara," 161.
591. Vogel (*American Indian*, 304) has compiled the various uses ascribed to *Pteridium aquilinum* by several North American Indian groups: the Delawares as a diuretic; the Menominees as a tea for women with caked breasts; and the Ojibwas for stomach cramps in women.
592. Bye, "Ethnoecology of the Tarahumara," 134, 161, 229.
593. Vogel, *American Indian Medicine*, 304; Lewis and Elvin-Lewis, *Medical Botany*, 290.
594. Lewis and Elvin-Lewis, *Medical Botany*, 26–27.
595. Pennington, *The Tarahumar*, 189.
596. Bye, "Ethnoecology of the Tarahumara," 252.
597. Urbina, Antonio de, "Relación de Serocahue," in Del Paso y Troncoso, *Relaciones*, 21.
598. Bennett and Zingg, *The Tarahumara*, 174.
599. Irigoyen-Rascón, *Cerocahui*, 90.
600. Bennett and Zingg, *The Tarahumara*, 174.
601. Pennington, *The Tarahumar*, 186.
602. Bye, "Ethnoecology of the Tarahumara," 176.
603. Deimel, "Pflanzen zwischen den Kulturen," 61.
604. Cabrera, *Plantas curativas*, 136.
605. Formerly *Chamomilla sueveolens* (Pursh) Rydb. or *Chamomilla nudum odoratum* (Lewis and Elvin-Lewis, *Medical Botany*, 339).
606. Irigoyen-Rascón, *Cerocahui*, 90.
607. Cabrera, *Plantas curativas*, 142.
608. Grinspoon, *Marihuana Reconsidered*, 37.
609. Irigoyen-Rascón, *Cerocahui*, 91.
610. Gaoni and Mechoulam, "A Total Synthesis," 3273–3275, in Grinspoon, *Marihuana Reconsidered*, 46; Lewis and Elvin-Lewis, *Medical Botany*, 405; Lisansky et al., "Drug-induced Psychoses," 88; Villacís, *Plantas medicinales*, 79.
611. Such as the non-nitrogenous, non-alkaloid substances Δ 1- 3, 4 trans-tetrahydrocannabinol and Δ6- 3,4 trans-tetrahydrocannabinol; cannabichromene, cannabinodiol, cannabinodiolic acid, and cannabinogerol (Grinspoon, *Marihuana Reconsidered*, 46; Lewis and Elvin-Lewis, *Medical Botany*, 427).
612. Pacher, Bátkai and Kunos, "The Endocannabinoid System, 389–462.
613. Cabrera, *Plantas curativas*, 131.

614. Ibid.
615. Synonyms: *Cacalia decomposita* A. Gray and *Psacalium decompositum* (Gray) Rob & Brett. Although *matarí* is unambiguously the name of *Cacalia* or *Odontotrichum*, Bye ("Ethnoecology," 181) claims it is also given to one other plant: *Conioselinum mexicanum* Coult. & Rose. He claims that the roots of that plant of the family Umbelliferae are used either in a poultice or in preparing a tea for rheumatism.
616. Brambila, *Diccionario rarámuri-castellano*, 306.
617. This is the form commonly used in Norogachi. The song by Erasmo Palma, "*Matirí bureba re*," praises the economic importance of the herb for the Tarahumaras who sell it in Chihuahua City.
618. Bennett and Zingg, *The Tarahumara*, 175.
619. Bennett and Zingg, *The Tarahumara*, 175–176; Pennington, *The Tarahumar*, 191; Brambila, *Diccionario rarámuri-castellano*, 205; Bye, "Ethnoecology of the Tarahumara," 142, 168, 277.
620. Bye, "Ethnoecology of the Tarahumara," 168.
621. Pennington, *The Tarahumar*, 13–14; Bennett and Zingg, *The Tarahumara*, 176; Brondo Whitt, *Nuevo León*, 165; Preciado et al. *Informe Kwechi*, appendix 2; Bye, "Ethnoecology of the Tarahumara," 168; Cardenal, *Remedios y prácticas*, 95; Cabrera, *Plantas curativas*, 148.
622. Brambila, *Diccionario rarámuri-castellano*, 566.
623. Pennington, *The Tarahumar*, 191.
624. Bennett and Zingg, *The Tarahumara*, 175.
625. Pennington, *The Tarahumar*, 192; Bye, "Ethnoecology of the Tarahumara," 168.
626. Deimel, "Pflanzen zwischen den Kulturen," 61.
627. Wagner, *Plantas medicinales*, 106.
628. Cardenal, *Remedios y prácticas*, 95; Salmón, *Sharing Breath with Our Relatives*, 356.
629. Bennett and Zingg, *The Tarahumara*, 175.
630. Bennett and Zingg, *The Tarahumara*, 176; Pennington, *The Tarahumar*, 110.
631. Bye ("Ethnoecology," 142) quotes Becerril Buitrón, "Contribución al estudio," stating that *Odontotrichum decompositum* has no value in treating diabetes.
632. Altamirano, "Estudio comparativo," 84; Cabrera, *Plantas curativas*, 148; Villacís, *Plantas medicinales*, 81.
633. Pennington, *The Tarahumar*, 193.
634. Pennington, *The Tarahumar*, 193; Brambila, *Diccionario rarámuri-castellano*, 307.
635. The Tarahumara name of *pinabete* (*Abies durangensis*) is also *mateó* (see "*okó*" entry in chapter 10).
636. Bye, "Ethnoecology of the Tarahumara," 175.
637. Bennett and Zingg, *The Tarahumara*, 144.
638. The name *tepozán* is given in southern Mexico to trees of the family Loganiaceae, including *Buddleia americana* L., *B. cordata* (*B. humboldtiana* Roem. & Schult.), *B. floccosa* Kunth, *B. parviflora* H.B.K., *B. sessiliflora*, *B. tomentella* Stand., *B. wrightii*

Rob., *B. lanceolata* Benth., and *B. microphylla* H.B.K. (*Enciclopedia de México*, 7:130). Wagner (*Plantas medicinales*, 115) identifies *tepozán* only as *Buddleia americana* and reports that in central Mexico its leaves and roots are used internally for hydrops and rheumatism, and externally for the treatment of wounds.

639. Pennington, *The Tarahumar*, 187; Cardenal, *Remedios y prácticas*, 96.
640. Bennett and Zingg, *The Tarahumara*, 144.
641. Villacís, *Plantas medicinales*, 113.
642. Preciado et al., *Informe Kwechi*, appendix 2.
643. Salmón, *Sharing Breath with Our Relatives*, 370.
644. Bye, "Ethnoecology of the Tarahumara," 171, 280.
645. Bye, "Ethnoecology of the Tarahumara," 171.
646. Pennington, *The Tarahumar*, 188; Bye, "Ethnoecology of the Tarahumara," 140.
647. Bennett and Zingg, *The Tarahumara*, 173.
648. Ibid.
649. Pennington, *The Tarahumar*, 212.
650. Cardenal, *Remedios y prácticas*, 101.
651. Bye, "Ethnoecology of the Tarahumara," 139.
652. Pennington, *The Tarahumar*, 129.
653. Bye, "Ethnoecology of the Tarahumara," 162.
654. Bye ("Ethnoecology," 271) also records *Agave americana* L. var. *expansa* (Jacobi) H. S. Gentry, *A. applanata* Koch; *A. multifilifera* H. S. Gentry, *A. pacifica* Trel., *A. parryi* Engelm., and *A. polianthiflora* H. S. Gentry.
655. Deimel, "Pflanzen zwischen den Kulturen," 62.
656. Pennington, *The Tarahumar*, 32, 35, 105.
657. Di Peso, *Casas Grandes*, 221.
658. The Aztecs identified the maguey plant with the goddess Mayaguel and other gods, collectively called the *tzentzontotochtin* ("400 rabbits"). These represented some of the behavioral quirks of people who were drunk. Mayaguel is commonly regarded as the goddess of *pulque*, a beverage obtained from *Agave pulcherrima* Hort. ex C. Koch. (Ortiz de Montellano, *Aztec Medicine*, 110).
659. Lumholtz, *Unknown Mexico*, 2:256.
660. Molina, *Vocabulario*, 55. The name "México" seems to derive from one or both of these Náhuatl words.
661. Di Peso, *Casas Grandes*, 221.
662. Beals, *The Comparative Ethnology*, 163–165; Driver and Creel, *Ethnography and Acculturation*, 19, in Di Peso, *Casas Grandes*, 227.
663. Pennington, *The Tarahumar*, 129, 150.
664. Tequila, the Mexican national beverage, is made from *Agave tequilana* Weber. *Mezcales* other than tequila, particularly those from Jalisco and Oaxaca, are also held in high repute by connoisseurs. *Bacanora* of Sonora, *raicilla* of coastal Jalisco, and *charanda* of Michoacán are other popular mezcales distilled from *Agave* sp. (Martínez

Limón, *Tequila*, 34–38). In the state of Chihuahua, *sotol* is prepared from *Dasylirion* sp. and, in the heartland of Tarahumara Country, the famous *lechuguilla* from Batopilas and *temorense* from Témoris are made, mostly by mestizos, from *Agave lechuguilla*.

665. Lumholtz, *Unknown Mexico*, 1:438–439; Irigoyen-Rascón, *Cerocahui*, 55.
666. Bennett and Zingg, *The Tarahumara*, 163; Pennington, *The Tarahumar*, 211. The use of *amole* (soap) from the roots of several plants of the family Agavaceae is a pan-Mexican Indian trait. Vogel (*American Indian*, 177) identifies the *amolli* mentioned by Sahagún—the Mesoamerican soap root—as *Sapindus saponarious* L., but Robalo (*Diccionario*, 23) prefers *Sapindus frutescens* Aubl. for *amole de bolita* or *xopalxocotl* and invokes a reference of Bustamante to Sahagún's writings referring to its use by the ancient Mexicans to treat rabies. Tarahumaras know several amoles—*awé* and *a*pora*, which is also used as a piscicide.
667. Pennington, *The Tarahumar*, 180.
668. Cardenal, *Remedios y prácticas*, 96.
669. Bye, "Ethnoecology of the Tarahumara," 162.
670. Cardenal, *Remedios y prácticas*, 123.
671. Cervantes in Standley, *Trees and Shrubs*, 90; Bennett and Zingg, *The Tarahumara*, 164.
672. Bye, "Ethnoecology of the Tarahumara," 162.
673. Pennington, *The Tarahumar*, 180.
674. Campos and Gaeta, *El rutuburi*, 6.
675. Bennett and Zingg, *The Tarahumara*, 164, 279; Cardenal, *Remedios y prácticas*, 96.
676. For example, extracts from *Agave sisalana* constitute a major source of starting materials (steroidal sapogenins) for hormone synthesis (Lewis and Elvin-Lewis, *Medical Botany*, 318). Among the agaves growing in western Tarahumara Country, *A. vilmoriniana* is rich in the sapogenin smilagenin (Bye, Burgess McGuire, and Mares Trías, "Ethnobotany," 91).
677. Lewis and Elvin-Lewis, *Medical Botany*, 19.
678. Wall et al., "Steroidal Sapogenins: VII," 1–7, in Pennington, *The Tarahumar*, 106n129.
679. Ortiz de Montellano, *Aztec Medicine*, 194.
680. Bye, "Ethnoecology of the Tarahumara," 174.
681. *Tepeguaje* comes from the Náhuatl words *tépetl*=mountain or hill and *huaxin*=guaje, or gourd. Robalo (*Diccionario*, 152) applies it to a tree of the family Leguminosae, *Acacia acapulamis* Kunth., and claims that the bark is used in medicine as an astringent, the gum as an alternative to arabic gum, and industrially the wood to make fine furniture.
682. Pennington, *The Tarahumar*, 183.
683. Bye, "Ethnoecology of the Tarahumara," 174. Bye ("Ethnoecology," 174, 281) reports another *Lysiloma*, *L. divaricatum* (Jacq.) McBride, which Tarahumaras call *wapakawe* or *wichawi*, with similar uses as a dental remedy and in tea for gastrointestinal disorders.

684. Pennington, *The Tarahumar*, 196.
685. Bye, "Ethnoecology of the Tarahumara," 140.
686. Irigoyen-Rascón, *Cerocahui*, 90.
687. Cabrera, *Plantas curativas*, 342.
688. Ibid.
689. Brambila, *Diccionario rarámuri-castellano*, 330.
690. Bye, "Ethnoecology of the Tarahumara," 134.
691. Ibid.
692. Bennett and Zingg, *The Tarahumara*, 158.
693. Bye, "Ethnoecology of the Tarahumara," 134.
694. Bennett and Zingg, *The Tarahumara*, 129, 158, 263.
695. Cardenal, *Remedios y prácticas*, 57.
696. Bennett and Zingg, *The Tarahumara*, 158, 263.
697. Bye, "Ethnoecology of the Tarahumara," 165.
698. Irigoyen-Rascón, *Cerocahui*, 90; Cardenal, *Remedios y prácticas*, 142.
699. Bye, "Ethnoecology of the Tarahumara," 281.
700. Pennington, *The Tarahumar*, 183.
701. Bennett and Zingg, *The Tarahumara*, 144.
702. Brambila, *Diccionario rarámuri-castellano*, 123.
703. Bennett and Zingg, *The Tarahumara*, 176.
704. Brambila, *Diccionario rarámuri-castellano*, 123.
705. Deimel, "Pflanzen zwischen den Kulturen," 61.
706. Brambila, *Diccionario rarámuri-castellano*, 123.
707. Bennett and Zingg, *The Tarahumara*, 176.
708. Deimel, "Pflanzen zwischen den Kulturen," 61.
709. Bennett and Zingg, *The Tarahumara*, 176.
710. Pennington, *The Tarahumar*, 188; Brambila, *Diccionario rarámuri-castellano*, 123, 348.
711. Vogel (*American Indian*, 336) identifies the whorled or dwarf milkweed as *Asclepias verticillata* L., the swamp milkweed as *A. incarnata* L., and the common milkweed as *A. cornuti* Decaisne or *A. syriaca* L. Bye ("Ethnoecology," 164) mentions another *Asclepias* sp. (*ecoranilla*), whose roots are used by the Tarahumaras to prepare a general medicinal tea. He also denounces *patito* (*Asclepias linaria* Cav.) as a dangerous purgative, which should be used with care and in small amounts.
712. Bye ("Ethnoecology," 168) also identifies *nakáruri* as *Odontotrichum globosum* (Robins & Fern.) Rydb. and claims that the root of this plant is used by Tarahumaras as a wash or in a poultice for rheumatism.
713. Brambila, *Diccionario rarámuri-castellano*, 349.
714. Bye, "Ethnoecology of the Tarahumara," 169.
715. Pennington, "Tarahumar Fish," 100, and *The Tarahumar*, 110. Note that *Stevia* (*roninowa*) is also known as *hierba de la mula*.
716. Pennington, "Tarahumar Fish," 100–101.

717. Pennington, *The Tarahumar*, 191.
718. Bye, "Ethnoecology of the Tarahumara," 169.
719. Bye ("Ethnoecology," 165, 275) identifies this *Tabebuia* as *T. chrysantha* (Jacq.) Nich. The genus *Lippia* belongs to the family Verbenaceae, *Monarda* and *Prunella* to the Verbenaceae, and *Tabebuia* to the Bignoniacea.
720. Pennington, *The Tarahumar*, 129.
721. Pennington, *The Tarahumar*, 127, 190.
722. Urbina, Antonio de, "Relación de Serocahue (Cerocahui)" in Del Paso y Troncoso, *Relaciones del siglo XVIII*, 20.
723. Bye, "Ethnoecology of the Tarahumara," 138, 175; Pennington, *The Tarahumar*, 190. *Lippia berlandieri* is highly appreciated in central Mexico as a stomach tonic (Wagner, *Plantas medicinales*, 107). Gracia Alcover (*Vitaminas y medicina*, 197), in Chile, identifies *orégano* as *Origanum vulgare* L., and reports for it medicinal virtues similar to those of Wagner's *Lippia*. Villacís (*Plantas medicinales*, 90) also endorses *O. vulgare* as the taxonomic identity of Mexican *orégano*. In addition to intestinal problems, he recommends it for menstrual disorders.
724. Merrill, *Rarámuri Souls*, 122.
725. Bennett and Zingg, *The Tarahumara*, 140.
726. Pennington, *The Tarahumar*, 107; Brambila, *Diccionario rarámuri-castellano*, 371.
727. *Nawé* is the Tarahumara piscicide par excellence; Rarámuri knowledge of piscicides is impressive: Pennington ("Tarahumar Fish," 95–102) lists thirty-one herbs used for this purpose. Bye ("Ethnoecology," 270–298) designates eleven herbs as piscicides. Gajdusek ("Tarahumara Indian Piscicide," 436) described *matésuwa* (*Gilia Macombii* Torrey [SYN. *Ipomoea thurberi* Torrey]).
728. Pennington, *The Tarahumar*, 182; Brambila, *Diccionario rarámuri-castellano*, 371.
729. Cardenal, *Remedios y prácticas*, 100.
730. Pammel, *A Manual of Poisonous Plants*, 558; Pennington, *The Tarahumar*, 107. Cuca ("Principios activos," 370) reported the isolation of two rotenoids, tephrosine and degualine, and a flavonone, 5-methoxy-isoloncocarpine, from this South American *Tephrosia*.
731. Formerly *Pinus pinea*. Also *Pinus pinceana* Gordon. The fruit, *piñón*, gathering of which is of great commercial importance for the Tarahumaras, is known as *isíkuri*.
732. Brambila (*Diccionario rarámuri-castellano*, 307) states that the Rarámuri names *mateyó*, *mategó*, or *mateó* designate the tree known in Spanish as *pinabete*. This tree has been classically identified as *Pseudotsuga mucronata* (Raf.) Sudw. Bye ("Ethnoecology," 271) applies the name *matioa* to *Pseudotsuga menziensii* (Mirb.) Franco and *mateyó* to both *Picea chihuahuana* Martínez and *Abies durangensis* Martínez.
733. Bye, "Ethnoecology of the Tarahumara," 271.
734. Brambila, *Diccionario rarámuri-castellano*, 509.
735. French internists of the turn of the twentieth century studied and used extensively turpentine (French *térébenthine*) for a variety of conditions, including pleurisy (De-

bove, Bouchet, and Sallard, *Aide-mémoire*, 831), nephritis arising from scarlet fever, typhoid fever, and gonorrhea.(Debove and Achard, *Manuel de thérapeutique*, 461). In America, both the Indians and later learned physicians regarded turpentine as an effective remedy for wounds and burns and as a balsam for throat and chest ailments.(Vogel, *American Indian*, 347; Parke, Davis, *Physician Manual*, 491). Kock-Weser ("Venenos," 733) reports that the ingestion of as little as fifteen grams of turpentine has been fatal.

736. Cabrera, *Plantas curativas*, 179; Villacís, *Plantas medicinales*, 88–89.
737. Lumholtz, *Unknown Mexico*, 1:409.
738. Pennington, *The Tarahumar*, 178.
739. Bye, "Ethnoecology of the Tarahumara," 161.
740. Bennett and Zingg, *The Tarahumara*, 128; Pennington, *The Tarahumar*, 135.
741. Pennington, *The Tarahumar*, 135.
742. Salmón, *Sharing Breath with Our Relatives*, 375.
743. Lumholtz, *Unknown Mexico*, 1:245.
744. The name *ortiga* is applied in Mexico to plants of the genera *Urtica* and *Urera* of the family Urticacea; *Cnidoscolus, Jatropha* (*Jumete*), and *Tragia* of the family Euphorbiaceae; and *Wiganda* of the Hydrophylaceae. *Urtica dioica* is commonly found in central Mexico while *U. subincisa* Benth. and *U. urens* L. are abundant throughout Mexico (*Enciclopedia de México*, 10:18).
745. Brambila, *Diccionario rarámuri-castellano*, 456.
746. Bye, "Ethnoecology of the Tarahumara," 171.
747. Deimel, "Pflanzen zwischen den Kulturen," 62.
748. Pennington, *The Tarahumar*, 125, 184.
749. Pennington, *The Tarahumar*, 184; Brambila, *Diccionario rarámuri-castellano*, 456.
750. Bye, "Ethnoecology of the Tarahumara," 171.
751. SEMARNAP, *Diagnóstico de productos*, 4.
752. Bye, "Ethnoecology of the Tarahumara," 240.
753. Bye, "Ethnoecology of the Tarahumara" 168, 240, 277.
754. Pennington, *The Tarahumar*, 192.
755. Bye, "Ethnoecology of the Tarahumara," 168.
756. Cardenal, *Remedios y prácticas*, 102.
757. Náhuatl: "medicine colored like blood (Cabrera, *Diccionario*, 149).
758. *Enciclopedia de México*, 10:206–207.
759. Pennington, *The Tarahumar*, 183.
760. Bennett and Zingg, *The Tarahumara*, 171.
761. Bye, "Ethnoecology of the Tarahumara" 165.
762. Cardenal, *Remedios y prácticas*, 118.
763. Bennett and Zingg, *The Tarahumara*, 171.
764. Irigoyen-Rascón, *Cerocahui*, 91. Bye ("Ethnoecology," 165) claims that the resin from *B. bipinnata* (Sessé & Moc) Engl. and *B. pennicillata* Engl. (both called *torote*), *B. lan-*

cifolia Engl. (*rusíwari*), and *Bursera* sp. (*muruwáka*) are used by the Tarahumaras in teas, washes, and infusions for a number of ailments.
765. Cabrera, *Plantas curativas*, 175.
766. Bennett and Zingg, *The Tarahumara*, 171. Bye, "Ethnoecology of the Tarahumara," 165; Cardenal, *Remedios y prácticas*, 118.
767. Bennett and Zingg, *The Tarahumara*, 169; Pennington, *The Tarahumar*, 121.
768. Identifications given by Bye to Burgess, in Burgess McGuire and Mares Trías, *Ralámuli nu'tugala go'ame*, 501.
769. Pennington (*The Tarahumar*, 32, 121, 150–151, 154–155) identifies *kakwara* as *Randia echinocarpa* and *kapuchí* as *Randia laevigata*.
770. Bennett and Zingg, *The Tarahumara*, 169.
771. Burgess McGuire and Mares Trías, *Ralámuli nu'tugala go'ame*, 297.
772. Bye, "Ethnoecology of the Tarahumara," 253.
773. Brambila, *Diccionario rarámuri-castellano*, 237.
774. Burgess McGuire and Mares Trías, *Ralámuli nu'tugala go'ame*, 303, 501.
775. Bennett and Zingg, *The Tarahumara*, 169.
776. Brambila, *Diccionario rarámuri-castellano*, 8.
777. Pennington, *The Tarahumar*, 150–151.
778. Identifications Bye gave to Burgess, in Burgess McGuire and Mares Trías, *Ralámuli nu'tugala go'ame*, 303, 501.
779. "Relación de Guazapares," "Relación de Chínipas," and "Relación de Batopilas," and Palmer in Gray, "Contributions," 380, in Pennington, *The Tarahumar*, 121.
780. Burgess McGuire and Mares Trías, *Ralámuli nu'tugala go'ame*, 297–308.
781. Bennett and Zingg, *The Tarahumara*, 169; Irigoyen-Rascón, *Cerocahui*, 83.
782. Bennett and Zingg, *The Tarahumara*, 169; Pennington, *The Tarahumar*, 157.
783. Bye, "Ethnoecology of the Tarahumara," 254; Brambila, *Diccionario rarámuri-castellano*, 235.
784. Irigoyen-Rascón, *Cerocahui*, 79.
785. Irigoyen-Rascón, *Cerocahui*, 91.
786. Hewitt in Bye, "Ethnoecology of the Tarahumara," 174; Pennington, *The Tepehuan of Chihuahua*, 182.
787. *Enciclopedia de México* (10:777) states that the name *poleo* is given also to other plants of the family Labiatae (or Lamiaceae), which share medicinal virtues with the European *poleo*, among them the Sonoran/Sinaloan bush *Clinopodium laevigatum* Stand., which may reach western Tarahumara Country. American pennyroyal (*Hedeoma pulegioides* [L.] Pers.) and the related *H. reverchoni* A. Gray and *H. hispida* Pursh are extensively discussed by Vogel (*American Indian*, 338–339) and Lewis and Elvin-Lewis (*Medical Botany*, 160, 294, 329). The antispasmodic, antiflatulent, and diaphoretic properties hailed by other Indian groups resemble those claimed for *poleo* by my own informants. Lewis and Elvin-Lewis report the presence of a volatile oil, pulegone, in *Hedeoma pulegioides*. Deimel ("Pflanzen zwischen," 62) calls *Agastache*

micrantha (A. Gray) Woot. & Standl., which is used by mestizos as a hypnotic and a gastrointestinal ailment remedy, *poleo* or *yerbena*.
788. Irigoyen-Rascón, *Cerocahui*, 91.
789. Brambila, *Diccionario rrámuri-castellano*, 600.
790. Cardenal, *Remedios y prácticas*, 108.
791. Brambila, *Diccionario rarámuri-castellano*, 125.
792. Bye, "Ethnoecology of the Tarahumara," 277.
793. The term *gordolobo* is applied through Mexico to plants from at least three genera, *Verbascum*, *Gnaphalium*, and *Senecio*. This is thought to have led to poisonings of some persons purchasing gordolobo in Mexican herbal stores in the United States (Kay, "Health and Illness," 42).
794. Deimel, "Pflanzen zwischen den Kulturen," 62.
795. Cabrera, *Plantas curativas*, 97.
796. Bennett and Zingg, *The Tarahumara*, 176.
797. Pennington, *The Tarahumar*, 192.
798. Bye, "Ethnoecology of the Tarahumara," 167.
799. Irigoyen-Rascón, *Cerocahui*, 90.
800. Bennett and Zingg, *The Tarahumara*, 176.
801. Pennington, *The Tarahumar*, 192.
802. Salmón, *Sharing Breath with Our Relatives*, 356.
803. Cabrera, *Plantas curativas*, 97.
804. Pennington, *The Tarahumar*, 185.
805. Irigoyen-Rascón, *Cerocahui*, 90.
806. Brambila, *Diccionario rarámuri-castellano*, 467.
807. Bennett and Zingg, *The Tarahumara*, 74.
808. Pennington (*The Tarahumar*, 32) identifies *repogá* as *aliso*; however, *aliso* (alder) is, of course, *Alnus* sp. Complicating matters, *Platanus* sp.—sycamores or plane trees—are also called, by the mestizos, *aliso* and by the Tarahumaras, *repogá*.
809. Brambila, *Diccionario rarámuri-castellano*, 174. Note that in Mexico the name *álamo* is given to species of a number of botanical families, including the Salicaceae (*Populus*, *Salix*), Platanaceae, Moraceae (*Ficus*), and Solanaceae (*Enciclopedia de México*, 1:377–379).
810. Pennington, *The Tarahumar*, 168, 210.
811. Pennington, *The Tarahumar*, 181; Bye, "Ethnoecology of the Tarahumara," 177.
812. Salmón, *Sharing Breath with Our Relatives*, 376; Bennett and Zingg, *The Tarahumara*, 174.
813. Bye, "Ethnoecology of the Tarahumara," 134, 164.
814. Bye, "Ethnoecology of the Tarahumara," 160.
815. Pennington, *The Tarahumar*, 193.
816. Bye, "Ethnoecology of the Tarahumara," 160; Salmón, *Sharing Breath with Our Relatives*, 383.

817. Preciado et al., *Informe Kwechi*, appendix 3.
818. Cabrera (*Diccionario*, 110) reports that *pipitzahua* or *pipizahua* are terms used in Mexico to name several plants of the family Compositae—i.e., *Perezia adnata* Gray, *Stevia viscida* H.B.K.—characterized by being strong laxatives. Since the nineteenth century, scientists have analyzed their contents, particularly pipitzahoic acid (after *pipitzahua*) or riolozic acid (after its discoverer, Dr. Río de la Loza), a mild purgative, found in *Perezia* sp. (Mohr, "On the Presence," 56). Cabrera proposed the Náhuatl etymology: *pipitzahuac*=reduplicated *pitzahuac*=thin shaft.
819. Brondo Whitt, *Nuevo León*, 28.
820. Bye, "Ethnoecology of the Tarahumara," 168.
821. Bye, "Ethnoecology of the Tarahumara," 168, 185.
822. Preciado et al., *Informe Kwechi*, appendix 3.
823. Bye, "Ethnoecology of the Tarahumara," 167, 277.
824. Irigoyen-Rascón, *Cerocahui*, 89.
825. Preciado et al., *Informe Kwechi*, appendix 4.
826. Bye, "Ethnoecology of the Tarahumara," 167, 236–237.
827. This is "shameful." Paradoxically, the name *sin vergüenza* recorded by Bennett means exactly the opposite; i.e., "shameless."
828. Bennett and Zingg, *The Tarahumara*, 338.
829. Bennett and Zingg, *The Tarahumara*, 338; Pennington, *The Tarahumar*, 170–171; Irigoyen-Rascón and Jesús Manuel Palma, *Rarajípari: La Carrera*, 146.
830. Pennington, *The Tarahumar*, 183.
831. SEMARNAP, *Diagnóstico de productos*, 5.
832. Burgess McGuire and Mares Trías, *Ralámuli nu'tugala go'ame*, 245.
833. Bye, "Ethnoecology of the Tarahumara," 228.
834. Burgess McGuire and Mares Trías, *Ralámuli nu'tugala go'ame*, 245.
835. Deimel, "Pflanzen zwischen den Kulturen," 62.
836. Pennington, *The Tarahumar*, 67.
837. Burgess McGuire and Mares Trías (*Ralámuli nu'tugala go'ame*, 245) claim that *ro'chíwari* grows amidst the cultured lands and on the hills but is never planted.
838. Bye, "Ethnoecology of the Tarahumara," 111.
839. Pennington, *The Tarahumar*, 57, 162.
840. Cardenal, *Remedios y prácticas*, 113.
841. Pennington, *The Tarahumar*, 51, 78, 114, 196.
842. Bye, "Ethnoecology of the Tarahumara," 279.
843. Brambila, *Diccionario rarámuri-castellano*, 571.
844. Bye, "Ethnoecology of the Tarahumara," 279.
845. Brambila, *Diccionario rarámuri-castellano*, 485; Pennington, *The Tarahumar*, 180.
846. Bye, "Ethnoecology of the Tarahumara," 279.
847. Brambila, *Diccionario rarámuri-castellano*, 64.
848. Brambila, *Diccionario rarámuri-castellano*, 485. Pennington (*The Tarahumar*, 168)

stated that carving *gomákari* from *Quercus arizonica* was a thing of the past. According to him, the best source of material for fashioning balls for the modern Tarahumaras is *urúbisi* (*Arbutus arizonica* or *A. glandulosa*), or, alternatively, *wisaró*.
849. Brambila, *Diccionario rarámuri-castellano*, 485.
850. Cabrera, *Plantas curativas*, 83.
851. Pennington, *The Tarahumar*, 180.
852. Salmón, *Sharing Breath with Our Relatives*, 369.
853. Pennington, *The Tarahumar*, 180; Brambila, *Diccionario rarámuri-castellano*, 485.
854. Burgess McGuire and Mares Trías, *Tase suhuíboa huijtabúnite*, 17.
855. Pennington, *The Tarahumar*, 180.
856. Mull and Mull, "Infanticide," 124.
857. Pennington, *The Tarahumar*, 115.
858. Bye, "Ethnoecology of the Tarahumara," 282; Deimel, "Pflanzen zwischen den Kulturen," 61.
859. Bennett and Zingg, *The Tarahumara*, 177.
860. Brambila, *Diccionario rarámuri-castellano*, 76, 279.
861. Cardenal, *Remedios y prácticas*, 78.
862. Bennett and Zingg, *The Tarahumara*, 177.
863. Salmón, "Cures of the Copper Canyon," 359.
864. Brondo Whitt, *Nuevo León*, 28.
865. Salmón, *Sharing Breath with Our Relatives*, 359.
866. Brambila, *Diccionario rarámuri-castellano*, 76, 279.
867. Bye, "Ethnoecology of the Tarahumara," 176.
868. Deimel, "Pflanzen zwischen den Kulturen," 61. Pennington (*The Tarahumar*, 170) describes other epiphytic plants of the family Bromeliacea, which are also used medicinally by the Tarahumaras: *retesíwara* (*Tillandsia karwinsyana* Schultes) that grows on mesquites and scrub oak, in a tea to relieve constipation, and *wechira* (*TIllandsia* sp.), in a wash to alleviate rheumatic pains. Bye ("Ethnoecology," 209) describes another *Tillandsia*, *T. mooreana* L. B. Smith, claiming that it is considered by the Tarahumaras living in the Batopilas canyon as the companion plant of *jíkuri*, inspiring respect and a sense of dread. Harming the plant may lead to disgrace for the offender.
869. Bennett and Zingg, *The Tarahumara*, 140.
870. Ibid.
871. Pennington, *The Tarahumar*, 186; Bye, "Ethnoecology of the Tarahumara," 166.
872. Cardenal, *Remedios y prácticas*, 114.
873. Deimel, "Pflanzen zwischen den Kulturen," 62.
874. Cardenal, *Remedios y prácticas*, 112.
875. Brambila, *Diccionario rarámuri-castellano*, 490.
876. Bennett and Zingg, *The Tarahumara*, 142.
877. Pennington, *The Tarahumar*, 110.
878. Deimel endorses these uses, but he contends that Tarahumaras use *ronínowa* exclu-

sively as a toothache remedy, while mestizos use it only for animal bites. In addition, he lists *ramíchi, ramiguchi, ramiojí*, and *ramúguri* (all conspicuously derived from the Tarahumara *rame*=tooth), which he identifies as *Callandria* sp., being used as Tarahumara dental remedies ("Pflanzen," 62).

879. Pennington, *The Tarahumar*, 91; Bennett and Zingg, *The Tarahumara*, 142.
880. Bye, "Ethnoecology of the Tarahumara," 169.
881. Bye, "Ethnoecology of the Tarahumara," 284.
882. Brambila, *Diccionario rarámuri-castellano*, 493.
883. Pennington, *The Tarahumar*, 190.
884. In Europe, *llantén* (*Plantago media* L.) is used for its emollient and astringent properties in ophthalmic poultices and gargling (Carvajal, *Plantas que curan*, 84). In Mexico, authors report more uses: externally as a cicatrizing agent for wounds, internally for liver, lung, and blood conditions (Wagner, *Plantas medicinales*, 32; García Rivas, *Plantas medicinales*, 87, and Pompa, *Medicamentos indígenas*, 148).
885. Brambila, *Diccionario rarámuri-castellano*, 493.
886. "Relación de Guazapares," in Del Paso y Troncoso, *Relaciones del siglo XVIII.*
887. Pennington, *The Tarahumar*, 190.
888. SEMARNAP, *Diagnóstico de productos*, 5.
889. Cardenal, *Remedios y prácticas*, 89.
890. Pennington, *The Tarahumar*, 192; Bye, "Ethnoecology of the Tarahumara," 276.
891. Pennington, *The Tarahumar*, 193.
892. The word *estafiate* proceeds, according to Aguirre Beltrán (*Medicina y magia*, 125) from the Náhuatl voice Iztauhyatl (water of the salt deity). He relates that during the feasts of the Aztec salt goddess, dancers and priests wore garlands made of it.
893. Aguirre Beltrán (*Medicina y magia*, 126) claims that the *techichinami*—the equivalent of the Tarahumaras' *waníame*—before sucking out a foreign body from the flesh of an individual, chewed estafiate as a prophylactic precaution. Its uses by the Aztecs include for weakness of the hands and anal and feet ailments. An extract of estafiate flowers was used in colonial Mexico to treat excessive heat and lightning stroke, intractable neuralgic pain, epilepsy, and amorous discord (De la Cruz, *Libellus*, 329–343, in Aguirre Beltrán, *Medicina y magia*, 126). Ortiz de Montellano (*Aztec Medicine*, 194–204) claims that *yauhtli* (*basigó*) and *iztauhyatl* were the Tlaloc's (rain god's) herbs, implying that they were used for diseases caused by the rain god.
894. Falcón Mariano, "Relación de Guaguachique (Wawachiki)," in Sheridan and Naylor, *Rarámuri*, 111.
895. Brambila, *Diccionario rarámuri-castellano*, 493–494.
896. Pennington, *The Tarahumar*, 192.
897. Bye, "Ethnoecology of the Tarahumara," 166.
898. Deimel, "Pflanzen zwischen den Kulturen," 62.
899. Rodríguez López, *Supersticiones de Galicia*, 25, in Aguirre Beltrán, *Medicina y magia*, 29.

900. Debove and Achard (*Manuel de thérapeutique*, 738) described also its toxic effects when ingested in high doses: vertigo, tremors, and convulsions.
901. Pennington, *The Tarahumar*, 184; Bye, "Ethnoecology of the Tarahumara," 179.
902. Cabrera, *Plantas curativas*, 207.
903. Debove and Achard, *Manuel de thérapeutique*, 738.
904. Ibid.
905. Bye, "Ethnoecology of the Tarahumara," 285.
906. Pennington, *The Tarahumar*, 190.
907. Bennett and Zingg, *The Tarahumara*, 142.
908. Ibid.
909. Schmutz and Hamilton, *Plants that Poison*, 73, in Salmón, "Cures of the Copper Canyon," 49.
910. Bye, "Ethnoecology of the Tarahumara," 163.
911. Also known as "Tarahumara sage."
912. Vogel, *American Indian Medicine*, 238, 359.
913. Bye, "Ethnoecology of the Tarahumara," 172.
914. Bennett and Zingg, *The Tarahumara*, 173; Bye, "Ethnoecology of the Tarahumara," 172.
915. Bye, "Ethnoecology of the Tarahumara," 173, 280. Bye reports that the *salvias* sold in Chihuahua City markets are *Buddleia scorpioides* and *Buddleia* sp. According to Wagner (*Plantas medicinales*, 112), *salvia real* from central Mexico is *Buddleia perfoliata* H.B.K. See Viesca (*Estudios*, 29–43) on identifying the *salvia* mentioned by López de Hinojosos (1578) as an antispasmodic and Pérez de Zárate (*Acerca de la medicina folklórica*, 120) for its medicinal use in rural Panama.
916. Bennett and Zingg, *The Tarahumara*, 173.
917. Pennington, *The Tarahumar*, 189.
918. SEMARNAP, *Diagnóstico de productos*, 5.
919. Bye, "Ethnoecology of the Tarahumara," 172.
920. The term *sangre de grado* or *sangre de drago* is used throughout the Americas to name a very diverse species of plants. For example, in Peru the name is given to *Croton palanostigma* Klotzsch. and to *Croton draconoides* Klotzsch. Latex from the former has been shown to speed up wound cicatrization (Zapata Ortíz, "Los medicamentos," 108); the latter is used to treat whooping cough, gastric ulcers, and tonsillitis (Scarpati, "Alcaloidi," 309). In Venezuela, the name is given to *Pterocarpus podocarpus* Blake whose bark and resin are used as a gum and as a hemostatic (Rodríguez and Machado, *Algunas plantas*, 411). In Arizona, *Jatropha cardiophylla* Muell. Arg. is used for weak blood (Kay, "Health and Illness," 141).
921. Bye, "Ethnoecology of the Tarahumara," 170.
922. Watson, "List of Plants Collected by Palmer," 439, in Pennington, *The Tarahumar*, 184.
923. *Saparí*, without the glottal occlusion, is the name of the green bean. Bye ("Ethno-

ecology," 163) identifies a *saparí*, a plant of the family Gramineae, as *Elyonorus barbiculmis* Hack. He states that the roots of that plant are boiled into a tea and taken in the treatment of gastrointestinal disorders and bites of poisonous animals.
924. Bye, "Ethnoecology of the Tarahumara," 163.
925. Pennington, "Tarahumar Fish," 97, and *The Tarahumar*, 106.
926. Brambila, *Diccionario rarámuri-castellano*, 507.
927. Bye, "Ethnoecology of the Tarahumara," 163.
928. Cardenal, *Remedios y prácticas*, 163.
929. Pennington, *The Tarahumar*, 186; Brambila, *Diccionario rarámuri-castellano*, 507. Scappert and Shore ("Cyanogenesis," 337–352) studied cyanogenesis in *Turnera ulmifolia* L. The aphrodisiac plant of the Mayas, *Damiana* (*Turnera diffusa* Wild. var. *aphrodisiaca*), also belongs to the family Turneracea.
930. Brambila, *Diccionario rarámuri-castellano*, 508.
931. Bye, "Ethnoecology of the Tarahumara," 264.
932. Irigoyen-Rascón, *Cerocahui*, 91–92.
933. Cardenal, *Remedios y prácticas*, 121.
934. Bennett and Zingg, *The Tarahumara*, 128.
935. *Sewáchari* is actually the name of the sunflower (*Helianthus annus* L.), surely applied to *Tithonia* because of its likeness to it. Pennington, *The Tarahumar*, 128, Brambila, *Diccionario rarámuri-castellano*, 517, and Bye, "Ethnoecology of the Tarahumara," 277, identify *sewáchari* as *Vigueria heliathoides* H.B.K.
936. Pennington, *The Tarahumar*, 192.
937. Wagner, *Plantas medicinales*, 112.
938. Villacís, *Plantas medicinales*, 103.
939. Deimel, "Pflanzen zwischen den Kulturen," 62.
940. *Wirote* means *enredadera*, a vine, and may therefore, be applied to a large number of species. Pennington (*The Tarahumar*, 185) reports that the milky excrescence on stems of a *güirote de leche* is used by the Rarámuri to cure insect bites and as a salve for sprains. Pennington seems to believe that this wirote is the same plant reported in several eighteenth century *Relaciones* as *gareque, guarec, guareque*, or *guarica* as an important medicinal plant among the Rarámuri. Nevertheless, it is more likely that *guareque* corresponds to the modern Tarahumara *waré*. The uses of the plant reported by the Franciscan friars are also more consistent with those reported by Zingg for *waré*.
941. Bennett and Zingg, *The Tarahumara*, 173.
942. Pennington, *The Tarahumar*, 185.
943. Bye, "Ethnoecology of the Tarahumara," 179.
944. Also known as *Haematoxilon boreale* S. Watson or *Caesalpinia echinata* Lamarck. A related species, *Haematoxilon campechianum* L., is used to obtain dyes and by several Indian groups for medicinal purposes (Vogel, *American Indian*, 408, 411).
945. Bye, "Ethnoecology of the Tarahumara," 174; Cardenal, *Remedios y prácticas*, 124.

946. Bennett and Zingg, *The Tarahumara*, 171; Pennington, *The Tarahumar*, 191.
947. Brambila, *Diccionario rarámuri-castellano*, 528.
948. Cardenal, *Remedios y prácticas*, 79.
949. Deimel, "Pflanzen zwischen den Kulturen," 62.
950. Bye, "Ethnoecology of the Tarahumara," 181.
951. Bye ("Ethnoecology," 123, 167, 181, 287) also calls *chiná* the plant *Cirsium mexicanum* DC., which is chewed to tighten lose teeth.
952. Cardenal, *Remedios y prácticas*, 62.
953. Pennington, *The Tarahumar*, 182; Bye, "Ethnoecology of the Tarahumara," 181, 260, 287.
954. Bye, "Ethnoecology of the Tarahumara," 287.
955. Pennington, *The Tarahumar*, 77, 126.
956. "Relación de Guazapares," 1777, in Del Paso y Troncoso, *Relaciones del siglo XVIII*.
957. Preciado et al., *Informe Kwechi*, appendix 4.
958. Pennington, *The Tarahumar*, 187.
959. Bye, "Ethnoecology of the Tarahumara," 181, 259.
960. Pennington, *The Tarahumar*, 181.
961. Bye, "Ethnoecology of the Tarahumara," 181, 260.
962. Cardenal, *Remedios y prácticas*, 79.
963. Bye, "Ethnoecology of the Tarahumara," 261.
964. Cardenal, *Remedios y prácticas*, 63.
965. Irigoyen-Rascón, *Cerocahui*, 92.
966. Pennington, *The Tarahumar*, 191; Brambila, *Diccionario rarámuri-castellano*, 533; Cardenal, *Remedios y prácticas*, 125; Bye, "Ethnoecology of the Tarahumara," 168.
967. Salmón, "Cures of the Copper Canyon," 50.
968. Corn styles are still used as a diuretic and urinary antiseptic throughout Mexico. This application extends as far as Panama (Pérez de Zárate, *Medicina folklórica*, 88).
969. Cabrera, *Plantas curativas*, 89.
970. Cardenal, *Remedios y prácticas*, 90.
971. Lumholtz, *Unknown Mexico*, 1:272.
972. Bennett and Zingg, *The Tarahumara*, 280.
973. Lumholtz, *Unknown Mexico*, 1:253.
974. Kennedy, "Tesguino Complex," 623.
975. Bye, "Ethnoecology of the Tarahumara," 173.
976. Pennington, *The Tarahumar*, 183.
977. Bye, "Ethnoecology of the Tarahumara," 173.
978. Cardenal, *Remedios y prácticas*, 126.
979. Salmón, "Cures of the Copper Canyon," 48.
980. Bennett and Zingg, *The Tarahumara*, 138.
981. Bye, "Ethnoecology of the Tarahumara," 180, 296.
982. Salmón, "Cures of the Copper Canyon," 48.

983. Cardenal, *Remedios y prácticas*, 111.
984. References to the use of *Datura* sp. as a remedy date back to the Badianus Manuscript. Vogel (*American Indian*, 165) states that Sahagún in 1577 described two main Aztec medicines derived from seeds of *Datura* sp.—*mixitl* and *tlapatl*—used for gout and rheumatic conditions. Sahagún remarked that *toloache* decreases appetite, then inebriates the subject and, lastly, causes insanity, which may become permanent. Lumholtz (*Unknown Mexico*, 1:4) reported that the Navajos considered the roots of *Datura meteloides* a powerful stimulant, although the better class among the tribe discouraged using it because it often led to madness and death. Yaquis in Sonora use a potion prepared from it for analgesia, particularly during childbirth. The psychoactive properties, particularly of the seeds, were well known and used in ritual curing. These uses reached Europe, where physicians tried seeds and extracts of *Datura* for conditions ranging from mania and epilepsy to ulcers and cancer (Vogel, *American Indian*, 165). From old times, the similarity of the effects from *Datura* and its derivates with those of *Atropa belladona* L. has been highlighted by medical writers.
985. "Relación de Guazapares," 1777, in Del Paso y Troncoso, *Relaciones del siglo XVIII*. De Ramón (*Diccionario popular*, 4:137) defines the archaic term *flucción* or *fluxión* as "a morbid accumulation of liquids or humors in any—generally overexcited—place of the body." The most common flucciones were those of the molar teeth and *fluxion de pecho*, an acute febrile affliction with labored breathing, cough, and difficult, frequently painful, expectoration of viscous sputa, often tinged with blood.
986. Lumholtz, *Unknown Mexico*, 1:356.
987. Bennett and Zingg, *The Tarahumara*, 138.
988. Ibid.
989. Pennington, *The Tarahumar*, 189.
990. Brambila, *Diccionario rarámuri-castellano*, 555.
991. Pennington, *The Tarahumar*, 167.
992. Bye, "Ethnoecology of the Tarahumara," 205; Cardenal, *Remedios y prácticas*, 111; Salmón, "Cures of the Copper Canyon," 48.
993. Salmón, "Cures of the Copper Canyon," 48.
994. Lewis and Elvin-Lewis, *Medical Botany*, 98, 419.
995. In Spain, the flower of the European *tilo* or *tila* has been traditionally used in preparing an anxiolytic tea; the leaves are reputedly used as an antispasmodic and for palpitations (Carvajal, *Plantas que curan*, 122). Wagner (*Plantas medicinales*, 43) adds cough, stomach, and kidney conditions to the list of indications; he claims tilo (*Tilia grandifolia* Ehrh.) is the "best diaphoretic agent known to man." The *te de tila* used in Tarahumara Country is obtained from Mexican herbalists. The flowers of the tilo tree may proceed from a number of different species: *Tilia americana* L., *T. vulgaris* Hayne, *T. tomentosa* Moench., *T. cordata* Mill., *T. mexicana* Schltdl., *T. platyphyllos* Scop., *T. heterophylla* Vent., or *T. grandifolia*.

996. Also known as *Agastache mexicana* (Kenth.) Lint. & Epling. or *Cedronella mexicana* (Kunth) Briq.
997. Cardenal, *Remedios y prácticas*, 130.
998. Cabrera, *Plantas curativas*, 245.
999. The Spanish word *trébol* (clover, shamrock) is applied throughout Mexico to herbaceous plants of the genera *Trifolium* and *Melilotus*, indicating that the leaves of these plants are composed of three leaflets (*Enciclopedia de México*, 12:427).
1000. Pennington, *The Tarahumar*, 183.
1001. Cardenal, *Remedios y prácticas*, 130.
1002. Deimel, "Pflanzen zwischen den Kulturen," 61.
1003. Pennington, *The Tarahumar*, 126.
1004. Bye, "Ethnoecology of the Tarahumara," 288.
1005. Lumholtz, *Unknown Mexico*, in Bye, "Ethnoecology of the Tarahumara," 176.
1006. Preciado et al., *Informe Kwechi*, appendix 5.
1007. Deimel, "Pflanzen zwischen den Kulturen," 61.
1008. Irigoyen-Rascón, *Cerocahui*, 92.
1009. Vogel, *American Indian Medicine*, 221, 276.
1010. Bennett and Zingg, *The Tarahumara*, 62.
1011. Pennington, *The Tarahumar*, 165.
1012. Cardenal, *Remedios y prácticas*, 71.
1013. Pennington, *The Tarahumar*, 117.
1014. Salmón, *Sharing Breath with Our Relatives*, 385; Cardenal, *Remedios y prácticas*, 132.
1015. Bye, "Ethnoecology of the Tarahumara," 285.
1016. Brambila, *Diccionario rarámuri-castellano*, 574.
1017. Bennett and Zingg, *The Tarahumara*, 175.
1018. Bennett and Zingg, *The Tarahumara*, 175.
1019. Irigoyen-Rascón, *Cerocahui*, 89.
1020. Pennington, *The Tarahumar*, 182.
1021. Brambila, *Diccionario rarámuri-castellano*, 574.
1022. Cardenal, *Remedios y prácticas*, 58, 87.
1023. Deimel, "Pflanzen zwischen den Kulturen," 62.
1024. Cabrera, *Plantas curativas*, 45.
1025. In many parts of Mexico, catnip (*Nepeta cataria* L.) is also called *valeriana*. For the uses of catnip as a remedy by North American Indians, see Vogel's *American Indian Medicine*, 330.
1026. Bye, "Ethnoecology of the Tarahumara," 182, 288.
1027. Although the active therapeutic components responsible for valerian's therapeutic effects have not been identified precisely (Krystal and Ressler, "The Use of Valerian," 842). Cabrera (*Plantas curativas*, 115) believed that the sedative and antispasmodic properties of valeriana are explained by the actions of the essential oil and

valerianic acid (also known as valeric or n-pentanoic acid.). Lapedes (*McGraw-Hill Dictionary*, 1703) reports that valepotriates are unlikely to play an important role in the sedative effects of valerian because they are highly unstable.

1028. Mercedes Herrera, personal communication, Norogachi, June 1, 1975.
1029. More precisely, *Verbena elegans* H.B.K. var. *asperata* Perry (Bye, *Ethnoecology*, 298). The classic European verbena is, of course, *Verbena officinalis* L., a plant reputed for its diaphoretic, diuretic, and antiseptic properties.
1030. Brambila, *Diccionario rarámuri-castellano*, 416. Actually *owáame* means "medicine, or remedy" (from *ówima*=to cure, heal).
1031. Bye, "Ethnoecology of the Tarahumara," 182, 298.
1032. Bennett and Zingg, *The Tarahumara*, 143; Cardenal, *Remedios y prácticas*, 98.
1033. "Relación de Guazapares," in Pennington, *The Tarahumar*, 188.
1034. Bennett and Zingg, *The Tarahumara*, 143.
1035. Pennington, *The Tarahumar*, 188.
1036. Bye, "Ethnoecology of the Tarahumara," 136, 182.
1037. In Del Paso y Troncoso, *Relaciones del siglo XVIII*. Also Pennington, *The Tarahumar*, 185. Falcón Mariano, "Relación de Guaguachique (Wawachiki)," in Sheridan and Naylor, *Rarámuri*, 112.
1038. Bennett and Zingg, *The Tarahumara*, 176.
1039. Bennett and Zingg, *The Tarahumara*, 172.
1040. Brambila, *Diccionario rrámuri-castellano*, 584.
1041. Bye, "Ethnoecology of the Tarahumara," 285.
1042. Bennett and Zingg, *The Tarahumara*, 172.
1043. Pennington, *The Tarahumar*, 117, 185.
1044. Bye, "Ethnoecology of the Tarahumara," 168.
1045. Bennett and Zingg, *The Tarahumara*, 172.
1046. Lumholtz, *Unknown Mexico*, 1:247–248.
1047. Campos and Gaeta, *El rutuburi*, 8.
1048. Lumholtz, *Unknown Mexico*, 1:273.
1049. Bennett and Zingg, *The Tarahumara*, 124.
1050. Pennington, *The Tarahumar*, 185, and *The Tepehuan of Chihuahua*, 182; Watson, "List of Plants Collected by Palmer," 439, in Bye," Ethnoecology of the Tarahumara," 178; Bye, "Ethnoecology of the Tarahumara" 168.
1051. Bennett and Zingg, *The Tarahumara*, 172.
1052. Lewis and Elvin-Lewis, *Medical Botany*, 39–40.
1053. Pennington, *The Tarahumar*, 186.
1054. Brambila, *Diccionario rarámuri-castellano*, 585.
1055. Deimel, "Pflanzen zwischen den Kulturen," 62.
1056. Pennington, *The Tarahumar*, 108.
1057. Bye ("Ethnoecology," 261) claims that *Ligusticum* grows in partially shaded, moist sites in the pine-oak forest where it is eagerly sought by the Tarahumaras for domestic use and for sale in the herb markets of Chihuahua City. Actually, chuchupaste's

reputation as a panacea reaches all of Mexico: Wagner (*Plantas medicinales*, 99) reports on its use against diabetes, headache, and insomnia.
1058. "Relación de Guazapares," "Relación de Nabogame," "Relación de Chínipas," quoted by Pennington, *The Tarahumar*, 187.
1059. Lumholtz, *Unknown Mexico*, 1:273.
1060. Salmón, *Sharing Breath with Our Relatives*, 354.
1061. Bye, "Ethnoecology of the Tarahumara," 181, 261.
1062. Salmón, *Sharing Breath with Our Relatives*, 354.
1063. Pennington, *The Tarahumar*, 187.
1064. Cardenal, *Remedios y prácticas*, 68.
1065. Brambila, *Diccionario rarámuri-castellano*, 585.
1066. Bye, "Ethnoecology of the Tarahumara," 181.
1067. Pennington, *The Tarahumar*, 165; Brambila, *Diccionario rarámuri-castellano*, 585.
1068. Bye, "Ethnoecology of the Tarahumara," 139, 147.
1069. From the Náhuatl *tetl*=rock, *ixcalli*=boiled, cooked (therefore *boiled stone*=lava) + *amatl*=the paper tree; that is, *tezcalama* is the "tree that grows on the lava" (Robalo, *Diccionario*, 472).
1070. Bennett and Zingg, *The Tarahumara*, 166.
1071. Pennington, *The Tarahumar*, 180; Bye, "Ethnoecology of the Tarahumara," 176.
1072. Bye, "Ethnoecology of the Tarahumara," 140.
1073. SEMARNAP, *Diagnóstico de productos*, 4.
1074. Brambila, *Diccionario rarámuri-castellano*, 587–588.
1075. Pennington, *The Tarahumar*, 193.
1076. Salmón, *Sharing Breath with Our Relatives*, 371.
1077. Bennett and Zingg, *The Tarahumara*, 175.
1078. Pennington, *The Tarahumar*, 181.
1079. Salmón, *Sharing Breath with Our Relatives*, 371.
1080. Also *Prosopis juliflora* (Swartz) DC. and *Prosopis glandulosa* Torr. or *Prosopis juliflora* var. *glandulosa* Cock (Vogel, *American Indian*, 336). Bye ("Ethnoecology," 281) reports among his specimens *Prosopis glandulosa* Torr. var. *torreyana* (L. Benson) M. C. Johnst.
1081. Brambila, *Diccionario rarámuri-castellano*, 169.
1082. Bye, "Ethnoecology of the Tarahumara," 177.
1083. Burgess McGuire and Mares Trías, *Ralámuli nu'tugala go'ame*, 280.
1084. Bennett and Zingg, *The Tarahumara*, 175.
1085. Bye, "Ethnoecology of the Tarahumara," 281.
1086. Brambila, *Diccionario rarámuri-castellano*, 67.
1087. Bennett and Zingg, *The Tarahumara*, 135; Pennington, *The Tarahumar*, 181.
1088. Pennington, *The Tarahumar*, 182.
1089. Brambila, *Diccionario rarámuri-castellano*, 169, 588.
1090. Pennington, *The Tarahumar*, 107.
1091. Ibid.

1092. Ibid.
1093. Villacís, *Plantas medicinales*, 82.
1094. Pennington, *The Tarahumar*, 182.
1095. Bye, "Ethnoecology of the Tarahumara," 173.
1096. Bennett and Zingg, *The Tarahumara*, 167.
1097. Bye, "Ethnoecology of the Tarahumara," 178.
1098. Bennett and Zingg, *The Tarahumara*, 175.
1099. Bye, "Ethnoecology of the Tarahumara," 175. Vogel (*American Indian*, 336), based on the Badianus Manuscript and Hrdlička, reports the use of mesquite as an ophthalmologic panacea among the Aztecs, Mescalero Apaches, Maricopas, and Pimas. Villacís (*Plantas medicinales*, 82) reports the use in central Mexico of infusions of *mezquite* seeds and bark for gastritis, diarrhea, dysentery, pharyngitis, and laryngitis.
1100. Brambila, *Diccionario rarámuri-castellano*, 597.
1101. Ibid.
1102. Pennington, *The Tarahumar*, 178.
1103. Bye ("Ethnoecology," 198) claims that the *Mammillaria heyderi* identified by Zingg was actually *M. craigii*, a closely related species. Dawson (*How to Know the Cacti*, 144) describes two varieties of *M. heyderi*: *M. heyderi* var. *macdougali* (Rose) L. Benson and *M. heyderi Muhlenpfordt* var. *heyderi*.
1104. Bye, "Ethnoecology of the Tarahumara," 200.
1105. Brambila, *Diccionario rarámuri-castellano*, 599.
1106. Cardenal, *Remedios y prácticas*, 112.
1107. From the Náhuatl *huiztli*=thorn, or spine and *náhuac*=around (Robalo, *Diccionario*, 160).
1108. Bruhn and Bruhn, "Alkaloids and Ehtnobotany," 245; Bye, "Ethnoecology of the Tarahumara," 199.
1109. Bennett and Zingg, *The Tarahumara*, 137.
1110. Cardenal, *Remedios y prácticas*, 112, 137.
1111. Salmón, *Sharing Breath with Our Relatives*, 361.
1112. Bye, "Ethnoecology of the Tarahumara," 199.
1113. Raffauf, *A Handbook of Alkaloids*, in Bye, "Ethnoecology of the Tarahumara," 199.
1114. Salmón, *Sharing Breath with Our Relatives*, 383.
1115. Pennington, *The Tarahumar*, 189.
1116. Salmón, *Sharing Breath with Our Relatives*, 383.
1117. Brambila (*Diccionario rarámuri-castellano*, 603) states that in Norogachi the term *wipá* is also given to a plant (other than tobacco) characterized by alternate leaves, a single stem, and bundled small yellow flowers. He stresses that a decoction prepared from the leaves has hemostatic action and is given to women with menstrual disorders.
1118. Vogel, *American Indian Medicine*, 386.
1119. Pennington, *The Tarahumar*, 68.

1120. Bennett and Zingg, *The Tarahumara*, 139.
1121. Aguirre Beltrán, *Medicina y magia*, 127–130; Sahagún (1577) and Ortiz de Montellano, *Aztec Medicine*, 154–156.
1122. Kock-Weser (in Wintrobe et al., *Harrison medicina interna*, 733) points out that the adult lethal dose of nicotine is only of 50 milligrams. Smoked tobacco is actually less toxic because nicotine is poorly absorbed from the whole tobacco and most of the alkaloid is burned. Literature on the noxious effects of nicotine and tobacco is too extensive and well known to be quoted here. *Nicotiana tabacum* also contains malic, oxalic, acetic, and pyridin-carbolic acids; asparagine, gum, resinoid substances, sulfates, chlorides, and phosphates of calcium, sodium, and magnesium; potassium nitrate, nicotein, nicocyamine, colidine, and pyridine. Lewis and Elvin-Lewis (*Medical Botany*, 392) report that the leaves of *Nicotiana tabacum* contain 0.6 to 9 percent of nicotine, as well as lesser amounts of nornicotine and an aromatic principle known as tobacco camphor. The roots contain the additional pyridine alkaloids anabasine and anabatine. Salmón ("Cures of the Copper Canyon," 49) claims that the medicinally active compounds in *N. trigonophylla* are eugenol, quercitrin, guaiacol, and rutin.
1123. Bye, "Ethnoecology of the Tarahumara," 180.
1124. Bennett and Zingg, *The Tarahumara*, 169.
1125. Pennington, *The Tarahumar*, 190.
1126. Bennett and Zingg, *The Tarahumara*, 138.
1127. Bye, "Ethnoecology of the Tarahumara," 179, 286.
1128. Bye, "Ethnoecology of the Tarahumara," 179.
1129. Deimel, "Pflanzen zwischen den Kulturen," 62.
1130. Cardenal, *Remedios y prácticas*, 41.
1131. Robalo, *Diccionario de aztequismos*, 329.
1132. Bye, "Ethnoecology of the Tarahumara," 272.
1133. Burgess McGuire and Mares Trías, *Ralámuli nu'tugala go'ame*, 475.
1134. Ibid.
1135. Bye, "Ethnoecology of the Tarahumara," 272.
1136. Cabrera, *Plantas curativas*, 362.
1137. Literally "flower of the dead."
1138. Cardenal, *Remedios y prácticas*, 56. Anzúres (*Religión, magia*, 29) claims that in central Mexico the leaves of *cempoalxóchitl* are used externally to alleviate "cold" illnesses, while in infusion they are drunk to correct and template a "cold stomach," as a diuretic, emmenagogue, and for empacho, flatulent colic, and "Miserere colic."
1139. Morton, *Atlas of Medicinal Plants*, 969.
1140. Martínez, Maximo, *Las plantas medicinales*, 69, in Orellana, *Indian Medicine*, 244.
1141. *Enciclopedia de México* (2:899) identifies *cempaxuchilito* as *Bidens angustissima* H.B.K., which is found in Jalisco, San Luis Potosí, Hidalgo, and Chiapas. The description of that plant resembles the one of the Chihuahuan *zempoalillo*.

11. The Tarahumaras

1. Balke and Snow, "Anthropological and Physiological Observations," 293–302; Groom, "Cardiovascular Observations," 304–314.
2. McMurry, Cerqueira, Connor, and Connor, "Changes in lipid and lipoprotein levels," 1704–1708.
3. Lumholtz, *Unknown Mexico*, 1:241.
4. Monárrez Espino and Martínez Salgado, "Condiciones de vida," 1.
5. See Daugherty, "Railroad Log," 61; Irigoyen-Rascón, *Cerocahui*, 196; Kennedy, *Tarahumara*, 196; González Rodríguez, *Crónicas de la Sierra*, 99; and Mull and Mull, "Differential Use of a Clinic," 257, and "Infanticide," 126.
6. Paredes, West, and Snow, "Biosocial Adaptation," 172.
7. Herrera Beltrán ("Aumentaron 30%," 1) reported 219 deaths of children under five in 1995 and 288 in 1996 in Tarahumara Country.
8. Irigoyen-Rascón, *Cerocahui*, 182.
9. From the original attempt of quantifying mortality by the second year of life, I learned that fifteen mestizos of the sample (twelve of them born at the hospital, two by C-section) and nine Tarahumaras lived past this landmark.
10. In 1974, Murray ("The Tarahumara Project," 207) determined by sampling families in Norogachi a general child mortality (child deaths/population) of 38 percent for Tarahumaras and 11 percent for mestizos. Monárrez Espino and Martínez Salgado's ("Condiciones de vida," 7) study in Guachochi reports that almost half of the Tarahumara children in their sample did not have a dead five-year-old or younger sibling, 24.8 percent only one, and 3.6 percent up to the age of five. In a later study ("Prevalencia de desnutrición," 9), the same authors calculated an adjusted infant mortality of 95.3 per 1,000 live births.
11. Mull and Mull, "Differential Use of a Clinic," 255; Monárrez Espino and Martinez Salgado, "Condiciones de vida," 6.
12. A fact reported since colonial times, for example by Father Figueroa, S.J. (in Dunne, *Early Jesuit Missions*, 94).
13. I found a remarkable exception at ranchería Borachiki, ten miles from Samachique, a twelve- or thirteen-year-old, with stigmata from severe mental retardation. The child had survived that far, never developing language and walking with very poor coordination. He was always inside the house totally naked and mostly ignored by the rest of the family whom he neither helped nor bothered.
14. Bennett and Zingg, *The Tarahumara*, 348.
15. Decorme, in 1608, in *La obra de los Jesuitas*, 2:252.
16. Lumholtz, *Unknown Mexico*, 1:243.
17. Bennett and Zingg, *The Tarahumara*, 229.
18. Mull and Mull, "Infanticide," 113–132.
19. See also Connor et al., "The Plasma Lipids," 113–142.

20. Cerqueira, "The Foods and Nutrient Intake of the Tarahumara Indians of Mexico," 1–100.
21. Monárrez Espino and Martínez Salgado, "Prevalencia de desnutrición," 8.
22. Laguna and Carpenter, "Raw Versus Processed Corn," 3–8; Laguna, *Bioquímica*, 565.
23. Ratkay, *An Account of the Tarahumara Missions*, 495–505.
24. Fonte, *Annua, 1611*, in González-Rodríguez, *Tarahumara*, 154–158, and in Sheridan and Naylor, *Rarámuri*, 9–11; Neumann, Letters to an Unknown Priest, September 15, 1693, p. 92.
25. Passin, "Place of Kinship," pt. 2, 480n40.
26. Gajdusek, "The Sierra Tarahumara," 37.
27. Brambila (*Hojas de un diario*, 66) describes how in the late 1940s an epidemic decimated Narárachi and its surroundings.
28. Mull and Mull, "Differential Use of a Clinic," 262.
29. Gajdusek and Rogers, "Specific Serum Antibodies," 831–832.
30. Gajdusek, "The Sierra Tarahumara," 37; Irigoyen-Rascón, *Cerocahui*, 188.
31. Mull and Mull, "Differential Use of a Clinic," 262.
32. Gajdusek and Rogers, "Specific Serum Antibodies," 831.
33. Mull and Mull, "Differential use of a Clinic," 252.
34. Gajdusek and Rogers, "Specific Serum Antibodies," 827.
35. Lumholtz, *Unknown Mexico*, 1:242; Gajdusek and Rogers, "Specific Serum Antibodies," 830.
36. Gajdusek and Rogers, "Specific Serum Antibodies," 820.
37. Irigoyen-Rascón, *Cerocahui*, 188–189.
38. Murray, "The Tarahumara Project," 210.
39. Irigoyen-Rascón, *Cerocahui*, 189.
40. Gajdusek, "The Sierra Tarahumara," 38.
41. Lumholtz, *Unknown Mexico*, 1:229. According to Bye (in Burgess McGuire and Mares Trías, *Ralámuli nu'tugala go'ame*, 500), *Amanita caesarea*. But also *Boletus edulis*. Lumholtz (*Unknown Mexico*, 2:200–201) reported that in the summer, especially in July, Tarahumaras from Guachochi, aided by dogs, looked for a false truffle (*Melanogaster variegatus* Farley var. *mexicanus*) that was esteemed as food. Bye ("Ethnoecology," 219) reports on *morochiki*, an edible mushroom, and *chimónowi*, a truffle. He also singles out *rarachaka* as a poisonous species.

Sister Aquina Hittman, a German nun from the Egerland (now in the Czech Republic), could find species of delicious edible mushrooms in the neighborhood of Norogachi: *herrenpilsen*, *eipilsen*, and *steinpilsen*. Tarahumara informants consulted reported that the mushrooms collected by the nun were not edible; they were afraid they might be poisonous.
42. Irigoyen-Rascón, *Cerocahui*, 155–160; Irigoyen-Rascón and Jesús Manuel Palma, "Rarajípari: The Kick-ball Race," 82–83; Palma, Erasmo, *Donde cantan los pájaros*, 107–110.

43. Deimel, *Tarahumara*, 70.
44. West, Paredes, and Snow, *Sanity in the Sierra Madre*, 1–3, in Yamamoto, "Psychohistorical View," 96–97; Paredes and Irigoyen-Rascón, "Jíkuri, the Tarahumara Peyote Cult," 121.
45. Kennedy, "Tesguino Complex," 630–647.
46. Lumholtz, *Unknown Mexico*, 1:237.
47. Lumholtz, *Unknown Mexico*, 1:243.
48. Mull and Mull, "Differential use of a Clinic," 245–264; Irigoyen-Rascón, *Cerocahui*, 169–183.
49. Aguirre Beltrán, "Medicina y salubridad," 29–31; Mull and Mull, "Differential Use of a Clinic," 255, 257.
50. Other barriers found by Tarahumaras to access health care are analyzed by Mull and Mull ("Differential Use of a Clinic," 260). A complaint repeatedly expressed to the press is the cost of treatment. Private and public facilities usually charge a very low recuperation fee, which is however excessive for Tarahumaras (Bustillos, "Indígenas, marginados de servicios médicos," 1).

Bibliography

Abbé, Juan Isidro Fernández. "Report on the Founding and Progress of this Mission of Jesús Carichí, from November 8, 1675." Translated by Daniel S. Matson. In Sheridan and Naylor, *Rarámuri*, 78–86.
Aguirre Beltrán, Gonzalo. *Medicina y magia: El proceso de aculturación en la estructura colonial.* Mexico City: Instituto Nacional Indigenista, 1963.
———. "Medicina y salubridad en las comunidades indígenas, o el encuentro de las dos culturas médicas mexicanas." *Mundo Médico* 2 (1974): 13.
Agurell, Stig. "Cactacea Alkaloids, 1." *Lloydia: The Journal of Natural Products* 32 (1969): 213–214. In Anderson, *Peyote*, 122.
Alegre, Francisco Javier, S.J. *Historia de la compañía de Jesús en Nueva España hasta el tiempo de su expulsión.* Edited by Ernest J. Burrus and Félix Zubillaga. Rome: Institutum Historicum S.I., 1960. First published 1780.
Almada, Francisco R. *Diccionario de historia, geografía y biografía chihuahuenses.* 2nd ed. Chihuahua, Mexico: Universidad de Chihuahua, 1968.
———. *Geografía del estado de Chihuahua.* Chihuahua, Mexico: Ruíz Sandoval, 1945.
———. *Resumen de historia del estado de Chihuahua.* Mexico City: Libros Mexicanos, 1955.
Altamirano, R. "Estudio comparativo de la planta llamada Matarique, *Cacalia decomposita* (Compositae)." *Estudio* 3:81–86. Mexico City: Instituto Médico Nacional, 1890.
Álvarez Gómez, Ángel. *Boletín metereológico del estado de Chihuahua: Compendio, 1957–1969.* Chihuahua, Mexico: Gobierno del Estado de Chihuahua / Secretaría de Agricultura y Ganadería, Unión Ganadera Regional, 1970.
Anderson, Edward F. *Peyote: The Divine Cactus.* Tucson: University of Arizona Press, 1980.
Anzúres y Bolaños, María del Cármen. "Religión, magia y medicina indígenas." *Estudios Indígenas* 2, no. 1 (1972): 25–34.
Aréchiga, Antonio Xavier. "Relación de Thomochic (Tomochi), 1777." In Del Paso y Troncoso, *Relaciones del siglo XVIII relativas a Chihuahua*.
Arnozan, X. *Précis de thérapeutique.* Paris: [Octave Doin], 1903.

Artaud, Antonin. *Los Tarahumaras*. Barcelona: Barral, 1972.
———. *México y viaje al país de los Tarahumaras*. Mexico City: Fondo de Cultura Económica, 1984.
Ascher, Robert, and Francis J. Clune. "Waterfall Cave, Southern Chihuahua, Mexico." *American Antiquity* 26 (1960): 2: 270–274.
Balke, Bruno, and Clyde Snow. "Anthropological and Physiological Observations on Tarahumara Endurance Runners." *American Journal of Physical Anthropology* 23 (1965): 293–302.
Bancroft, Hubert Howe. *History of the North Mexican States and Texas*. San Francisco: A. L. Bancroft, 1884.
Bannerman, R. H. *Traditional Medicine and Health Care Coverage: A Reader for Health Administrators and Practitioners*. Geneva: World Health Organization, 1983.
Basauri, Carlos. *Monografía de los Tarahumaras*. Mexico City: Talleres Gráficos de la Nación, 1929.
Bassols, Angel Batalla. *El noroeste de México: Un estudio geográfico económico*. Mexico City: Universidad Autónoma de México, Instituto de Investigaciones Económicas, 1972.
Beals, Ralph Leon. *The Comparative Ethnology of Northern Mexico before 1750*. Ibero-Americana, 2. Berkeley: University of California Press, 1932.
Becerril Buitrón, Ana María. "Contribución al estudio farmacológico de la raíz de matarique." Thesis, Universidad Nacional Autónoma de México, 1950. In Bye, "Ethno-ecology of the Tarahumara," 142.
Bennett, Wendell C., and Robert M. Zingg. *The Tarahumara: An Indian Tribe of Northern Mexico*. Chicago: University of Chicago Press, 1935. Facsimile reprint. Glorieta, N.M.: Rio Grande, 1976.
Blackwell, Will H. *Poisonous and Medicinal Plants*. Englewood Cliffs, N.J.: Prentice Hall, 1990.
Blanco Madrid, Elco, Irma D. Enríquez Anchondo, and María Elena Siqueiros Delgado. *Manual de plantas tóxicas del estado de Chihuahua*. Chihuahua, Mexico: Centro Librero la Prensa, 1983.
Bradley, C. E. "Yerba de la flecha: Arrow and Fish Posion of the American Southwest." *Economic Botany* 10, no. 4 (1956): 362–366.
Brambila, David. *Bosquejos del alma tarahumara*. Mimeograph. Sisoguichi, Mexico: published by author, 1967.
———. *Diccionario castellano-rarámuri*. Mexico City: Obra Nacional de la Buena Prensa, 1983.
———. *Diccionario rarámuri-castellano*. Mexico City: Obra Nacional de la Buena Prensa, 1976.
———. *Gramática rarámuri*. Mexico City: Buena Prensa, 1953.
———. *Hojas de un diario*. Mexico City: Jus, 1950.
———. "Psicologia y educación del Tarahumar." *América Indigena* 19, no. 3 (1959): 199–208.

———. *Supersticiones y costumbres de los Rarámuri.* Edited by Ricardo Robles. Mimeograph. Sisoguichi, Mexico: R. Robles, 1960.
Brondo Whitt, E. *Nuevo León: novela de costumbres, 1896–1903.* Mexico City: Lumen, 1935.
Brouzès, Françoise. "De la tortilla aux corn-flakes : L'alimentation dans une communauté indigene de la basse Tarahumara (Mexique)." Thesis, Université de Bordeaux, 1979.
———. "La nourriture partagée: Introduction à l'étude de la kórima chez les Tarahumaras (Mexique)." Ph.D. diss., Université de Bordeaux, 1980.
Bruhn, Jan G., and Catarina Bruhn. "Alkaloids and Ethnobotany of Mexican Peyote Cacti and Related Species." *Economic Botany* 27 (1973): 241–251. In Bye, "Ethnoecology of the Tarahumara," 212.
Bruhn, Jan G., and Jan Erik Lindgren. "Cactacea Alkaloids, 23: Alkaloids of *Pachycereus pecten-aboriginum* and *Cereus jamacaru*." *Lloydia: The Journal of Natural Products* 39 (1976): 175–177.
Burgess McGuire, Donald. "Leyendas tarahumaras." In González Rodríguez, Luis, and Don Burgess, *Tarahumara,* 70–177. Mexico City: Chrysler de México, 1985.
Burgess McGuire, Donald. *¿Podrías vivir como un Tarahumara?* Edited by Bob Schalwijk. Mexico City: Abeja, 1975.
Burgess McGuire, Donald, Ramón López Batista, Ignacio León Pacheco, Luis Castro Jiménez, and Albino Mares Trías. *Rarámuri riěcuara: Juegos y deportes de los Tarahumaras.* Chihuahua, Mexico: Don Burgess McGuire, 1981.
Burgess McGuire, Donald, and Albino Mares Trías. *Ralámuli nu'tugala go'ame: Comida de los Tarahumaras.* Chihuahua, Mexico: Don Burgess McGuire, 1982.
———. *Tase suhuíboa huijtabúnite — No moriremos de diarrea.* 2nd ed. Chihuahua, Mexico: Don Burgess McGuire, 1993.
Burrus, Ernest J., and Félix Zubillaga. *El noroeste de México: Documentos sobre las misiones jesuíticas, 1600–1769.* Mexico City: Universidad Autónoma de México, Instituto de Investigaciones Históricas. 1986
Bustillos, Blanca. "Indígenas, marginados de servicios médicos: No tienen para pagar lo poco que les cobran." *El Heraldo de Chihuahua,* January 1, 1995.
Bye, Robert A. "Ethnoecology of the Tarahumara of Chihuahua, Mexico." Ph.D. diss., Harvard University, 1976.
———. "Hallucinogenic Plants of the Tarahumara." *Journal of Ethnopharmacology* 1 (1979): 23–48.
———. "Medicinal Plants of the Sierra Madre: Comparative Study of Tarahumara and Mexican Market Plants." *Economic Botany* 40, no. 1 (1986): 103–124.
———. "Plantas psicotrópicas de la Tarahumara." *Cuadernos Científicos, CEMEF* 4 (1975), 49–72.
Bye, Robert A., Donald Burgess McGuire, and Albino Mares Trías. "Ethnobotany of the Western Tarahumara of Chihuahua, Mexico: 1. Notes on the Genus *Agave*." Harvard University, *Botanical Museum Leaflets* 24, no. 5 (1975): 85–112.

Bye, Robert A., and Lincoln Constance. "A New Species of *Tauschia* (Umbelliferae) from Chihuahua, Mexico." *Madroño* 26, no. 1 (1979): 44–47.

Cabrera, Luis G. *Diccionario de aztequismos*. 2nd ed. Mexico City: Oasis, 1975.

———. *Plantas curativas de México, propiedades medicinales de las más conocidas plantas de México, su aplicación correcta y eficaz*. 5th ed. Mexico City: Cicerón, 1958.

Cabrera, Luis G., Lyle Campbell, and Marianne Mithun, eds. *The Languages of Native America: Historical and Comparative Assessment*. Austin: University of Texas Press, 1979.

Campos, Javier M., and Ascención Gaeta. *El rutuburi: Una danza tarahumara*. Norogachi, Mexico: Javier Campos, 1970.

Caraveo, Carlos. "Cambio demográfico y etno-histórico en la región de Casas Grandes." In Irigoyen-Rascón, *Memoria del seminario de lenguas*, 17–31.

Cardenal, Francisco Fernández. *Remedios y prácticas curativas en la Sierra Tarahumara*. Chihuahua, Mexico: Camino, 1993.

Carvajal, Pío Arias. *Plantas que curan, plantas que matan*. Colección de Medicina Popular. Mexico City: Galve, 1960.

Castetter Edward F., and Willis H. Bell. *The Aboriginal Utilization of the Tall Cacti in the American Southwest*. University of New Mexico Bulletin, no. 307. Biological series 5, no. 1. Ethnobiological Studies in the American Southwest, 4. Albuquerque: University of New Mexico, 1937.

Castini, Pier Gian. *Carta anua, 1621*. In González Rodríguez, *Crónicas de la Sierra Tarahumara*, 47, and in Decorme, *La obra de los Jesuitas mexicanos*, 2:216.

———. *Carta annua, 1623*. In González Rodríguez, *Crónicas de la Sierra Tarahumara*, 51–52.

Cavo, Andrés. *Los tres siglos de México durante el gobierno español*. Mexico City: Luis Abadiano y Valdés, 1838.

Cerqueira, María Teresa. "The Foods and Nutrient Intake of the Tarahumara Indians of Mexico." Ph.D. diss., University of Iowa, 1975.

Champion, Jean R. "Acculturation among the Tarahumaras of Northwest México since 1890." *Transactions of the New York Academy of Sciences*, 17 (1955): 560–566.

Ching Vega, Oscar W. "Cascada de Basaséachi, reina de la Sierra Madre." *El Heraldo de Chihuahua*, August 11, 1976.

———. "La Tarahumara: El último refugio de una cultura que resiste la incomprensión." *El Heraldo de Chihuahua*, February 29, 1976.

Christelow, Allen. Introduction to *The Letters of Father Joseph Neumann* and his *Historia seditionum*. TS, Bolton Collection, Bancroft Library, University of California at Berkeley, 1936.

Claeys, M., L. Pieters, J. Courthort, D. Vanden Berghe, and A. J. Vlietnick. "Umuhengerin: A New Antimicrobially Active Flavonoid from *Lantana trifolia*." *Journal of Natural Products* 51, no. 5 (1988): 966–968.

Clavijero, Francisco Javier. *Historia antigua de México con cisertaciones sobre la tierra, animales y habitantes de México.* Mexico City: Juan Navarro, 1853.
Comisión del Río Fuerte. "Registros termo-pluviométricos de Norogachi." Unpublished manuscript. Mexico City: Secretaría de Recursos Hidráulicos Oficina de Climatología, 1976.
Comisión Nacional para la Erradicación del Paludismo. *Boletín Zonal,* no. 2 (1973). Navojoa, Mexico: Comisión Nacional para la Erradicación del Paludismo.
Committee on Form and Style of the Council of Biology Editors. *CBE Style Manual.* 3rd ed. Washington, D.C.: American Institute of Biological Sciences, 1972.
Connor, William E., Maria Teresa Cerqueira, Rodney W. Connor, Robert B. Wallace, M. René Malinow, and H. Richard Casdorph. "The Plasma Lipids, Lipoproteins and Diet of the Tarahumara Indians of Mexico." *American Journal of Clinical Nutrition* 31 (1978): 113–142.
Cortés, Hernan. *Cartas de relacíon.* 2nd ed. Mexico City: Mexicano Unidos, 1985.
Cramaussel, Chantal, and Salvador Álvarez Suarez. "La creación de un espacio politico simbolico: La conquista de la Nueva Vizcaya." Ph.D. diss., Escuela Nacional de Antropología, Mexico City, 1982.
Cuca, Luis Enrique Suárez. "Principios activos de *Lonchocarpus* y *Tefrosias.*" In IILA / CISO, *Simposio internazionale,* 367–371.
Cuéllar Zazueta, Rina. "Ulama, la supervivencia de un juego prehispánico en Sinaloa." *Tips de Aeroméxico* 15 (2000): 60–63.
Cutler, Hugh. "Cultivated Plant Remains from Waterfall Cave, Chihuahua." *American Antiquity* 26, no. 2 (1958): 277–279.
Daugherty, Franklin W. "Railroad Log: From Chihuahua to Los Mochis." In West Texas Geological Society, *Geologic Field Trip Guidebook,* 45–72.
Dawson, E. Yale. *How to Know the Cacti.* Dubuque, Ia.: Wm. C. Brown, 1963.
Debove, G. M., and Achard, Ch. *Manuel de thérapeutique.* Vol. 3. Paris: J. Rueff, 1902.
Debove, G. M., G. Pouchet, and A. Sallard. *Aide-mémoire de thérapeutique.* Paris: Masson / Libraires de l'Académie de Médecine, 1910.
Decorme, Gerard. *La obra de los Jesuitas mexicanos durante la epoca colonial, 1572–1767.* 2 vols. Mexico City: Antigua Libreria Robredo de José Porrúa e Hijos, 1941.
Deeds, Susan M. "Resistencia indígena y vida cotidiana en la Nueva Vizcaya: Trastomos y cambios étnico-culturales en la época colonial." In Molinari Medina and Porras Carrillo, *Identidad y cultura,* 55–69.
Deimel, Claus. "Die Peyoteheilung der Tarahumara." In *Schreibheft: Zeitschrift für Literatur.* Festschrift zum 50 Geburtstag von H. Fichte 25 (1985): 155–163.
———. "Narárachi, Zwischen Traditionalismus und Integration." Aufstatz zur Vorlage bei der Stiftung Studienkreis, Hamburg, West Germany, 1976.
———. "Pflanzen zwischen den Kulturen: Tarahumaras und Mestizen der Sierra Madre im Noroeste de México; Ethnobotanische Vergleiche." *Curare* 12 (1989): 41–64.
———. *Tarahumara: Indianer im Norden Mexikos.* Frankfurt am Main, Germany: Syndikat, 1980.

De la Cruz, Martín. *Libellus de medicinalibus indorum herbis*. Edited by Efrén C. del Pozo, 329–343. Mexico: Instituto Mexicano de Seguridad Social (IMSS), 1964. First published 1552.

De la Peña, Moisés T. *Chihuahua económico*. Chihuahua, Mexico: Adrián Morales, 1948.

Del Castillo, Gabriel. "Sentence on Convicted Tarahumara Rebels." Translated by C. Villa-Preselzky. In Sheridan and Naylor, *Rarámuri*, 65. First published 1697.

Del Castillo, Rodrigo. "La vida cotidiana en San Miguel de las Bocas del rio Florido en 1662." In González Rodríguez, *Crónicas de la Sierra Tarahumara*, 249–260.

Del Paso y Troncoso, Francisco. *Relaciones del siglo XVIII relativas a Chihuahua; Mss encontrados y coleccionados por F. Del Paso y Troncoso en los Archivos de la Real Academia de Historia de Madrid y del Archivo de Indias de Sevilla*. Biblioteca de Historiadores Mexicanos. Mexico City: Vargas Rea, 1950.

De María y Campos, Teresa. "Los animales en la medicina tradicional mesoamericana." *Anales de Antropología* 16 (1979): 183–223.

De Ramón, Luis P. *Diccionario popular universal de la lengua española*. Barcelona: Establecimiento Tipográfico-Editorial de Pablo Riera y Sans, 1896.

Der Marderosian, A., and L. Liberti. *Natural Product Medicine: A Scientific Guide to Foods, Drugs, Cosmetics*. Philadelphia: G. F. Stickley, 1988.

De Velasco, Pedro R. *Danzar o morir: Religión y resistencia a la dominación en la cultura Tarahumara*. Mexico City: 1983.

De Vries, Fred. "El caso de la ciudad de Chihuahua: Migrantes y flujos migratorios en el norte de México." Estudios Urbanos y Regionales. Chihuahua, Mexico: Desarrollo Económico del Estado de Chihuahua, 1985.

Díaz Infante, Carlos. *100,000 kilómetros misioneros en la Nueva Tarahumara*. Sisoguichi, Mexico: published by author, 1969.

———. "Dialecto tarahumara de Chinatú," 92–108. *Primer seminario sobre lenguas habladas en el estado de Chihuahua*. Chihuahua: Universidad Autónoma de Chihuahua (UACH), 1981.

———. "El Tarahumar y su cultura." *Renovación* 1 (1973): 4–5.

———. "La cultura tarahumara reflejada en su lengua." In Irigoyen-Rascon, *Memoria del seminario de lenguas*, 60–67.

Dibildox Martínez, José Luis, Celestino Villa Ayala, Gabriel Parga Terrazas, Javier Avila Aguirre, Francisco Chávez Acosta, Javier Campos Morales, and Rosendo Martínez Flores. *Declaration of the Diocese of the Sierra Tarahumara on the Exploitation of the Forest, 2000*. Translated by Cyrus Reed. Austin: Border Trade and Environment Project, Texas Center for Policy Studies, 2000.

Dick, Herbert W. *Bat Cave*. School of American Research, no. 27. Santa Fe, N.M.: School of American Research, 1965.

Dieulafoy, Georges. *Manuel de pathologie interne*. Paris, 1901.

Di Peso, Charles C. *Casas Grandes: A Fallen Trading Center of the Gran Chichimeca*. Dragoon, Ariz.: Amerind Foundation; Flagstaff, Ariz.: Northland, 1974.

———. "Prehistory: O'otam." In Ortiz and Sturtevant, *Handbook of North American Indians*, 9:91–99.

———. "Prehistory: Southern Periphery." In Ortiz and Sturtevant, *Handbook of North American Indians*, 9:152–161.

Dirección General de Estadística. *Censos de población, 1970*. Mexico City: Secretaría de Industria y Comercio, 1970.

Documentos para la historia de México. 4th ser., vol. 3. Mexico City: Vicente García Torres, 1857.

Domínguez, Silverio, José Peón y Contreras, Domingo Orvañanos, Benito Bordas, William B Atkinson, Henry M. Lyman, Christian Fenger, W. T. Belfield, Buchanan Burr, Morris L King, Webster H. Jones, and Consejo Superior de Salubridad of Mexico City. *El médico práctico doméstico y enciclopedia de la medicina*. Guelph, Canada: World / Mexico City: Griffin and Campbell, 1889.

Driver, Harold E., and Wihelmine Creel. *Ethnography and Acculturation of the Chichimeca-Jonaz of Northeast Mexico*. Indiana University Research Center in Anthropology, Folklore, and Linguistics, Publication no. 26. *International Journal of American Linguistics* 29, no. 2, part 2 (1963).

Duke, James A. *Handbook of Medicinal Herbs*. Boca Raton, Fla.: CRC, 1985.

———. "Jatropha curcas L." In *Handbook of Energy Crops*, 1983. www.hort.purdue.edu/newcrop/duke_energy/Jatropha_curcas.html.#sthash.a4Tps8wb.dpuf.

Duke, James A., and Edward Ayensu. *Medicinal Plants of China*. Vol. 1. Algonac, Mich.: Reference Publications, 1985.

Dunlay, Thomas W. M. "Indian Allies in the Armies of New Spain and the United States: A Comparative Study." *New Mexico Historical Review* 53, no. 3 (1981): 239–258.

Dunne, Peter Masten. *Early Jesuit Missions in Tarahumara*. Berkeley: University of California Press, 1948.

Enciclopedia de las hierbas. Barcelona: Jaimes Libros, 1979.

Enciclopedia de México. Mexico City: Enciclopedia de México, 1978.

Enciclopedia Monitor de Salvat ilustrada. Mexico City: Salvat Editores de México; Novara, Italy: Istituto Geográfico di Agostini, 1971.

Engelmann, George J. *Labor among Primitive Peoples: Showing the Development of the Obstetric Science of To-day from the Natural and Instinctive Customs of all Races, Civilized and Savage, Past and Present*. 3rd ed. Saint Louis, Mo.: J. H. Chambers, 1884. Electronic Text Center, University of Virginia Library.

Escobedo Chávez, Emma. *Bibliografía básica del estado de Chihuahua*. Chihuahua, Mexico: Centro de Estudios Regionales de la Universidad Autónoma de Chihuahua / Programa Cultural de las Fronteras, Secretaría de Educacíon Pública (SEP), 1983.

Escribanía de cámara: Minutes of Governor Guajardo Fajardo Trial of Residence, February 17, 1656. Archivo General de Indias, Seville, Spain. Commentary in Porras Muñoz, *Iglesia y estado en la Nueva Vizcaya*, 564.

Esteyneffer, Juan de. *Florilegio medicinal*. Edited by María del Cármen Anzúres y Bolaños. Mexico City: Academia Nacional de Medicina, 1978.

Estrada, Ignacio Javier. Letter to the Provincial F. Juan Antonio de Oviedo, S.J., November 23, 1730. Translated by Daniel Matson. In Sheridan and Naylor, *Rarámuri*, 73–78.

Evans, Richard Schultes. "Botanical Sources of the New World Narcotics." In Weil, Metzner, and Leary, *The Psychedelic Reader*, 89–110.

———. *The Plant Kingdom and Hallucinogens*. Part 2. Cambridge, Mass.: Botanical Museum of Harvard University, 1969.

Evía y Valdés, Diego. Letter to the Governor, April 4, 1652. Archivo General de Indias (AGI), Audíencia de Guadalajara 63, Seville, Spain.

———. Letters to the King, February 16, 1642; April 2, 1652; April 6, 1652; and March 3, 1653. Archivo General de Indias (AGI), Audíencia de Guadalajara 63, Seville, Spain.

Falcón Mariano, Joseph Augustin. "Relación de Nuestra Señora del Pópulo de Guaguachique (Wawachiki), Dec. 4, 1777." Biblioteca Nacional de Madrid. Mss 2449. Translated by Patrick Hayes. In Sheridan and Naylor, *Rarámuri*, and in Pennington, *Tarahumar of Mexico*, 116.

Fernald, Lyndon Merritt, ed. *Gray's Manual of Botany*. 8th ed. New York: American Book, 1950.

Ferrari, Giorgio. "Ricerche sulle Laureacee del genere *Ocotea* dell'America latina." In IILA / CISO, *Simposio internazionale*, 373–386.

Fisher, M. G. "Some Comments Concerning Dosage Levels of Psychedelic Compounds for Psychotherapeutic Experiences." In Weil, Metzner, and Leary, *The Psychedelic Reader*, 149–163.

Fisher, Richard D., and Luis G. Verplancken. *Chihuahua, Mexico: Copper Canyon, Sierra de Tarahumara, Canyon Train Adventure and Colonial Pueblos*. Tucson, Ariz.: Sunracer, 1989.

Fontana, Bernard L., Edmond J. B Faubert, and Barney T. Burns. *The Other Southwest: Indian Arts and Crafts of Northwestern Mexico*. Phoenix: Heard Museum, 1977.

Fontana, Bernard L., and John P. Schaefer. *Tarahumara: Where Night is the Day of the Moon*. Flagstaff, Ariz.: Northland, 1979.

Fonte (or Font), Juan. *Annua, 1607*. In González Rodríguez, *Crónicas de la Sierra Tarahumara*: 156–160.

———. *Annua, 1608*. In González Rodríguez, *Tarahumara: La sierra y el hombre*, 154–158, and in Sheridan and Naylor, *Rarámuri*, 9–11.

———. *Annua, 1611*. In González Rodríguez, *Tarahumara: La sierra y el hombre*, 159–162, and *Crónicas de la Sierra Tarahumara*, 186–193.

Foster, George M. "El legado Hipocrático latinoamericano: 'Caliente' y 'frío' en la medicina popular contemporánea." Translated by Xavier Lozoya. *Medicina Tradicional* 2, no. 6 (1979): 5–24.

Foster, Michael S. "Loma San Gabriel: Una cultura del noroeste de mesoamérica." 16th Mesa Redonda de la Sociedad Mexicana de Antropología, Saltillo, Coahuila, 1979. In Lazalde, *Durango*, 38.

Fried, Jacob. "An Interpretation of Tarahumara Interpersonal Relations." *Anthropological Quarterly* 34, no. 2 (1961): 110–120.

---. "The Relation of Ideal Norms to Actual Behavior in Tarahumara Society." *Southwestern Journal of Anthropology* 9, no. 3 (1953): 286–295.

Fuller, Thomas C., and Elizabeth McClintock, *Poisonous Plants of California*. California Natural History Guides, 53. Berkeley: University of California Press, 1986.

Gajdusek, Carleton D. "The Sierra Tarahumara." *Geographical Review* 43, no. 1 (1953): 15–38.

---. "Tarahumara Indian Piscicide: *Gilia macombii* Torrey." *Science* 120, no. 3115 (1954): 436.

Gajdusek, Carleton D., and Nancy C. Rogers. "Specific Serum Antibodies to Infectious Disease Agents in Tarahumara Indian Adolescents of Northern Mexico." *Pediatrics* 16, no. 6 (1955): 819–835.

Galeffi, Corrado. "Recenti acquisizioni nel campo dei curare indigeni." In IILA / CISO, *Simposio internazionale*, 561–576.

Gamboa, Eduardo Carrera. "Arqueología en la Sierra Tarahumara: Elementos para la interpretación de los orígenes de los pueblos indios del norte de México." In Molinari Medina and Porras Carrillo, *Identidad y cultura*, 35–53.

Gamio, Manuel. "Papel de la antropología." (Summary of the second proposition issued to the 2nd Pan-American Scientific Congress)" In INI, *INI, 30 años después*, 26–38.

Gaoni, Y., and Raphael Mechoulam. "A Total Synthesis of dl-Δ1-tetrahydrocannabinol, the Acitve Constituent of Hashish." *Journal of the American Chemical Society* 87 (1965): 3273–3275.

García Gutiérrez, Carlos, and Luis García Gutiérrez. "Geologic Setting of Ore Deposits in the State of Chihuahua, Mexico." In West Texas Geological Society, *Geological Field Trip Guidebook*.

García Manzanedo, Héctor. "Notas sobre la medicina tradicional en una zona de la Sierra Tarahumara." *América Indígena* 23 (1963): 61–70.

García Rivas, Heriberto. *Plantas medicinales de México*. Mexico City: Panorama, 1988.

Garrett, W. E. "South to Mexico City." (Tarahumara infant mortality data from Luis Verplancken, S.J.). *National Geographic* 134, no. 2 (August 1968): 145–153.

Giner Rey, Miguel Angel. *Apuntes para la historia del real de minas de Santa Rosa de Uruachic*. Chihuahua, Mexico: Talleres Gráficos del Gobierno del Estado de Chihuahua, 1995.

---. *Uruachic: 250 años de historia*. Chihuahua, Mexico: Centro Librero la Prensa, 1986.

Gold, Barry S., Richard C. Dart, and Robert A. Barish. "Bites of Venomous Snakes." *New England Journal of Medicine* 347, no. 5 (2002): 347–356.

Gómez González, Filiberto. *Rarámuri: Mi diario tarahumara*. Mexico City: Excélsior, 1948.

González Reyna, Jenaro. *Memoria geológica minera del estado de Chihuahua*. Chihuahua, Mexico: Stylo, 1956.

González Rodríguez, Luis. *Crónicas de la Sierra Tarahumara*. Colección Centenario. Chihuahua, Mexico: Camino, 1992.

———. *El noroeste novohispano en la época colonial*. Instituto de Investigaciones Antropológicas, Universidad Nacional Autónoma de México. Mexico City: Porrúa, 1997.

———. *Etnología y misión en la Pimería Alta, 1715–1740*. Mexico City: Instituto Nacional de Investigaciones Históricas, Universidad Nacional Autónoma de México, 1977.

———. "Jesús Hielo Vega: Primer sacerdote tarahumar." CENAMI. *Estudios Indígenas* 4, no. 4 (1975): 487–488.

———. "La etnografía acaxee de Hernando de Santarén." *Tlalocan* 8:355–394.

———. "Mitos étnicos y mitos de la iglesia." *Christus* 37, no. 443 (1972): 32–35.

———. *Révoltes des indiens tarahumars*. From the Latin text of Joseph Neumann's *Historia seditionum*. . . . Translation, introduction, notes, and analytical indexes. Paris: Institut des Hautes Études de l'Amerique Latine, 1971.

———. *Tarahumara: La sierra y el hombre*. SEP/80. Mexico City: SEP and Fondo de Cultura Económica, 1982.

Good, Byron J., and Mary-Jo Del Vecchio Good. "The Meaning of Symptoms: A Cultural Hermeneutic Model for Clinical Practice." In *The Relevance of Social Science for Medicine*, edited by Leon Eisenberg and Arthur Kleinman, 165–196. Dordrecht, Holland: D. Reidel, 1980.

Götzl, E., and O. Schimmer. "Mutagenicity of Aristolochic Acids (I, II) and Aristolic Acid in New YG Strains in *Salmonella Typhimurium* Highly Sensitive to Certain Mutagenic Nitroarenes." *Mutagenesis* 8, no. 1 (1993): 17–22.

Gracia Alcover, Blas. *Vitaminas y medicina herbaria: Medicina herbaria chilena; Plantas, hierbas y frutas medicinales*. Mexico City: Vida Naturista, 1950.

Gray, A. "Contributions to American Botany: A Revision of the North American Ranunculi." Proceedings of the American Academy of Arts and Sciences, 21 (1886): 363–370.

Greenblatt, Milton. "*LJ West's Place in Social and Community Psychiatry*." In *The Mosaic of Contemporary Psychiatry*, edited by Kales, Pierce, and Greenblatt, 3–13.

Griffen, William B. "Culture Change and Shifting Populations in Central Northern Mexico." Anthropological Papers of the University of Arizona, no. 13. Tucson: University of Arizona Press, 1969.

———. "Indian Assimilation in the Franciscan Area of Nueva Vizcaya." Anthropological Papers of the University of Arizona, no. 33. Tucson: University of Arizona Press, 1979.

———. "Procesos de extinción y continuidad social y cultural en el norte de México durante la colonia." *América Indígena* 30, no. 3 (1970): 679–725.

———. "Some Problems in the Analysis of the Native Indian Population of Northern Nueva Vizcaya during the Spanish Colonial Period." In *Themes of Indigenous Acculturation in Northwestern Mexico*, edited by Thomas B. Hinton and Phil C. Weigand, 50–56. Anthropological Papers of the University of Arizona, no. 38 (Tucson: University of Arizona Press, 1981).

Grinspoon, Lester. *Marihuana Reconsidered*. Cambridge. Mass.: Harvard University Press, 1971.

Groom, Dale. "Cardiovascular Observations on Tarahumara Indian Runners—the Modern Spartans." *American Heart Journal* 81, no. 3 (1971): 304–314.
Guadalajara, Tomás, and José Tardá. "Relation of the Tarahumara Indians, 1675." In González Rodríguez, *Tarahumara: La sierra y el hombre*, 168–179.
Guevara Sánchez, Arturo. *Un sitio arqueológico de la ciudad de Chihuahua*. Chihuahua, Mexico: Centro Regional de Chihuahua, Instituto Nacional de Antropología e Historia, 1985.
Gummerman, George J., and Emil W. Haury. "Prehistory: Hohokam." In Ortiz and Sturtevant, *Handbook of North American Indians*, 9:75–90.
Gurría Lacroix, José, and Miguel León Portilla, eds. *Historia de México*. Mexico City: Juan Salvat, 1978.
Gutiérrez, Tonatiúh, and Electra Gutiérrez. "50,000 Tarahumaras en la sierra, un escenario majestuoso." *Excélsior* (Mexico City), April 21, 1974.
Hackett, Charles Wilson. *Historical Documents Relating to New Mexico, Nueva Vizcaya and Approaches thereto, to 1773*. Vol. 2. Carnegie Institution of Washington Publication, no. 330. Washington, D.C.: Carnegie Institution of Washington, 1926. In Pennington, *The Tarahumar of Mexico*, 3, 19.
Hammouda, Y., and M. M. Motawi. "Principal alkaloid isolated from *Tecoma stans* (L.) H.B.K. (*Bignonia stans* L.) Bignoniaceae." *Egyptian Pharmacology Bulletin* 41 (1959): 73.
Hancock, Richard Humphris. *Chihuahua: A Guide to the Wonderful Country*. International Training Programs. Norman: University of Oklahoma Press, 1978.
Hargreaves, R. T., R. D. Johnson, D. S. Millington, M. H. Mondal, W. Beavers, L. Becker, C. Young, and K. L. Rinehart. "Alkaloids of American Species of *Erythrina*." *Lloydia: The Journal of Natural Products* 37, no. 4 (1974): 569–580. In Bye, "Ethnoecology of the Tarahumara."
Harris, James C. "The Starry Night (*La nuit etoilée*)." *Archives of General Psychiatry* 39 (2002): 978–979.
Haughton, P. J. "The Scientific Basis for the Reputed Activity of Valerian." *Journal of Pharmacology and Pharmacotherapeutics* 51 (1999): 505–512.
Hayden, Julian D. "La arqueología de la Sierra del Pinacate, Sonora, México." In *Sonora: Antropología del desierto*, edited by Beatriz Braniff C. and Richard Stephen Felger, 27:281–305. Colección Científica Diversa. Mexico City: Instituto Nacional de Antropología e Historia (INAH) / SEP, Centro Regional del Noroeste, 1976.
Hegnauer, R. *Chemotaxonomie der Pflanzen*. 3 vols. Lehrbücher und Monographien aus dem Gebiete der Exakten Wissenschaften. Basel, Switzerland: Birkhäuser, 1962–1967.
Hernández, Francisco. *Historia natural de Nueva España*. Vol. 3, book 17, *Historia de las plantas*. Mexico City: Universidad Nacional Autónoma de México, 1959. First published 1577.
Herrera Beltrán, Claudia. "Aumentaron 30% las muertes de niños menores de 5 años en la Tarahumara." *La Jornada* (Mexico City), April 15, 1997.

Herring, Hubert Clinton, and Helen Baldwin Herring. *A History of Latin American from the Beginnings to the Present.* New York: Knopf, 1968.
Herzen, V. *Guide-formulaire de thérapeutique.* 7th ed. Paris: Libraire J. B. Bailliére et Fils, 1913.
Hester, James J. "Late Pleistocene Extinction and Radiocarbon Dating." *American Antiquity* 26, no. 1 (1960): 58–77. In Di Peso, *Casas Grandes,* 68.
Hewitt, W. P. "Chihuahua Orchid Trails." *American Orchid Society Bulletin* 14 (1946): 398–406.
Hippocrates. *Nature of Man.* Translated by W. H. S. Jones. Loeb Classical Library. Cambridge, Mass.: Harvard University Press, 1931.
Hoffman, Albert. "Investigaciones sobre los hongos alucinógenos mexicanos y la importancia que tienen en la medicina sus substancias activas." *Artes de México* 124, no. 16 (1969): 24–31.
Hold, K. M., N. S Sirisoma, T. Ikeda, T. Narahashi, and J. E. Casida. "Alfa-thujone (the Active Component of Absinthe): Gamma-aminobutyric Acid Type A Receptor Modulation and Metabolic Detoxification." *Proceedings of the National Academy of Sciences of the United States of America* 97 (2000): 3826–3831. In Harris, "The Starry Night": 978–979.
Hrdlička, Aleš. *Physiological and Medical Observations among Indians of Southwestern U.S. and Mexico.* Bureau of American Ethnology, Bulletin 34. Washington, D.C.: Smithsonian Institution, 1908.
IILA (Istituto Italo-Latino Americano) / CISO (Centro Italiano di Storia Ospitaliera). *Simposio internazionale sulla medicina indigena e popolare dell'America latina.* Rome: IILA / CISO, 1977.
INEGI (Instituto Nacional de Estadística, Geografía e Informática). *XI censo general de población y vivienda.* Mexico City: INEGI, 1990.
———. *XII censo general de población y vivienda.* Tabulados de la muestra censal. Mexico City: INEGI, 2000. www.inegi.gob.mx/.
INI (Instituto Nacional Indigenista). *Atlas de las lenguas indígenas.* Mexico City: INI, 1995.
———. *Características de la población indígena de México.* Mexico City: INI, 2001. www.cdi.ini.gob.mx.
———. "Censo indígena de México." *Memorias del Instituto Nacional Indigenista,* 101. Mexico City: INI, 1945.
———. *Flora medicinal indígena de México, I: Treinta y cinco monografías del atlas de las plantas de la medicina tradicional mexicana.* Mexico City: INI, 1994.
———. *INI, 30 años después: Revision crítica.* México Indígena. Mexico City: INI, 1978.
———. "Tendencias migratorias de la población indígena de México." Subdirección de Investigación, Departamento de Investigación Básica para la Acción Indigenista, 4–48. Mexico City: INI, 1995.
Instituto Chihuahuense de la Cultura. "Basaséachi: Portal del hombre, el tiempo y la cultura." *El Heraldo de Chihuahua,* July 23, 2000.

Irigoyen-Rascón, Fructuoso R. *Cerocahui: Una comunidad en la Tarahumara.* Mexico City: Universidad Nacional Autónoma de México, 1974.

———, ed. *Memoria del seminario de lenguas habladas en el estado de Chihuahua: Ra'itzari, la palabra.* Chihuahua, Mexico: Centro de Estudios Regionales de la Universidad Autónoma de Chihuahua / Dirección de Culturas Populares, 1982.

———. "Migración de indígenas mejicanos a los Estados Unidos: Una realidad invisible." *Psychline* 4, no. 2 (2002): 29–31.

———. "Psychiatric Disorders among the Tarahumara Indians of Northern Mexico." *Curare* 12 (1988) 169–173.

Irigoyen-Rascón, Fructuoso, Gilberto P. Carrillo, and Ignacio León Pacheco. "Pesca tradicional en la Tarahumara." In *Memories of the First International Symposium of Fishing and Education* 4. Cancún, Mérida, El Carmen, Mexico, 1979.

Irigoyen-Rascón, Fructuoso, and Erasmo Palma. *Chá okó: Manual de propedéutica en Rarámuri.* Mexico City: Buena Prensa, 1977.

Irigoyen-Rascón, Fructuoso, and Jesús Manuel Palma. "Rarajípari: The Kick-ball Race of the Tarahumara Indians." *Annals of Sports Medicine* 2, no. 2 (1985): 79–94.

———. *Rarajípari: La Carrera de Bola Tarahumara* [*The Kick-ball Race of the Tarahumara Indians*]. Chihuahua, Mexico: Centro Librero la Prensa / Ayuntamiento de Chihuahua, 1995.

Irigoyen-Rascón, Fructuoso, and Alfonso Paredes, "Biosocial Adaptation in the Tarahumara Ecosystem: The Ekistic Principles of the Tarahumara." Unpublished manuscript, 1977.

Irwin-Williams, Cynthia. "Post-pleistocene Archaeology, 7000–2000 B.C." In Ortiz and Sturtevant, *Handbook of North American Indians*, 9:31–42.

Jones, G. H., M. Fales, and W. C. Wildman. "The Structure of Tecomanine." *Tetrahedron Letters* 6 (1963): 397–400.

Jordán, Fernando. *Crónica de un país bárbaro.* Mexico City: Asociación de Periodistas Mexicanos, 1956.

Kales, Anthony, Chester M. Pierce, and Milton Greenblatt, eds. *The Mosaic of Contemporary Psychiatry in Perspective.* New York: Springer-Verlag, 1992.

Kay, Margarita Artschwager. "Health and Illness in a Mexican Barrio." In *Ethnic Medicine in the Southwest*, edited by E. H. Spicer, 99–166. Tucson: University of Arizona Press, 1977.

———. "Poisoning by Gordolobo." *HerbalGram: The Journal of the American Botanical Council* 32 (1994): 42.

Kelley, John Charles. "Reconnaissance and Excavation in Durango and Southern Chihuahua, Mexico." In *Year Book of the American Philosophical Society, 1953.* New York: American Philosophical Society, 1953.

Kennedy, John G. *Inápuchi: Una comunidad tarahumara gentil.* Inter-American Indian Institute 58. Mexico City: Instituto Indigenista Interamericano, Departamento de Antropología, 1970.

———. *Tarahumara of the Sierra Madre: Beer, Ecology, and Social Organization.* Arlington Heights, Ill.: AHM, 1978.

———. "Tesguino Complex: The Role of Beer in Tarahumara Culture." *American Anthropologist* 65 (1963): 620–640.

Kimberleigh, J. Field, Michael J. Sredl, Roy C. Averill-Murray, and Terry B. Johnson. "A Proposal to Re-establish Tarahumara Frogs (*Rana tarahumarae*) into Southcentral Arizona." Phoenix: Nongame and Endangered Wildlife Program, Arizona Game and Fish Department, 2003. www.gf.state.az.us/pdfs/wc/ratastep10.pdf.

King, Robert E. "Geological Reconnaissance in Northern Sierra Madre Occidental of Mexico." *Geological Society of America Bulletin* 50, no. 11 (1939): 1625–1722.

Kock-Weser, Jan. "Venenos más comunes." In Wintrobe et al., *Harrison medicina interna*, 722–737.

Koehler, Jack. "Crossing the Divide." *Continental* 7, no. 9 (2003): 34–37.

Kroeber, A. L. *Anthropology: Race, Language, Culture, Psychology, Prehistory.* New York: Harcourt, Brace, 1948.

———. *Uto-Aztecan Languages of Mexico.* Ibero-Americana, no. 8. Berkeley: University of California Press, 1934.

Krystal, Andrew D., and Ilana Ressler. "The Use of Valerian in Neuropsychiatry." *CNS Spectrums* 6, no. 10 (2001): 841–847.

La Barre, Weston. "Primitive Psychotherapy in Native American cultures: Peyotism and Confession." *Journal of Abnormal and Social Psychology* 42, no. 3 (1947): 294–309.

Laguna, José. *Bioquímica.* 2nd ed. Mexico City: Médica Mexicana, 1967.

Laguna, José, and KJ Carpenter. "Raw Versus Processed Corn in Niacin Deficient Diets" *Nutrition* 45, no. 1 (1951): 3–8.

La Llave, Pablo de. "Materia Medica of the New Mexican Pharmacopoeia, Part 8." *American Journal of Pharmacy* 57, no. 12 (1885). https://archive.org/details/americanjourna157phil.

———. "Sobre el axin, especie nueva de *Coccus* y de la grasa que de é se extrae." *Registro Trimestre* (April 2, 1832): 147–152.

Lapedes, Daniel L., ed. *McGraw-Hill Dictionary of Scientific and Technical Terms.* 2nd ed. New York: McGraw-Hill, 1978.

Lara, Andres. "Sepelios," 5–6. In Brambila, *Supersticiones y costumbres.*

Lartigue, François. *Indios y bosques: Políticas forestales y comunales en la Sierra Tarahumara.* Ediciones de la Casa Chata, 19. Mexico City: Centro de Investigaciones y Estudios Superiores en Antropología Social, 1983.

Laurencich Minelli, Laura. "Iconografie del peyotl nel Messico precolombiano." In IILA / CISO, *Simposio internazionale*, 255–265.

Lazalde, Jesús F. *Durango indígena: Panorámica cultural de un pueblo prehispánico en el noroeste de México.* Durango, Mexico: Impresiones Gráficas de México, Gómez Palacio, 1987.

Lazcano Sahagún, Carlos, *Barrancas del Cobre.* Guía México Desconocido. Edición especial no. 26. Mexico City: Jilguero, 1996.

Le Blanc, Steven A. "The Dating of Casas Grandes." *American Antiquity Reports* 45, no. 4 (1980): 799–806.
León, Carlos A. "De cómo llegar a ser shaman." *Revista Colombiana de Psiquiatría* 7 (1978): 11–31.
———. "Mental Health and Culture: Some Vignettes." In *Mental Health, Cultural Values and Social Development*, edited by R. C. Nann, S. D. Butt, and Ignacio Ladrido, 41–51. Dordrecht: Holland: D. Reidel, 1984.
Lévi, Jerôme M. "La flecha y la cobija: Codificación de la identidad y la resistencia en la cultura material rarámuri." In Molinari and Porras, *Identidad y cultura*, 127–153.
Lewis, Walter H., and Memory P. F. Elvin-Lewis. *Medical Botany: Plants Affecting Man's Health*. New York: Wiley-Interscience, John Wiley and Sons, 1977.
Lionnet, Andrés. *Los elementos de la lengua tarahumara*. Mexico City: Universidad Nacional Autónoma de México, Instituto de Investigaciones Históricas, 1972.
Lisansky, Jonathan, Rick J. Strassman, David Janowsky, and Craig S. Rich. "Drug-induced Psychoses." In *Transient Psychosis, Diagnosis, Management and Evaluation*, edited by P. J. Tupin, U. Halbreich, and J. J. Peña. New York: Brunner/Mazel, 1984.
Lookout, Morris. "Alcohol and the Native American." *Alcohol Technical Reports* 4, no. 4 (1975): 30–37.
López Austin, Alfredo. "Cosmovisión y medicina náhuatl." In Viesca Treviño, *Estudios sobre etnobotánica y antropología médica*, 13–27.
———. "La dualidad 'frío-caliente' como elemento de la concepción médica prehispánica." *Medicina Tradicional* 2, no. 6 (1979): 22–24.
López de Hinojosos, Alonso. *Suma y recopilación de cirugia con un arte para sangrar muy útil y provechosa*. Mexico City: Antonio Ricardo, 1578.
López, Gregorio. *Tesoro de medicinas para diversas enfermedades*. Mexico City: Francisco Rodríguez Lupercio, 1674.
Lozoya, Mariana. "*Tronadora* (*Tecoma stans* [L.] H.B.K.)." *Medicina Tradicional* 3, no. 10 (1980): 1–4.
Ludwig, Arnold M. "Altered States of Consciousness." In *Altered States of Consciousness*, edited by Charles C. Tart, 10–13. New York: John Wiley, 1969.
Lumholtz, Carl Sofus. *Unknown Mexico: Explorations in the Sierra Madre and Other Regions, 1890–1898*. 2 vols. New York: C. Scribners, 1902. Facsimile reprint. New York: Dover, 1987.
Lyon, Gaston. *Traité élémentaire de clinique thérapeutique*. 4th ed. Libraires de l'Académie de Médecine. Paris: Masson, 1902.
Maderey, Laura Elena. "Aspectos hidrológicos de la cuenca del Río Conchos." Universidad Nacional Autónoma de México. *Boletín del Instituto de Geografía* 2 (1969): 109–138.
———. "Características hidroclimáticas en la Barranca del Cobre y su influencia en la vegetación." In *Memorias de la primera reunión nacional de agroclimatología*, 116–129. Mexico City: Universidad Autónoma de México, 1985.
———. "Hydroclimatic Contrasts and their Influence on the Vegetation, Barranca del

Cobre, Mexico—A Lengthening of the Tropics in Temperate Zone." *GeoJournal* 19, no. 1 (1989): 87–92.

Maisch, John M. "Useful Plants of the Genus *Psoralea*." *American Journal of Pharmacy* 61, no. 7 (1889): 345–352.

Mandell, Arnold J., and Mark A. Geyer. "The Euphorohallucinogens." In *Comprehensive Textbook of Psychiatry*, edited by Benjamin L. Sadock, Harold I. Kaplan, and Alfred M. Freedman, 586–589. 3rd. ed. Baltimore: Williams and Wilkins, 1983.

Márquez Terrazas, Zacarías. "Notas sobre historia de la Tarahumara [1600–1767]," 34–42. In Irigoyen-Rascón, *Memoria del seminario de lenguas*.

———. *Origen de la iglesia en Chihuahua*. Chihuahua, Mexico: Camino, 1991.

———. *Satevó: Período colonial*. Chihuahua, Mexico: Gobierno del Estado de Chihuahua, 1990.

Marsella, Anthony J., and Geoffrey M. White, eds. *Cultural Conceptions of Mental Health and Therapy*. Dordrecht, Holland: D. Reidel, 1982.

Marshall, Richard, ed. *Great Events of the 20th Century*. Pleasantville, N.Y.: Reader's Digest, 1977.

Martínez, J. Rosario C. "El Tarahumara." In Sariego, *El indigenismo en Chihuahua*, 95–113.

Martínez, José Bonifacio, et al. *Mi Chihuahua hoy: Monografía del estado*. Chihuahua, Mexico: Centro Librero la Prensa, 1996.

Martínez, Maximino. *Las plantas medicinales de México*. Mexico City: Botas, 1969.

Martínez Caraza, Leopoldo. *El norte bárbaro de México*. Mexico City: Panorama, 1983.

Martínez del Río, Pablo. "A Preliminary Report on the Mortuary Cave of Candelaria, Coahuila, Mexico." *Bulletin of the Texas Archaeological Society* 24 (1953): 206–253.

Martínez Limón, Enrique. *Tequila: Tradición y destino*. Mexico City: Revimundo México, 1998.

Martínez Marín, Carlos "El reparto de la riqueza," 1101–1114. In Gurría Lacroix and León Portilla, *Historia de Mexico*.

Mateos, Agustín M. *Etimologías griegas del español*. 8th ed. Mexico City: Esfinge, 1964.

———. *Etimologías latinas del español*. 12th ed. Mexico City: Esfinge, 1968.

McGregor, L. M., and Cármen M. Meza. "Los insectos dentro de la farmacopea Mexicana." Paper presented at the First Seminar of Traditional Medicine, Oaxaca, Mexico, August 29–31, 1979.

McLaughlin, J. L. "Cactus Alkaloids, 6: Identification of Hordenine and N-methyltyramine in *Ariocarpus fissuratus* varieties *fissuratus* and *lloydia*." *Lloydia: The Journal of Natural Products* 32 (1969): 392–394.

McMurry, Martha P., Maria Teresa Cerqueira, Sonja L. Connor, and William E. Connor. "Changes in Lipid and Lipoprotein Levels and Body Weight in Tarahumara Indians after Consumption of an Affluent Diet." *New England Journal of Medicine* 325, no. 24 (1991): 1704–1708.

Merino, Manuel Alcántara, Josefina Clemente Lelin et al., eds. *Farmacopea nacional de los Estados Unidos Mexicanos*. 4th ed. Mexico City: Dirección General de Control de Alimentos, Bebidas y Medicamentos, 1974.

Merrill, William L. "The Concept of Soul among the Rarámuri in Chihuahua, Mexico: A Study on Worldview." Ph.D. diss., University of Michigan, 1981.
———. "God's Saviours in the Sierra Madre." *Natural History* 3 (1983): 58–66.
———. "La identidad ralámuli, un perspectiva histórico." In Molinari Medina and Porras Carrillo, *Identidad y cultura*.
———. *Rarámuri Souls: Knowledge and Social Process in Northern Mexico*. Washington, D.C.: Smithsonian Institution Press, 1988.
———. "Tarahumara Social Organization, Political Organization and Religion." In Ortiz and Sturtevant, *Handbook of North American Indians*, 10:290–305.
Miller, Wick R. "Nota sobre los lenguajes extintos del noroeste de México de supuesta filiación Uto-azteca." In Irigoyen-Rascón, *Memoria del seminario de lenguas*.
Mohr, Charles. "On the Presence of Pipitzahoic Acid in the *Perezia* . . ." *American Journal of Pharmacy* 56, no. 4 (1884): 185–193.
Molina, Fray Alonso de. *Vocabulario en lengua castellana y mexicana y mexicana y castellana*. Mexico City: Porrúa, 1970. First published 1571.
Molinari Medina, Claudia, and Eugeni Porras Carrillo, eds. *Identidad y cultura en la Sierra Tarahumara*. Colección Obra Diversa. Mexico City: Instituto Nacional de Antropología e Historia / Congreso del Gobierno del Estado de Chihuahua, 2001.
Monache, Franco Delle. "Aspetti chimici degli allucinogeni dell'America latina." In IILA / CISO, *Simposio internazionale*, 501–516.
Monárrez, Joel Espino, and Homero Martínez Salgado. "Cobertura de vacunación en niños Tarahumaras menores de cinco años en el municipio de Guachochi, Chihuahua. www.chi.itesm.mx/~investig/Salud_Indigena/7sintesisvacunacion.htm, 1998.
———. "Condiciones de vida de los Tarahumaras menores de 5 años y sus familias en el municipio de Guachochi, Chihuahua." Chihuahua, Mexico: Instituto Tecnológico de Monterrey, Unidad Chihuahua, 1977. www.chi.itesm.mx/~investig/Salud_Indigena /7sintesissocioeconomica.htm.
———. "Prevalencia de desnutrición en niños tarahumaras menores de cinco años en el municipio de Guachochi, Chihuahua." *Salud Pública de México* 42, no. 1 (2000): 8–16.
Mora, Mar de. *Censo de nombre de Dios, 1778*. Archivo Franciscano, roll 9, box 16, Archivos Microfilmados del Museo de Antropología e Historia, Mexico City.
Mörner, Magnus. *Race Mixture in the History of Latin America*. Boston, Mass.: Little, Brown, 1967.
Morton, Julia F. *Atlas of Medicinal Plants of Middle America: Bahamas to Yucatán*. Springfield, Ill.: Charles C. Thomas, 1981.
Mull, Dorothy S., and Dennis Mull. "Differential Use of a Clinic by Tarahumara Indians and Mestizos in the Mexican Sierra Madre." *Medical Anthropology* 9, no. 3 (1985): 245–264.
———. "Infanticide among the Tarahumara of the Mexican Sierra Madre." In *Child Survival: Anthropological Perspectives on the Treatment and Maltreatment of Children*, edited by Nancy Scheper-Hughes, 113–132. Dordrecht, Holland: D. Reidel, 1987.
Murray, William Breen. "The Tarahumara Project: An Experiment in Medical Anthro-

pological Field Teaching." In *The Relevance of Anthropology in Medical Education: A Mexican Case Study*. Ph.D. diss., McGill University, 1980.

Myerhoff, Barbara G. *Peyote Hunt: The Sacred Journey of the Huichol Indians*. Ithaca, N.Y.: Cornell University Press, 1974.

Naranjo, Plutarco. "Influencia de la medicina aborigen en la medicina popular actual." In IILA / CISO, *Simposio internazionale*, 235–247.

Neumann, Joseph. *Historia seditionum, quas adversus Societatis Iesu missionarios corumque auxiliares moverunt nationes indicae, at potissimum Tarahumara in America Septentrionali, Regnoque Novae Cantabriae, jam toto ad fidem catholicam propemodum redacto*. Prague: Typia Univers, Carolo-Ferd. Soc. Jesu ad S, Clem (Charles-Ferdinand Universal Press of the Society of Jesus), 1730. Microfilm, Bolton Collection, Bancroft Library, University of California at Berkeley. Translated by Marian L. Reynolds, 1936. Translated into Spanish by Joaquín Díaz Anchondo and Luis González Rodríguez, 1991.

———. Letter to Father Stowasser. July 29, 1686. Translated by Marian L. Reynolds, 1936. Bolton Collection, Bancroft Library, University of California at Berkeley.

———. Letters to an Unknown Priest in the Province of Bohemia. January 15, 1681; February 20, 1682; September 15, 1693, and April 23, 1698. Translated by Marian L. Reynolds, 1936, Bolton Collection, Bancroft Library, University of California at Berkeley.

Neumann, Joseph, and Luis González Rodríguez. *Historia de las rebeliones en la Sierra Tarahumara, 1626–1724*. Colección Centenario, no. 8. Chihuahua, Mexico: Camino, 1991.

New Encyclopaedia Britannica. 15th ed. 32 vols. Chicago: Encyclopaedia Britannica, 1995.

New Webster's Medical Dictionary. Springfield, Mass.: Merriam-Webster, 1986.

Norman, James, and David Hiser. "The Tarahumaras: Mexico's Long Distance Runners." *National Geographic* 149, no. 6 (1976): 712–718.

Oberti, Santiago. "*Tianguispepetla (Althernanthera repens* [L.] Kuntze)." *Medicina Tradicional* 2 (1980): 8.

Ocampo, Manuel. *Historia de la misión de la Tarahumara, 1900–1965*. 2nd ed. Colección México Heróico. Mexico City: JUS, 1966.

Oliverio, Alberto, and Castellano, Claudio. "Psicofarmacologia degli Allucinogeni dell'America Latina." In IILA / CISO, *Simposio internazionale*, 517–531.

Orellana, Sandra L. *Indian Medicine in Highland Guatemala*. Albuquerque: University of New Mexico Press, 1987.

Orozco, María Elena H. *Tarahumara: Una antigua sociedad futura*. Chihuahua, Mexico: Gobierno del Estado / Coplade, 1992.

Ortiz, Alfonso, and William C. Sturtevant, eds. *Handbook of North American Indians*. Vol. 9, *Southwest*. Washington, D.C.: Smithsonian Institution, 1979.

———. *Handbook of North American Indians*. Vol. 10, *Southwest*. Washington, D.C.: Smithsonian Institution, 1983.

Ortiz de Montellano, Bernard R. *Aztec Medicine, Health and Nutrition.* New Brunswick, N.J., 1990.

Ovalle, Ignacio Fernández. "Bases programáticas de la acción indigenista." In INI, *INI, 30 años después*, 9–21.

Pacher, P., S. Bátkai, and G. Kunos. "The Endocannabinoid Ssystem as an Emerging Target of Pharmacotherapy." Pharmacology Review 58, no. 3 (2006): 389–462.

Palacios, Prudencio Antonio de los. *Notas a la recopilación de las leyes de Indias.* Edited by Bernal de Buqueda. Mexico City: Universidad Nacional Autónoma de México, 1979.

Palma, Erasmo. *Cantos tarahumaras.* Edited by Carlos Preciado and Manuel Villalobos. Mimeograph. Norogachi, Mexico: Carlos Preciado and Manuel Villalobos, 1976.

———. *Donde cantan los pájaros chuyacos.* Chihuahua, Mexico: Gobierno del Estado de Chihuahua, 1989.

Pammel, L. H. *A Manual of Poisonous Plants.* Cedar Rapids, Ia.: Torch, 1911.

Paredes, Alfonso, and Fructuoso Irigoyen-Rascón. "Jíkuri, the Tarahumara Peyote Cult: An Interpretation." In Kales, Pierce, and Greenblatt, *The Mosaic of Contemporary Psychiatry*, 121–129.

Paredes, Alfonso, Louis J. West, and Clyde C. Snow. "Biosocial Adaptation and Correlates of Acculturation in the Tarahumara Ecosystem." *The International Journal of Social Psychiatry.* 16, no. 3 (1970): 164–174.

Parke, Davis, and Company. *Physician Manual of Therapeutics.* Detroit, Mich.: Parke, Davis, 1901.

Pascual, José. "An Account of the Missions, 1651." In Sheridan and Naylor, *Rarámuri*, 17–30. Translated by Patrick Hayes from *Documentos para la historia de México*, 179–209.

Passin, Herbert. "The Place of Kinship in Tarahumara Social Organization." Part 1. *Acta Americana* 1 (1943): 360–383.

———. "The Place of Kinship in Tarahumara Social Organization." Part 2. *Acta Americana* 1 (1943): 470–495.

———. "Tarahumara Prevarication: A Problem in Field Method." *American Anthropologist* 44 (1942): 235–247.

Pécoro, Fernando, and Nicolás de Prado (or Di Prato). "Historia de la nueva entrada a las naciones de Chínipas, Guailopos, Guazapares, Témoris y Otras, 1675–1681." In González Rodríguez, *Crónicas de la Sierra Tarahumara*, 81–107.

Peña, Moyrón Enrique. "Una Atlántida en la Tarahumara." *El Heraldo de Chihuahua*, June 6–7, 1976.

Pennington, Campbell W. "La carrera de bola entre los Tarahumaras de México: Un problema de difusión." *América Indígena* 30, no. 1 (1970).

———. *La cultura de los Eudeve del noroeste de México.* Noroeste de México 6. Mexico City: Centro Regional del Noroeste, INAH / SEP, 1982.

———. *The Pima Bajo of Central Sonora.* Salt Lake City: University of Utah Press, 1980.

———. "Plantas medicinales utilizadas por el Pima Montañés de Chihuahua." *América Indígena* 23, no. 1 (1973): 213–232.

———. *Plantas medicinales utilizadas por los Tarahumaras* [Medicinal Plants Utilized by the Tarahumara]. Chihuahua, Mexico: Esparza, 1973.

———. "Tarahumara." In Ortiz and Sturtevant, *Handbook of North American Indians*, 10:276–289.

———. "Tarahumar Fish Stupefaction Plants." *Economic Botany* 12, 1 (1958): 95–102.

———. *The Tarahumar of Mexico: Their Environment and Material Culture*. Salt Lake City: University of Utah Press, 1963. Reprint, 1974.

———. *The Tepehuan of Chihuahua: Their Material Culture*. Salt Lake City: University of Utah Press, 1969.

Peón y Contreras, José. "Empacho." In Domínguez et al., *El médico práctico y doméstico*, 211–216.

Pérez de Ribas, Andrés. *Historia de los triunfos de nuestra Santa Fé entre gentes las más bárbaras y fieras del nuevo orbe, 1645*. Partial translation by Patrick Hayes in Sheridan and Naylor, *Rarámuri*.

———. Relación de Sinaloa, 1635." In González Rodríguez, *El noroeste novohispano en la época colonial*, 204–217.

Pérez de Zárate, Dora. *Acerca de la medicina folklórica panameña*. Panama: Universitaria, 1996.

Pérez González, Gregoria. "Naturalia: La conservación de la cotorra serrana." *El Heraldo de Chihuahua*, August 17, 2003.

Pieris Derahiyagala, Paul E. "*Proboscidea*." In *New Encyclopaedia Britannica*. 15th ed. Encyclopaedia Britannica, 1984.

Pietrich, Blanche, "De cinco balazos fue abatido el rarámuri José Cruz Gardea." *La Jornada* (Mexico City), July 9, 1996.

Plancarte, Francisco M. *El problema indígena Tarahumara*. Instituto Nacional Indigenista: Memorias, 5. Mexico City: INI, 1954.

Pompa, Gerónimo. *Medicamentos indígenas*. 42nd ed. Madrid: América, 1975.

Porras Carrillo, Eugeni. "La Sierra Tarahumara: Una región multiétnica y pluricultural." In Molinari Medina and Porras Carrillo, *Identidad y cultura*.

Porras Muñoz, Guillermo. *Iglesia y estado en la Nueva Vizcaya, 1562–1821*. Pamplona, Spain: Universidad de Navarra, 1966.

Prado, Luis. "Peyote." In Brambila, *Supersticiones y costumbres de los Rarámuri*, 14–16, and in Irigoyen-Rascón, *Cerocahui*, 130–135.

Preciado, Carlos, Manuel Villalobos, Jorge Ibarra, Juan Arciniega, Mauricio Rivera, and Fructuoso Irigoyen-Rascón. *Informe Kwechi: Report of a Three-Year Research among the Tarahumaras of Kwechi, Chihuahua, on their Culture, Religion and Education*, edited by Fructuoso Irigoyen-Rascón. Privately published for the Tarahumara Mission, 1974.

Presidencia de la República, México. *Convenio IMSS-COPLAMR para el establecimiento de servicios de salud en el medio rural.* Mexico City: Solidaridad Social por Cooperación Comunitaria, May 25, 1979.

Pusztai, Arpad. *Plant Lectins.* Cambridge: Cambridge University Press, 1992.

Quillet, Aristide. *Diccionario enciclopédico ilustrado Quillet.* 8 vols. Mexico City: Cumbre, 1978

Raffauf, R. F. *A Handbook of Alkaloids and Alkaloid Containing Plants.* New York: Wiley-Science, 1970. In Bye, "Ethnoecology of the Tarahumara," 200.

Ramos, Martha Leticia, and Mónica Iturbide. "Migración tarahumara a la ciudad de Chihuahua." In Brouzès Pelissier, Françoise, coordinator, *Los Rarámuri hoy: Memorias.* Chihuahua, Mexico: Instituto Nacional Indigenista / Culturas Populares, Unidad Regional Chihuahua, 1991.

Ratkay, Johannes M. *An Account of the Tarahumara Missions and a Description of the Tarahumaras and of their Country.* 1683 Translated by Marian L. Reynolds, 1936. Microfilm, Bolton Collection, Bancroft Library, University of California at Berkeley.

Recopilación de las leyes de Indias. 4 vols. Madrid: Cultura Hispánica, 1973.

Retana, Juan Fernández de, Juan Fernández de la Fuente, and Martín Ugalde. "Declaración de Nicolás de Cocomórachi," 1697. Translated by Cármen Villa Prezelski. In Sheridan and Naylor, *Rarámuri,* 66–67.

Reyes Ochoa, Alfonso. "Yerbas del Tarahumara" (1934). In *Obras Completas,* 10:121. Mexico City: Fondo de Cultura Económica, 1959.

Reynolds, Marian L. *Some Letters of Father Joseph Neumann, S.J., Missionary to the Heathen Tarahumaras; Together with his "Historia Seditionum."* Collected by Herbert E. Bolton, translated by M. L. Reynolds and with an introduction by Allen Christelow. TS, Bolton Collection, Bancroft Library, University of California at Berkeley, 1936.

Rice, John. "Drought Adds to Number of Indian Children Hospitalized." Associated Press. *McAllen (Tex.) Monitor,* November 8, 1994.

Richards, Bill. "*Arundo* has 2 Lives: A Pest in California, but to Florida a Boon." *Wall Street Journal,* October 16, 2002, 76.

Riva Palacio, Vicente, Juan de Dios Arias, Alfredo Chavero, José María Vigil, and Julio Zárate. *México a través de los siglos.* 16 vols. Facsímile ed. Mexico City: Cumbre, 1984. First published 1889.

Robalo, Cecilio. *Diccionario de aztequismos, o sea jardín de las raíces aztecas, palabras del idioma náhuatl, azteca o mexica introducidas al idioma castellano bajo diversas formas.* Contribución al Diccionario Nacional. Colección Daniel. Mexico City: Fuente Cultural, 1912.

Robles, Ricardo J. "La actividad misionera: Un diálogo intercultural." *Christus* 37, no. 443 (1972).

———. *Mujé narí nurema sinéame pari naí wichimoba arewá suwibuma naí nurama. Consagración Episcopal de Monseñor José Llaguno Farías, S.J.; Ordenación sacerdotal de Jesús Hielo Vega y Luis Raúl Garibaldi V.* Program. Sisoguichi, Mexico, April 13, 1975.

Rodríguez López, Jesús. *Supersticiones de Galicia y costumbres vulgares*. Buenos Aires, Argentina: Nova, 1940.

Romano Pacheco, Arturo. "Deformación cefálica intencional." In *Antropología física, época prehispánica*, edited by J. Romero, 3:195–227. Colección México, Panorama Histórico y Cultural. Mexico City: Instituto Nacional de Antropología e Historia, 1974.

Rosales, Juan B., José María Ponce de León, Manuel Chico, and Eduardo Primero. *El estado de Chihuahua de 1821 a 1921: Centenario de la consumación de la independencia de México*. Chihuahua, Mexico: Talleres Tipográficos de la Escuela de Artes y Oficios, 1921.

Rubel, Arthur J. "Concepts of Disease in Mexican American Culture." *American Anthropologist* 62 (1960): 795–814.

Sahagún, Bernardino. *Historia general de las cosas de la Nueva España*, edited by Angel María Garibay. Mexico City: Porrúa, 1956.

Salmón, Enrique. "Cures of the Copper Canyon: Medicinal Plants of the Tarahumara with Potential Toxicity." *HerbalGram: The Journal of the American Botanical Council* 34 (1995): 44–55. In http://www.herbalgram.org/membersarea/hgarticles/hg34coppercanyon.html.

———. *Sharing Breath with Our Relatives: Rarámuri Plant Knowledge, Lexicon, and Cognition*. Ph.D. diss., Arizona State University, 1999. Ann Arbor, Mich.: UMI Dissertation Services.

Salvatierra, Juan María. "Texto de la entrada a Témoris, Cerocahui, Guazapares, Huisaromes y Cuiteco, 1680." In González Rodríguez, *Crónicas de la Sierra Tarahumara*, 113–126.

Sánchez, J. M. L., and L. E. Estrada. "Distribución geográfica de la *Tronadora Tecoma stans* (L.) H.B.K. en México, una planta medicinal." In *Plantas medicinales de México: Introducción a su estudio*, edited by L. E. Estrada. Chihuahua, Mexico: UACH, 1990.

Sánchez-Téllez, Maria Carmen, and Francisco Guerra. *Pestes y remedios en la conquista de América*. Madrid: Universidad de Alcalá de Henares, Departamento de Historia, Facultad de Filosofía y Letras, 1985.

Santarén, Hernando. "La etnografía acaxee de Hernando de Santarén." In González Rodríguez, *El noroeste novohispano en la época colonial*, 135–150.

Sariego, Juan Luis. *El indigenismo en Chihuahua: Antología de textos*. Chihuahua, Mexico: Escuela Nacional de Antropología e Historia / Instituto Nacional de Antropología e Historia, 1998.

Sauer, Carl. *The Distribution of Aboriginal Tribes and Languages in Northwest Mexico*. Iberoamericana, 5. Berkeley: University of California Press, 1934.

Sayer, Chlöe. *Costumes of Mexico*. Austin: University of Texas Press, 1985.

Scappert, Phillip J., and Joel S. Shore. "Cyanogenesis in *Turnera ulmifolia* L." *Evolutionary Ecology Research* 2 (2000): 337–352.

Scarpati, Maria Luisa. "Alcaloidi del *Croton draconoides*, una pianta medicinale del Perú." In IILA / CISO, *Simposio internazionale*, 309–311.

Schmidt, Robert H., Jr. *A Geographical Survey of Chihuahua*. Southwestern studies monograph no. 37. El Paso: University of Texas at El Paso, 1973.

———. *A Geographical Survey of Sinaloa*. Southwestern studies monograph no. 50. El Paso: University of Texas at El Paso, 1973.

Schmutz, Ervin M., and Lucretia Breazeale Hamilton. *Plants that Poison: An Illustrated Guide for the American Southwest*. Flagstaff, Ariz.: Northland, 1989.

Sejourné, Laurette. *The Burning Water: Thought and Religion in Ancient Mexico*. Translated by Irene Nicholson. London: Thames and Hudson, 1978.

SEMARNAP (Secretaría del Medio Ambiente, Recursos Naturales y Pesca). *Diagnóstico de productos no maderables en Chihuahua, Durango, Jalisco, Michoacán, Guerrero y Oaxaca*. Mexico City: Centro de Investigación y Docencia Económicas (CIDE), 2000. www.semarnat.gob.mx/pfnm.

Sheridan, Thomas E., and Thomas H. Naylor, eds. *Rarámuri: A Tarahumara Colonial Chronicle, 1607–1791*. With a foreword by Charles W. Polzer. Flagstaff, Ariz.: Northland, 1979.

Smith, Paul W. *Mis amigos indígenas / My Indian Friends*. Puebla, Mexico: Compañía Impresora de Libros y Revistas, 1960.

Späth, Ernst. "Über das Carnegin." *Berichte der Deutschen Chemischen Gesellschaft* 62, no. 4 (1929) 1021–1024.

Spicer, Edward. *Cycles of Conquest*. Tucson: University of Arizona Press, 1962.

Spier, Leslie. "The Distribution of Kinship Systems in North America." *University of Washington Publications in Anthropology* 1, no. 2 (August 1925): 69–88.

Standley, Paul C. *Trees and Shrubs of Mexico*. Contributions from the United States National Herbarium 23. Washington, D.C.: U.S. Government Printing Office, 1920–1926.

Steele, Susan. "Uto-Aztecan: An Assessment for Historical and Comparative Linguistics." In Cabrera, Campbell, and Mithun, *The Languages of Native America*, 444–544.

Steinegger, Ernst, and Rudolf Hänsel. *Lehrbuch der Allgemeine Pharmakognosie*. Berlin: Springer, 1963.

Swadesh, Mauricio. *Indian Linguistic Groups of Mexico*. Mexico City: INAH, 1959.

Terrazas Sánchez, Filiberto. *La guerra Apache en México (Viento de octubre)*. 2nd ed. Mexico City: B. Costa Amic, 1977.

Thord-Gray, I. *Tarahumara-English, English-Tarahumara Dictionary*. Coral Gables, Fla.: University of Miami Press, 1955.

Tibon, Gutierre. *Historia del nombre y de la fundación de México*. Mexico City: Fondo de Cultura Económica, 1981.

Torres Gaona, Miguel René. "Recientes investigaciones sobre algunas plantas chilenas." In IILA / CISO, *Simposio internazionale*, 347–352.

Trabulse, Elías. "Los hospitales de Nueva España en los siglos XVI y XVII." In *Historia de México*, edited by Jorge Gurría Lacroix and Miguel León Portilla, 1421–1428. Turner, Nancy, and Adam F. Szczawinski. *Common Poisonous Plants and Mushrooms of North America*. Portland, Ore.: Timber, 1991.

Turner, Victor W. *The Ritual Process: Structure and Anti-structure.* Chicago, Ill.: Aldine, 1969.
Tyler, Varro E., Lynn R. Brady, and James R. Robbers. *Pharmacognosy.* Philadelphia: Lea & Febiger, 1981.
Unger, M. S. "Bibliography on LSD and Psychotherapy," 241–249. In Weil, Metzner, and Leary, *The Psychedelic Reader.*
Uranga, Ernesto. *Uirichiki: El fuerte de los piel cobriza.* Sisoguichi, Mexico, 1963.
Urbina, Antonio de. "Relación de Serocahue (Cerocahui), October 27, 1777." In Del Paso y Troncoso, *Relaciones del siglo XVIII relativas a Chihuahua.*
Urbina, Manuel. "*Aje, Axin o Ajin.*" *Naturaleza* 7 (1902): 363–365.
Uribe, Beatriz, and Cármen León. *Estudio comparativo de niños tarahumares en dos escuelas de grados diferentes de aculturación.* Ph.D. diss., Universidad Iberoamericana, Mexico City, 1978.
Valdés, María. *Los indios en los censos de población.* Mexico City: Universidad Nacional Autónoma de México, 1996.
Valiñas, Leopoldo. "Lengua, dialectos e identidad étnica en la Sierra Tarahumara." In Molinari Medina and Porras Carrillo, *Identidad y cultura,* 105–125.
Van Gennep, Arnold. *The Rites of Passage.* Chicago: University of Chicago Press, 1960.
Velarde, Luis Javier. "La primera relación de la Pimería Alta." In González Rodríguez, *Etnología y misión,* 27–88.
Velasco, Juan Bautista. "Relación de la entrada a Chínipas, 1601." In González Rodríguez, *Crónicas de la Sierra Tarahumara,* 37–41.
Velázquez Moctezuma, Xavier. "Epazote (*Chenopodium ambrosioides* L.)." *Medicina Tradicional* 2 (1979): 6.
Venning, Frank D. *Cacti.* Racine, Wis.: Golden Press / Western, 1974.
Verplancken, Luis, Gilberto Chacón, Carlos Díaz Infante, and Ricardo Robles. Foreword to Bennett and Zingg, *The Tarahumara.*
Viesca Treviño, Carlos, ed. *Estudios sobre etnobotánica y antropología médica.* Mexico City: Instituto Mexicano para el Estudio de las Plantas Medicinales, 1976.
Villacís, Luis. *Plantas medicinales de México.* Mexico: Epoca, 1978.
Villa Rojas, Alfonso. *Los elegidos de Dios: Etnografía de los Mayas de Quintana Roo.* Mexico City: Instituto Nacional Indigenista, 1978.
Villa Rojas, Alfonso, and Robert Redfield. *Chan Kom: A Maya Village.* Carnegie Institution of Washington, 448. Washington, D.C.: Carnegie Institution, 1934.
Vivó, Jorge A. *Razas y lenguas indígenas de Mexico.* Instituto Panamericano de Geografía e Historia. Publicación 52. Mexico City: [Industrial Gráfica], 1941.
Vogel, Virgil J. *American Indian Medicine.* Norman: University of Oklahoma Press, 1970.
Wagner, Federico. *Plantas medicinales y remedios caseros.* Mexico City: Aurora, 1950.
Wall, Monroe E., C. Roland Eddy, J. J. Willaman, D. S. Correll, B. G. Schubert, and H. S. Gentry. "Steroidal Sapogenins: XII. Survey of Plants for Steroidal Sapogenins and Other Constituents." *Journal of the American Pharmaceutical Association* 43, no. 8 (1954): 503–505.

Wall, Monroe E., Merle M. Krider, C. F. Krewson, C. Roland Eddy, J. J. Willaman, D. S. Correll, B. G. Schubert, and H. S. Gentry. "Steroidal Sapogenins: VII. Survey of Plants for Steroidal Sapogenins and Other Constituents." *Journal of the American Pharmaceutical Association* 43, no. 1 (1954): 1–7.

Wampler, Joseph. *New Rails to Old Towns: The Region and Story of the Ferrocarriles Chihuahua al Pacífico*. Berkeley, Calif.: published by author, 1969.

Wasson, G. "Notes on the Present Status of *Ololiuhqui* and the other Hallucinogens in Mexico." *Botanical Museum Leaflets* 20, no. 6 (1963): 161–193. In Weil, Metzner, and Leary, *The Psychedelic Reader*, 153–175.

Waters, Michael R., and Thomas W. Stafford, Jr. "Redefining the Age of Clovis: Implications for the Peopling of the Americas." *Science* 315, no. 5815 (23 February 2007): 1122–1126.

Watson, Sereno. "List of Plants Collected by Dr. Edward Palmer in Southwestern Chihuahua, Mexico in 1885." *Proceedings of the American Academy of Arts and Sciences* 21 (1886): 414–455.

Weaver, Thomas. "Changes in Forestry Policy, Production and the Environment in Northern Mexico, 1960–2000." *Journal of Political Ecology* 7 (2000). http://jpe.library.arizona.edu/volume_7/Weaver00.pdf.

Weil, Gunther M., Ralph Metzner, and Timothy Leary. *The Psychedelic Reader: Selected from* The Psychedelic Review. New Hyde Park, N.Y.: University Books, 1965.

West, Louis Jolyon. *En un mundo convulsionado por crecientes índices delictivos, los Tarahumaras poseen la clave de una existencia pacífica*." Associated Press, January 1, 1975. *El Heraldo de Chihuahua*, February 29, 1976.

West, Louis J., Alfonso Paredes, and Clyde C. Snow. *Sanity in the Sierra Madre: The Tarahumara Indians*. Presented at the 122nd annual meeting, American Psychiatric Association. Miami Beach, Florida, 1969. In Yamamoto, Joe, "*Psychohistorical View*."

West Texas Geological Society. *Geologic Field Trip Guidebook thru the States of Chihuahua and Sinaloa*. Publication 74-63 (Midland, Tex.: West Texas Geological Society, 1974).

Willaman, J. J., and Bernice G. Schubert. *Alkaloid-bearing Plants and their Contained Alkaloids*. U.S. Department of Agriculture, Agricultural Research Service. Technical Bulletin 1234. Washington, D.C.: U.S. Department of Agriculture, 1961.

Wintrobe, Maxwell M. et al., eds. *Harrison medicina interna*. 4th Spanish ed. Mexico City: Blakiston / La Prensa Médica Mexicana, 1973.

Woodbury, Richard B., and Ezra B. W. Zubrow. "Agricultural Beginnings, 2000 B.C.–A.D. 500." In Ortiz and Sturtevant, *Handbook of North American Indians*, 9:43–60.

World Health Organization. *The Promotion and Development of Traditional Medicine*. Technical report series 622. Geneva: World Health Organization, 1978.

Ximénez, Francisco. *Cuatro libros de la naturaleza y virtudes de los plantas y animales de uso medicinal en la Nueva España*.... Mexico City: Oficina tip. de la Secretaría de Fomento, 1888.

Yamamoto, Joe. "Psychohistorical View of Transcultural Psychiatry." In Kales, Pierce, and Greenblatt, *The Mosaic of Contemporary Psychiatry*, 89–98.

Zapata Ortíz, Vicente. "Los medicamentos en la medicina indígena del Perú." In IILA / CISO, *Simposio internazionale*, 106–113.

Zeiner, Arthur R., Alfonso Paredes, and Lawrence Cowden. "Physiological Responses to Ethanol among the Tarahumara Indians." *Annals of the New York Academy of Sciences* 273 (1976): 151–158.

Zeiner, Arthur R., Alfonso Paredes, Robert A. Musicant, and Lawrence Cowden. "Racial Differences in Psychophysiological Responses to Ethanol and Placebo." *Currents in Alcoholism*, no. 1 (1977): 271–286.

Zepeda, Nicolás de. "Alzamiento y asiento de los Tarahumares en 1649." In González Rodríguez, *Crónicas de la Sierra Tarahumara*, 241–247.

Zingg, Robert Mowry. *Report on the Archeology of Southern Chihuahua*. Contributions of the University of Denver, 3. Denver: University of Denver, Center of Latin American Studies, 1940.

Index

Page numbers in italic type indicate illustrations. The abbreviation (HRC) at the end of a main entry indicates it appears in the herbal remedy compendium in chapter 10.

Abbé, Father, 294n45
abdominal pain. *See* gastrointestinal ailments
aborí (HRC), 155–159, 310n1, 310nn6–7, 311n23
abortion, 158–159, 270–271, 311n23
Abutilon trisulcatum (HRC), 215–216
Acacia spp. (HRC), 260–262, 336n681, 352n1099
Acalypha phlioides (HRC), 196–197
acculturation forces, Champion's model, 56
Achillea millefolium (HRC), 183
Acourtia spp. (HRC), 226–227, 337n711
Adaintum capilus-veneris (HRC), 214–215, 333n587, 333n591
adornments from plant materials, 124, 161, 162, 170, 315n124
afterbirth, care of, 99–100, 303n9
afterlife, in religious beliefs, 69, 300n23
Agastache pallida, 320n248
Agave spp. (HRC), 221–224, 335n658, 335n664, 336n666, 336n676
aggressive behavior, attitudes about, 67–68, 134, 186, 251
agitated states, mental disorders, 133–134
agriculture: in canyon floors, 25; colonial labor system, 45–47, 295nn57–58; Conchos River irrigation, 20; crop destruction during rebellions, 48, 296n70; in economic system, 70–71; historical expansion, 37, 290nn8–9
Aguirre Beltrán, 308n13, 344nn892–893
akáame (HRC), 159
alcohol consumption study, 8
alferesía, 133, 308n54
alguacils, in political structure, 78–79, 302n57
Almada, Francisco R., 18
Alnus oblongifolia (HRC), 235–236, 341nn808–809
Aloe vera (HRC), 244–245
Alternanthera spp. (HRC), 170–171, 316n136
Althea officinalis (HRC), 215–216
altitude of region, overview, 17
amapola (HRC), 160
Ambrosia spp. (HRC), 178–179, 319nn221–222
amphibians, in ecological habitat, 34, 289n49
Amsinckia sp. (HRC), 172
amulets, 124, 127, 137, 139, 306n37, 307n51
anará (HRC), 160, 312n30
Anazi Basketmaker culture, 37
Anderson, Edward F., 328n470
Andopogon citratus (HRC), 265

383

anger, 67–68, 134, 186, 251
animals. *See* fauna
Anoda spp. (HRC), 253
Antirrhinum spp. (HRC), 233–234
ápago (HRC), 160, 312n34
aposí (HRC), 160–161
Arbutus spp. (HRC), 200–201
Arciniega, Juan, 286n18
Arctostaphylos pungens (HRC), 209–211, 331n553
ardimientos, role in disease, 123
Aréchiga, Antonio Xavier, 332n568
arguasiris, in political structure, 78–79, 302n57
*a*rí* (HRC), 162–163, 312n50
Ariocarpus fissuratus (HRC), 201–202, 328nn467–471, 329n480, 329n484
arisí (HRC), 163–164
Aristolochia spp. (HRC), 197–199
armadillos, in ecological habitat, 31
Arnozan, X., 311n23
arrayanes (HRC), 164
arrowheads, as fishing evidence, 38
arrows, plants for, 167, 179, 197, 257, 320n266, 325n399
Arroyo de Basuchil, mammoth remains, 37, 290n3
Arroyo del Santísimo, route of, 21
Artaud, Antonin, 149
Artemisia spp. (HRC), 242–243, 244, 344nn892–893, 345n915
Arundo donax (HRC), 166–167
Asclepias spp. (HRC), 195, 200, 226–227, 327nn453–454, 337n711
Asplenium monanthes (HRC), 214–215, 333n587, 333n591
atole, 210, 219, 263, 270
authority notification, for Jíkuri ritual, 142–143
ax wounds, 279–280, 301n48
Ayensu, Edward, 314n109
Aztecs: healing practices, 201, 207, 250, 320n248, 324n359, 327n450, 344n893, 348n984, 352n1099; in language families, 38, 39, 40, 71, 301n37; rituals/belief systems, 289n42, 311n17, 321n273, 335n658, 344n892; smallpox impact, 43, 293n39

babárachi (HRC), 164
Baccharis spp. (HRC), 174–175
bachí (HRC), 164–165
bacierera (HRC), 165
badgers, in ecological habitat, 31
bajíachi celebrations, 6–7, 67–68, 88–89, 95
bakánowa, 121, 127, 134, 154, 165–166, 308n5
bakúi (HRC), 166–167
bakusí (HRC), 167–168
banagá (HRC), 168
Bancroft, Hubert Howe, 52
baptisms, 69, 82, 100, 157–158, 268–269, 294n41, 294n46
bariguchi (HRC), 168
barrancas region, physiographic overview, 17–18, 21, 288n12
Basaséaci waterfall, 21, 288n20
basigó (HRC), 168–170, 315n117
Básile, Jácome, 48, 294n41
Basketmaker culture, 37, 291n13
batagá (HRC), 170
batari: and body scent, 58–59; at celebrations and festivities, 7, 110; at death ceremonies, 101; herbal ingredients, 186, 204, 222, 232, 241, 333n587; for Jíkuri ritual, 142, 144, 146
Batopilas, temperatures, 22
Batopilas River, route of, 20
bats, 30, 120
ba'wí ówima, in Jíkuri ritual, 151–152, 310n13
begging, attitudes about, 71
bejoké (HRC), 260–262, 352n1099
belief systems. *See* disease, conceptual explanations; religious beliefs
Bennett, Wendell C., 129, 271, 301n48, 320n248, 332n574, 342n827
Berlandiera lyrata (HRC), 187
be'techókuri (HRC), 170–171, 316n136
Beudin, Cornelio, 48
bezoars, 307n51

bichinaba (HRC), 171
Bidens angustissima (HRC), 353n1141
bigamy trials, 76
biological agent variable, in health conceptual framework, 11
birds: in disease explanations, 119–120, 121, 124, 306n26; in ecological habitat, 31–33, 289n44; in legends, 40–41, 154, 289n42; in religious festivities, 109
birth registry system, 268
bladder ailments. *See* urinary ailments, herbal remedies
bluebird species, 31
boarding schools, 55
body characteristics, 59–60
body painting, 104–105, 109–110, 127
body scent, cultural group differences, 58–59
bola, cultural explanation, 128, 307n41
boldo (HRC), 171–172, 316n148
bones, taboos, 122, 157–158
Borago officinalis (HRC), 172
borraja (HRC), 172
Boulenger, G. A. (on frogs), 289n49
Bouvardia glaberrima (HRC), 244
Brambila, David, 218, 226, 239, 306n26, 306n31, 338n732, 352n1117
brasilillo (HRC), 172
breast-feeding practices, 270
brewing. *See* batari
Bruhn, Jan G., 263
Buddleia spp. (HRC), 219–220, 256–258, 334n638
Burgess McGuire, Donald, 265, 342n837
burial caves, 37–38
Bursera spp. (HRC), 231–232, 339n764
busichí okorá, 132
Bye, Robert A., reporting on herbs, 174, 176–181, 185, 187, 195, 197–198, 202, 204, 208, 230, 235, 256, 262–263, 265, 290n8, 312n30, 312n34, 319nn221–222, 322n295, 324n359, 327n450, 328n469, 331n542, 333n587, 334n615, 336n681, 338n732, 339n764, 343n868, 345n915, 345n923, 350n1057, 355n41

Cabrera, Luis G., 265, 316n136, 324n359, 330n512, 342n818, 349n1027
Cacalia decomposita (HRC), 218–219, 334n615
cacalosúchil (HRC), 173
Caesalpinia spp. (HRC), 164, 174, 251
Cáhita people, 39, 293n39
calahuala (HRC), 173, 317n161
Calliandra spp. (HRC), 172, 237–238, 342n827
calomeca de Los Bajos (HRC), 173–174
Campos, Indio de los, 139, 309n17
cancer, 132–133
Candameña Canyon, 21
canela (HRC), 174
Cannabis sativa (HRC), 216–217, 333n611
canyonlands, physiographic overview, 17–18, 21, 288n12
Capsicum spp. (HRC), 206–208, 330n528
Caraveo, Carlos, 49, 50, 297n79
Cardenal, Francisco Fernández, 195, 231, 250, 266, 332n574, 353n1138
cardinal species, 32
Cardiospermum halicacabum (HRC), 235
Carlowrightia sp. (HRC), 198
Carteria axin (HRC), 162–163, 312n50
Carya illinoensis (HRC), 211–212, 332n568
Casas Grandes culture, 38, 90, 292n27
cascalote (HRC), 174, 317n166
Castilleia sp. (HRC), 198
Castini, Juan, 44, 293n39
cattle, domestic, 29
Cave-Dweller phase, 37–38
cave shelters, 65–66
Ceanothus buxifolius (HRC), 260–262, 352n1099
Cedronella mexicana (HRC), 253
Celtis pallida (HRC), 260–262, 352n1099
Centaurea americana (HRC), 204
Cephalocereus leucocephalus (HRC), 175–177, 318n192
Cercocarpus montanus (HRC), 179, 319n225
Cerocahui (Irigoyen-Rascón), 7, 59, 279
Cerqueira, María Teresa, 271
cha'gusi or *cha'gúnari* (HRC), 174–175
Champion, Jean R., 56

chapeyó, during religious festivities, 72, 79, 111–112
character traits, 67–68
Charles III, 52
chawiró or *chawé* (HRC), 175–177, 318n192
Cheilantes spp. (HRC), 214–215, 333n587, 333n591
Cheilanthes spp. (HRC), 214–215, 262, 333n587, 333n591
Chenopodium spp. (HRC), 181–183, 187–189, 203–204, 321n273
chichiquelite (HRC), 177–178, 318n206
chicory, 319n219
chicura (HRC), 178–179, 319n219, 319nn221–222
Chihuahua, physiographic overview, 15–16
Chihuahua al Pacífico, 20–21
Chihuahua City, 41, 52, 291n13
chiká (HRC), 179, 319n225
chikí (HRC), 179
chikuri nakara (HRC), 179–180
childbirth and pregnancy: afterbirth care, 99–100, 303n9 (ch. 5); in disease conceptual framework, 13; herbal and animal remedies, 31, 156, 167, 205, 215–216, 230, 236, 254, 265, 348n984; before labor, 94–98; labor and delivery process, 98–99, 282
chimney styles, 58, 65
chiná or *chinaka* (HRC), 249–250
Ching Vega, Oscar W., 301n48
Chínipas Mission, 49
Chínipas people, 41, 44, 292n28, 294n46
Chínipas River, traverse route, 21
Chiricahua culture, remains, 37
cholera, 44
cholugos, in ecological habitat, 30
cho'péinari or *cho'pénara* (HRC), 180, 320n248
cho'rí (HRC), 180–181
Christelow, Allen, 294n52
Christianity. *See* missionary activity, Catholic; religious beliefs
chúcha or *chúchaka* (HRC), 181, 320n266

chu'í, role in disease, 122–123
chu'ká (HRC), 181–183
cientos (HRC), 183
Cinnamonum zeylanicus (HRC), 174
ciruelo del campo (HRC), 183
Cissus sp. (HRC), 256
Ciudad Guerrero, precipitation patterns, 22
Clínica San Carlos, 7–8, 286n16
clothing: cultural patterns, 5–6, 59–64, 299n7; death ceremonies, 101; during pregnancy and childbirth, 95, 97; during religious festivities, 104, 105, 111
Clovis culture, 5, 36–37
Cobre Canyon, 18
Coccus axin (HRC), 162–163, 312n50
cocolmeca (HRC), 183–184, 321n287
Coix lacryma (HRC), 170
cold-heat complex, disease, 114–115, 305n6
colds. *See* respiratory ailments
collective activities, 88–93
Cologania angustifolia (HRC), 233–234
communion phase, in Jíkuri ritual, 149–151, 310n11
Concho people, 39, 47, 49, 51, 294n50, 295n65
Conchos River, 18, 20, 288n16
Conferva sp. (HRC), 168
congenital disease, 126, 270, 354n13
Conioselinum mexicanum (HRC), 334n615
Conium maculatum (HRC), 258–259, 350n1057
Consejo Supremo de la Raza Tarahumara, 83–84
Constance, Lincoln, 202
contraceptive pill, herbal origins, 4
contrahierba (HRC), 184, 322n295
Contreras, Gaspar, 45, 46
Conyza spp. (HRC), 247
cooking, cultural group differences, 58–59
copalquín (HRC), 185–186
Copper Canyon, 288n12
Córdova de Gardea, Guadalupe, 9
corn beer. *See* batari
coronilla (HRC), 187

Coryphanta compacta (HRC), 262–263, 352n1107
cosa, cultural explanation, 128, 307n41
Cosmos spp. (HRC), 237
cough. *See* respiratory ailments
court trials, 74–77, 301n48
Coutarea spp. (HRC), 185–186
Cowden, Lawrence, 281
cranial deformations, 38, 291n17
crime, 67–68, 74–77, 301n48
Croix, Teodoro de, 46
crop ceremonies/protections, plants for, 157, 238, 254
crosses: during Jíkuri ritual, 144, 145–147, 150; during *noríruachi* festivities, 105; protective functions, 124, 306n37; on si'pírakas, 145–146; during *wikubema* ritual, 157; wood used for, 159, 161
Crotalaria ovalis (HRC), 179–180
Crotalus spp. (HRC), 247. *See also* snakes
Croton spp. (HRC), 185–186, 345n920
Cucurbita spp. (HRC), 163–165
cultural features variable, in health conceptual framework, 11
curanderas, 139, 308n6
curare, 4
Cusihuiríachi, mining activity, 49
Cydonia vulgaris (HRC), 224–225
Cymbopagon sp. (HRC), 265

Dalea wislizenii (HRC), 198
dances, 66, 91–93, 103–104, 109–112, 142, 149
Datura spp. (HRC), 251–252, 348nn984–985
Daugherty, Franklin W., 287n7
death ceremonies, 101–102, 157–158, 226
death taboos, 122–123
death toll, epidemics, 43, 293nn38–39
Decorme, Gerard, 49, 292n28, 293n38, 297n72
Deeds, Susan M., 294n45
deer, 30, 67
deforestation, 27
deformities, physical, 126, 270, 354n13
Deimel, Claus, 187, 218, 226, 281, 343n878

deities, 68–69, 91–92, 299n21
De la Peña, Moisés T., 288n23
demeanor, social, 59, 67–68
demons, 299n21
de Mora, Mariano, 295n65
De Ramón, Luis P., 308n54, 348n985
devitalization, in disease explanations, 116–117, 134
Deza y Ulloa, Antonio, 292n27
diarrhea. *See* gastrointestinal ailments
Diaz-Infante, Carlos, 55, 80, 302n63
Dichondra spp. (HRC), 179–180
digestive disorders. *See* gastrointestinal ailments
Di Peso, Charles, 37, 290n8, 291n13
disease, conceptual explanations, 13, 113–114; animals, 118–120, 121, 124, 305nn14–16, 306n26, 306n37; chu'í contact, 122–123; cold-heat complex, 114–115, 305n6; fear relationships, 125–127; iwigara loss, 115–119, 124–125, 305n7, 305n11; masses in body, 128; mental ailments, 133–134; miscellaneous conditions, 127–128, 130–133; progression of illness, 125; protective functions, 124, 306nn36–37; ritual/taboo violations, 120–123, 130–131, 132–133; sexually-transmitted types, 121, 123, 130, 132, 279; skin disorders, 128–129; whirlwinds, 123, 306n31; witchcraft, 117–118, 133, 138–139, 305n7; about worms/parasites, 129–130. *See also* health conditions, western medical perspective
diseases, colonial era, 43–44, 293nn38–39, 294n41, 294n45
dogs, domestic, 29
domestic animals, 29
Dorstenia spp. (HRC), 322n295
dreams, 67, 118, 123, 137–138, 150, 166
drums, 103, 106, 110, 146–147
Dryopteris normalis (HRC), 214–215, 333n587, 333n591
Duke, James A., 314n109
dwellings, 38, 58, 65–66, 291n15

dyes from plants, 162, 164, 174, 186, 190, 193, 216, 236, 248, 346n944
dysentery. *See* gastrointestinal ailments

eagles, in ecological habitat, 32–33
Echeverría, Father, 293n39
Echinocactus spp. (HRC), 318n192
Echinocereus spp. (HRC), 175–177, 262–263, 318n192, 352n1107
Eclipta alba (HRC), 166–167
ecological habitat, overview: climate, 22–27; fauna species, 29–35; flora species, 27–29; hydrographic features, 18, 20–22, 288n20; physiographic features, 15–18, 287nn7 8, 288n12; soil characteristics, 27
economic system, 70–71, 88
education systems, 52, 55, 295n58
ejido authorities, 82–83
Elvin-Lewis, Memory P. F., 169, 319n225, 324n363, 340n787, 353n1122
Elyonorus barbiculmis (HRC), 345n923
Elytraria spp. (HRC), 206
emic perspective, 4–5, 11
empacho: cultural explanations, 130–131, 307n50; herbal remedies, 163, 169, 171–172, 180, 181, 205, 227, 237, 242
end-of-the-year cycle, festivities, 111–112
environmental variable, in health conceptual framework, 10
epasote (HRC), 187–189
epidemics, colonial era, 43–44, 293nn38–39, 294n41, 294n45
Epithelanta micromeris (HRC), 201–202, 328nn467–471, 329n480, 329n484
equipatas, precipitation patterns, 22
Equisetum spp. (HRC), 166–167, 314n109
erá or *eráka* (HRC), 189–190, 324n359
Eriogonum spp. (HRC), 168
Erodium cicutarium (HRC), 194
erosion problems, 27
Eryngium spp. (HRC), 249–250
Erythrina spp. (HRC), 161–162
escorcionera (HRC), 190, 324nn363–364
Esteyneffer, Juan de, 52, 298n91

Estrada, Ignacio Javier, 47
estrenina (HRC), 190–191
etic perspective, 4–5
eucalipto (HRC), 191
Eucalyptus spp. (HRC), 191
Eucheria socialis (HRC), 200–201
Eupatorium spp. (HRC), 244
Euphorbia sp. (HRC), 195
Evía y Valdés, Diego de, 49, 297n73
eye ailments: frequency of, 132, *278*; herbal remedies, 162, 193, 203, 209, 216, 219, 223, 226, 230, 242, 246, 254, 262
Eyerúame cult, 68
Eysenhardtia polystachia (HRC), 231

Fajardo, Guajardo, 48, 295n59, 296n67, 296nn70–71, 297n73
falling diseases, 125, 132–133
falling down, beliefs about, 122
family-based healing practices, 135–136
family structure, 71–72
fauna: in disease explanations, 118–120, 121, 124, 305nn14–16, 306n26, 306n37; in ecological habitat, 29–35, 289n44, 289n49; for health remedies, 162–163, 200–201, 212, 225, 243, 332n571, 332n574; in legends, 40–41, 154, 289n42
fear. See *majawá*
federal government, Mexican, 55–56, 84–86
Ferdinand V, 295n57
Ficus petiolaris (HRC), 259–260, 351n1069
Figueroa, Jerónimo, 44
Fimbristylis sp. (HRC), 204
fire, during Jíkuri ritual, 144, 149, 309n7
fire baptism, 100, 157–158
fish, in ecological habitat, 34, 289n49. *See also* pisicides (HRC)
Fisher, Richard D., 288n20
fishing, archaeological evidence, 38
flight to the gentiles, 47, 51
flora, in ecological habitat, 27–29. See also specific plants, e.g., *Agave* spp.
Florilegio Medicinal (Esteyneffer), 52, 298n91
Foeniculum vulgare (HRC), 199

fontanel, depressed, 126, 127–128, 138, 233, 282
Fonte, Juan, 42, 43, 44, 48
food-offering ceremonies, 100–101, 142, 144, 309n8
footprint style, cultural group differences, 58
fossils, 17, 288n9
Fouquieria spp. (HRC), 220–221
foxes, in ecological habitat, 29, 289n33
Franciscans, *19*, 52–53, 294n50. *See also* missionary activity, Catholic
Franseria spp. (HRC), 174–175, 178–179
Fraxinus spp. (HRC), 254
fright. *See majawá*
frogs, 34, 289n49, 306n26
Fuerte River, 18, 20
funeral ceremonies, 101–102, 142, 149, 153, 157–158

Gaeta, Ascención, 310n13
Gajdusek, Carleton D., 18, 274, 279
Galván, Father, 54–55
gambling, 90
games with balls, varieties, 303n9. *See also* kickball races
Gamio, Manuel, 4
Ganó legend, 160–161, 312n40
García Gutiérrez, Carlos, 287n8
García Gutiérrez, Luis, 287n8
García Manzanedo, Héctor, 306n26
Gardea, Cruz, 302n68
Gardea, Juan, 302n68
Gardea, Marciano, 302n68
garó or *garóko* (HRC), 260–262, 352n1099
gastrointestinal ailments: cultural explanations, 130–131, 307n50; herbal remedies, 156, 159–161, 163, 164, 166–169, 171–175, 177–181, 183, 184, 186, 187, 192, 194, 197–198, 200, 202, 204, 215–217, 227, 229, 231, 235–237, 239–240, 242–244, 246–247, 250, 253, 255–256, 259, 265–266, 315n117, 319n222; western medical perspective, 272, 274, *275*
gender roles: childbirth assistance, 98–99;

clothing, 59–64, 299n7; family structure, 71; healers, 137, 139, 308n6; Jíkuri ritual, 145, 149–150, 152; kickball races, 90, 152–153; during *Nawésari* ceremony, 79; religious festivities, 107, 108, 304n2; *tutuguri* ceremony, 93
generaris, in political structure, 72, 77
Gentiana spp. (HRC), 220
geology of region, overview, 17, 287nn7–8
Geranium sp. (HRC), 180
gestures and greeting styles, 59, 115
Glandorff, Father, 44, 298n91
Gnaphalim spp. (HRC), 234–235, 341n793
goats, domestic, 29
gobernadora (HRC), 191–192, 324n376
"God and the Devil" legend, 40–41
gods, 68–69, 91–92, 299n21
González Rodríguez, Luis, 60, 187, 295n55, 325n399
gossip, 81, 302n60
go'tó (HRC), 192
Götzl, E., 198–199
granada (HRC), 192
Grand Canyon, 18, 288n12
grasshoppers, in ecological habitat, 35
Greeks, disease theory, 114, 304n2 (ch. 7)
greeting styles and gestures, 59, 115
Griffen, William B., 47, 294n50
Guachochi, temperatures, 22
Guadalajara, Tomás de, 49
Guailopos people, 292n28
guamúchil (HRC), 193
Guapalayna, in Urique Canyon, *21*
Guazapari people, 44
Gummerman, George J., 37
*gu*rú* (HRC), 193

Haematoxilon spp. (HRC), 248, 346n944
hallucinogens, 162, 165–166, 177, 216–217, 222, 251–252, 324n363, 328n428, 329n484, 348n984. *See also jíkuri* entries
hangovers, herbal remedies, 163, 207
Haury, Emil W., 37
hawks, in ecological habitat, 32–33

Hayden, Julian D., 38
headaches: cultural explanations, 131, 134; herbal remedies, 163, 165, 171, 178, 180, 188, 196, 198–201, 207, 212–214, 216, 223, 226, 229–234, 237, 242, 245, 249, 252–253, 255, 258–259, 262, 264, 320n248, 350n1057; and Jíkuri ritual, 142, 152
healers: gender roles, 137, 139, 308n6; levels of, 135–136; si'páames, 69, 136–137, 143–152, 308n5, 308n11; sukurúames, 117, 118, 119, 121, 138–139; towitas, 138; waniames, 121, 130, 138, 308nn10–13. See also *owirúame*
healing ceremonies, plants for, 221–225, 250–251, 254, 264
health conditions, western medical perspective: alcohol abuse, 280–281; children, 268–274, 354n7, 354n10, 354n13; chronic illness, 277–279; infectious diseases, 272–277; mental illness, 281–282; mushroom poisoning, 280, 355n41; nutrition, 271–272; trauma, 279–280. See also disease, conceptual explanations; western medical practices
heart ailments: herbal remedies, 152, 173, 181, 197–198, 207, 209, 214, 230, 239, 244–245; prevalence, 132, 281
heat-cold complex, disease, 114–115, 305n6
Hedeoma spp. (HRC), 180, 340n787
Heisenberg's uncertainty principle, 6–7, 285n14
Helianthemum glomeratum (HRC), 240
Hernández, Fernando Beltrán, 7
Herrera Beltrán, Claudia, 354n7
Hewitt, W. P., 234
Hielo Vega, Jesús, 81
Hieracium fendleri (HRC), 179–180
hierbabuena (HRC), 193–194
hierba de la chuparrosa (HRC), 194
hierba de la flecha (HRC), 194–195, 325n399
hierba de la golondrina (HRC), 195
hierba de la piedra (HRC), 198
hierba de la víbora (HRC), 195–196
hierba de la virgen (HRC), 196
hierba del burro (HRC), 196

hierba del cáncer (HRC), 196–197, 326n420
hierba del indio (HRC), 197
hierba del pastor (HRC), 197, 326n426
hierba del piojo (HRC), 198
hierba del toro (HRC), 198
hierba de Santa María (HRC), 198–199
higuerilla (HRC), 199, 327n445
hinojo (HRC), 199
Hintonia spp. (HRC), 185–186
Hío people, 292n28
Hippocrates, 114, 304n2 (ch. 7)
history overview: acculturation factors summarized, 56; before European arrivals, 36–42; impact of epidemics, 43–44, 293nn38–39, 294n41, 294n45; indigenous uprisings, 47–51, 296nn67–68, 296nn70–71, 297n72, 297n79; labor system, 45–47, 294n52, 295n55, 295nn57–58; after Mexican independence, 53–56; missionary activity, 44–45, 52–53, 295n65; passive resistance strategy, 51–52; Spanish arrival, 42–43; twentieth-century trends, 54–57
hitting diseases, 125
Hittman, Aquina, 355n41
Hrdlička, Aleš, 186, 332n574, 352n1099
humoral theory, disease, 114, 304n2 (ch. 7)
hunting, 67
hydrographic features, overview, 18, 20–22, 288n20
Hyptis spp. (HRC), 244, 345n915

ibagápitanos, 77, 78, 79, 105, 110
Ibarra, Jorge, 286n18
igualama (HRC), 199–200, 327n450
Indian Confederation, 49–50
infanticide, 270–271
infant mortality rates, 95, 268–269, 354n7, 354nn9–10
influenza: colonial era epidemics, 44, 293n39; herbal remedies, 169, 196, 239; modern prevalence, 272, 274, 276
INI (National Indigenous Institute), 9–10, 64, 85–86, 267
inmortal (HRC), 200, 327nn453–454

insecticides (HRC), 198, 228, 237
insects: in ecological habitat, 35; for health remedies, 162–163, 200–201, 225, 243; in legends, 40–41
invertebrates, in ecological habitat, 34–35. *See also* worm-based diseases
Iostephane heterophyla (HRC), 190, 324nn363–364
Ipomoea spp., 324n363
irrigation, Conchos River, 20
Irwin-Williams, Cynthia, 37
isotherms, 26
iwí (HRC), 200–201
iwigara loss, 115–119, 122, 137, 305n7, 305n11

jail punishments, 76–77
Jatropha spp. (HRC), 203, 244–245, 345n920
Jíkuri (HRC), 201–202, 263, 328nn467–471, 329n480, 329n484, 343n868; as deity, 69; in disease explanations, 120–121, 134; for dreamlike state, 118
Jíkuri ritual: behavior expectations, 143, 145, 309n6, 309n9; cactus collecting process, 140; closing activity, 151–153; communion phase, 149–151, 310n11; dance phase, 149; hiding of, 52; preamble phase, 141–143; preparation phase, 143–145; protective functions, 153–154, 310n15; rasping phase, 145–149; research methodology, 140–141; shaman role, 136–137
Jordán, Ricardo, 55
Jova people, 15, 39, 41, 50
juanita (HRC), 202
jube (HRC), 202–203
Juglans spp. (HRC), 211–212, 332n568
jumete (HRC), 203
Juniperus spp. (HRC), 155–159, 225, 310n1, 310nn6–7, 311n23; in ecological habitat, 28
júpachi (HRC), 203–204

kachana (HRC), 204
kapitanos, in political structure, 72, 74, 78
Karwinskia humboldtiana (HRC), 256–258

kasará or *kasaráka* (HRC), 204
Kennedy, John G., 67, 281, 299n7
keyóchuri (HRC), 205, 330n512
kichínowari (HRC), 206
kickball races: overview, 6, 66, 89–91; during ceremonies/rituals, 102, 106, 152–153; plant/herb uses, 156–157, 164, 166, 202, 219, 230, 237–238, 263; during pregnancy, 90, 94; in taboo stories, 118, 122
kidney ailments. *See* urinary ailments, herbal remedies
kinship terminology, 70, 301n37
kirí (HRC), 206
kobishi/kobisi, 58–59, 97, 176, 270, 272
Kock-Weser, Jan, 338n735, 353n1122
Koehler, Jack, 288n12
kokoyome, characteristics, 38, 291n15
komerachi, court trials, 74, 301n40
konéma ritual, 92, 93
koremá, beliefs about, 119–120
korí (HRC), 206–208, 330n528
kórima concept, 70–71
korísowa (HRC), 208
Krameria spp. (HRC), 260
kuchíwari. *See* worm-based diseases
kusarí (HRC), 208–209, 331n542
kusí urákame (HRC), 209, 209n548
ku'wí (HRC), 209–211, 331n553
Kwechi, Marist Brothers project, 8–9, 286n18

labor system, 45–47, 49, 54, 294n52, 295n55, 295nn57–58
lachi (HRC), 211–212, 332n568
láchimi (HRC), 212, 332n571, 332n574
land ownership, 82–83
Lantana spp. (HRC), 205, 330n512
Lapedes, Daniel L., 349n1027
lario (HRC), 212, 332n577
Larrea spp. (HRC), 191–192, 324n376
latido, cultural explanations, 128, 307n42
laurel (HRC), 212–213
Laws of the Indies, 42–43, 292n35
Lemairocereus thurberi (HRC), 175–177
Lepidium virginicum (HRC), 238, 342n837

Lewis, Walter H., 169, 319n225, 324n363, 340n787, 353n1122
life rhythms, 66–67
lightning: herbal protections, 154, 157, 259, 344n893; in religious beliefs, 69, 96, 306n36
Ligusticum porteri (HRC), 258–259, 350n1057
limoncillo (HRC), 213
linaza (HRC), 213
Linum usitassimum (HRC), 213
lions, in legends, 40–41
Lippia spp. (HRC), 227
Litsea spp. (HRC), 212–213
livestock, 29, 70, 91, 122
lizards: in ecological habitat, 34; remedies with, 212, 332n571, 332n574
Llaveia axin (HRC), 162–163, 312n50
Loma San Gabriel culture, 291n13
Lookout, Morris, 5
Lophophora williamsii (HRC), 201–202, 328nn467–471, 329n480, 329n484
lowíame, cultural explanations, 133–134
lumbering activity, 56, 83
Lumholtz, Carl Sofus, writings on: childbirth beliefs, 303n9; culture origins, 292n25; disease and health, 44, 268, 279, 307n40; fauna, 289n42, 289n49; gossip practices, 302n60; herbal remedies, 229, 320n248, 348n984; infanticide, 271; Jíkuri ritual, 69, 145–146, 153, 309n7, 310n11, 310n15; *Jíkuri rosápari*, 328n469; physical appearance, 60; plant foods, 176, 222, 355n41; seasonal residence patterns, 299n11; socioeconomic status, 54; suicide, 282; waterfalls, 288n20
Lysiloma spp. (HRC), 224, 336n681, 336n683

ma'achiri (HRC), 213–214
ma'chogá (HRC), 214–215, 333n587, 333n591
Madera, temperatures, 22
madrone trees, 28
maize, 37, 207, 264, 271–272, 290n9. See also *Zea mays* (HRC)
majawá: in disease explanations, 119, 125–127, 131, 305n7; herbal remedies, 200, 231, 258, 259–260
mala mujer (HRC), 215
malaria: and belief system, 124; herbal remedies, 156, 161, 184–185, 191, 198, 206, 216, 232; prevalence, 278–279
malnutrition and nutrition, 270, 271–272, 273
malva (HRC), 215–216
Mamillaria sp. (HRC), 201–202, 328nn467–471, 329n480, 329n484
mammals, 29–31, 40–41
Mammillaria spp. (HRC), 262–263, 352n1107
mammoth remains, 36–37, 290n3
manzanilla (HRC), 216
Mares Trías, Albino, 265, 342n837
maría (HRC), 216
Mariana Póbora, creation of, 84
Mariano, Falcón, 53, 205, 242, 257, 259, 312n50, 321n273
Marist Brothers, 8–9, 286n18
mariwana (HRC), 216–217, 333n611
Marker, Russell E., 4
Márquez Terrazas, Zacarías, 295n55, 297n76
marriage, 55, 95
Martínez, Manuel, 44
Martínez de Hurdaide expedition, 44, 294n46
Martinez Salgado, Homero, 271, 354n10
Martynia spp. (HRC), 180–181
Mascagnia macroptera, 160, 312n30
mastuerzo (HRC), 217–218
matachín dances/dancers, 101, 111–112, 142, 143–144, 152, 153
matáka stone, during Jíkuri ritual, 145, 149–150, 151
matarí (HRC), 218–219, 334n615
ma'tegochi (HRC), 219
Matelea sp. (HRC), 180–181
mateó (HRC), 219–220, 334n638
mateóchiri (HRC), 220
maternal mortality rates, 97
matesa (HRC), 220
Matricaria spp. (HRC), 216
mawirí (HRC), 220–221

mayora, in political structure, 72, 77–78
Mayo River, route of, 21
McLaughlin, J. L., 263
mé (HRC), 221–224, 335n658, 335n664, 336n666, 336n676; in Jíkuri ritual, 152
measles, 293n39
mechawí (HRC), 224, 336n681, 336n683
medical clinics, introduction/expansion, 283
megafauna remains, 36–37, 290n3
Melilotus spp. (HRC), 253, 353n999
membrillo (HRC), 224–225
Méndez, Pedro, 294n46
menses problems: in belief system, 95, 131; herbal remedies, 156, 188, 199, 204, 214–216, 234, 242–243, 250–251, 271, 328n471, 338n723, 352n1117; mental ailments, 117, 133–134, 281–282, 305n7
mental illness, 133–134, 281–282
mental retardation. *See* retardation
Mentha canadensis (HRC), 193–194
Merrill, William L., 69, 78, 300n23, 301n37, 302n59
mestizo population: childbirth, 97; chronic illness, 277–278; in disease/health framework, 114, 127, 130, 304n2 (ch. 7); growth of, 55–56; infant mortality rate, 269, 354n10; as municipal authorities, 302n68; Tarahumara relations summarized, 57; western medical care, 53–54
methodology, research, 7–11, 267–268
Mexican Revolution, 55, 82, 86
mice, in ecological habitat, 30
migratory movements, 6, 25, 51, 56, 66, 299n11
Miller, Wick R., 39
Mimosa spp. (HRC), 260–262, 352n1099
Mimulus guttatus (HRC), 242, 344n884
mining activity: colonial labor system, 45–46, 49, 294n52, 295n55; after Mexican independence, 53–54; sites of, 17, 18, *19*, 20, 21, 287n8
missionary activity, Catholic: arrival of, 42, 292n34, 295n55; bishop's feud, 297n73; and defensive colonization strategy, 47; during disease epidemics, 43–44, 294n41; indigenous rebellion impact, 48–49, 297n73; jurisdictional dispute, 45, 294n50; linguistics studies, 300n33; medical practices, 43–44, 52, 53, 54–55, 283, 294n41; overview, 44–45, 52–53, 295n65, 298n91; post-rebellion acceleration of, 49, 52–53, 297n76, 297n78; in reduccion system, 292n35; translation of "soul," 115–116; during the twentieth century, 8–9, 54–55, 81–82, 283, 302n63. *See also* religious beliefs
missions, map, *19*
Monarda austromontana (HRC), 227
Monárrez Espino, Joel, 271, 354n10
Moors, representations of during noríruachi festivities, 103–105, 109
Moranta, Gerónimo, 44
morema, in Jíkuri ritual, 151
Moreno Bakasórare, Luis, 9
*mo*riwá* or *mo*riwáka* (HRC), 225
mortality rates, 95, 268–269, 274, 354n7, 354nn9–10
Morton, Julia F., 266
mukí chiwara (HRC), 225–226
Mulatos Canyon, 21
Mülhenbergia porteri (HRC), 166–167
Mull, Dennis, 136, 271, 274, 282
Mull, Dorothy S., 136, 271, 274, 282
municipal authorities, role of, 84–86, 302n68
munísowa (HRC), 226
Murray, William Breen, 279, 299n11, 354n10
mushrooms, 280, 355n41
music, 66, 92, 111, 143, 145–148
Myrtus communis (HRC), 164

naká or *nakáka* (HRC), 226–227, 337n711
nakáruri (HRC), 227, 337n712
naming ceremonies, 100, 157–158
na'pákori (HRC), 227
Naranjo, Plutarco, 4
National Indigenous Institute (INI), 9–10, 64, 85–86, 267
nawá (HRC), 227

nawé (HRC), 228, 338n727, 338n730
nawésaris: by generaris, 77; during Jíkuri ritual, 143, 144–145, 147–148; by mayoras, 78; oratory style, 302n59; by siríames, 61, 73, 75, 79–81, 110
Naylor, Thomas H., 292n34, 297n76, 297n83, 299n11
Neuman, Joseph, writings on: alcohol abuse, 280–281; disease epidemics, 44, 294n41; dwellings, 299n11; physical appearance, 60; rebellions, 48, 49, 50, 51; reduccion system, 43
Nicotiana spp. (HRC), 264, 352n1117, 353n1122
Nolina spp. (HRC), 193
Nombre de Dios mission, 295n65
noríruachi festivities, 103–110, 304nn2–3
Norogachi, medical clinic, 7–8, 286n16
Notholaena spp. (HRC), 173
nutéa rite, 100–101, 122–123, 142
nutrition and malnutrition, 271–272

oak species. See Quercus spp.
Odontotrichum spp. (HRC), 218–219, 334n615, 337n712
okó (HRC), 228–229, 338nn731–732, 338n735
Olea europea (HRC), 229–230
olivo (HRC), 229–230
oná or onáka (HRC), 230
Onorúame, in belief system, 68–69
oparúame, cultural intolerance, 134
Opata people, 39, 42
opium (HRC), 160
Opuntia spp. (HRC), 189–190, 324n359
Orellana, Sandra L., 266
ortiga or ortiguilla (HRC), 230, 339n744
Ortiz de Foronda, Diego, 50
Ortiz de Montellano, 344n893
Osorio (Spanish commander), 42, 44
Oteros River, 21
otters, in ecological habitat, 31
ówina (HRC), 231
owirúame: childbirth assistance, 97–98; death ceremonies, 101; fire baptism, 100; healing practices, 117, 124–125, 137–138; in political structure, 72
owls, 33, 120

Pablo, Don, 297n76
Pachycereus pecten-aboriginum (HRC), 175–177
pain relief, herbal remedies, 157, 159–161, 166–167, 174, 178, 184, 187–188, 190–191, 197–198, 201, 207, 214, 217–218, 223, 231–232, 245, 248–249, 254, 258–260, 262–263, 343n868, 344n893. See also specific ailments, e.g., gastrointestinal ailments; headaches; toothache
Palma, Agustín, 80
Palma Tuchéachi, Erasmo, 9, 92
palo dulce (HRC), 231
palo mulato (HRC), 231–232, 339n764
Panicum bulbosum (HRC), 166–167
papache (HRC), 232–233
Papaver spp. (HRC), 160
Papigochi Mission, 48
parasites. See worm-based diseases
parches (HRC), 233
Paredes, Alfonso, 281
pariseos, during festivities, 103–110, 142, 304nn2–3
Parmelia caperata (HRC), 236
Parral mining district, 45
parrots, in ecological habitat, 32
Parthenium tomentosum (HRC), 256–258
Pascual, José (Joseph), 44, 325n399
pasmo, cultural explanation, 132
Pasquale, Julio, 44, 47, 48
Passin, Herbert, 301n37, 302n60
passive resistance strategy, Spanish intrusion, 3, 51–52
Pécoro, Francisco, 49, 297n78
Pectis spp. (HRC), 213
Pegüis Canyon, 288n16
Pellaea termifolia (HRC), 214–215, 333n587, 333n591
Pennington, Campbell W., writings on:

geology, 287n7; herbal remedies, 185, 190, 195, 196, 198, 201, 210, 226, 239, 244, 258–259, 264, 343n868, 346n940; migration patterns, 53; mining centers, 45; missionary activity, 295n55; non-medical plant uses, 159, 342n848; origins, 291n13, 292n25; plant classifications, 176, 312n30; territory boundaries, 41

Peón y Contreras, José, 307n50

Perea, Pedro de, 44

Pérez de Zárate, Dora, 307n50

Perezia thurberi (HRC), 236, 342n818

perritos (HRC), 233–234

Peumus boldus (HRC), 171–172, 316n148

Phaseolus metcalfei (HRC), 192

Phoradendron spp. (HRC), 239–240, 343n868

Phragmites communis (HRC), 166–167

Phrynosoma obiculare (HRC), 212, 332n571, 332n574

physical traits of Tarahumara, 59–64, 66–67, 90–91, 121–122, 268

physiographic features, overview, 15–18, 287nn7–8, 288n12

Phytolacca sp. (HRC), 212, 332n577

Piedra Volada waterfall, 288n20

pigeons, in ecological habitat, 32

pigs, domestic, 29

Pima people, 38–39

pinole *vs.* maize, 271–272

pintos, during noríruachi festivities, 103–104

Pinus spp., 28, 32, 228–229, 338nn731–732, 338n735

pinworms. *See* worm-based diseases

Pionocarpus madrensis (HRC), 204

Piper sp. (HRC), 183–184

pisicides (HRC), 227, 228, 241, 246, 250, 255, 258, 261, 336n666, 338n727, 338n730

Pisonia capitata, 260–262, 352n1099

Plantago spp. (HRC), 197, 242, 326n426, 344n884

Platanus spp. (HRC), 235–236, 341nn808–809

Plumbago spp. (HRC), 170–171, 190–191

Plumeria spp. (HRC), 173

pneumonia, 274

poleo (HRC), 234, 340n787

poliomyelitis viruses, 279

political organization, 72–79, 82–86, 110, 301n40, 301n43

Polygonum punctatum (HRC), 208

Polypodium spp. (HRC), 317n161, 333n587

population centers, map, *19*

population statistics, 5–6, 15, 52

Populus spp. (HRC), 265

porcupines, in ecological habitat, 31

Potentilla thurberi (HRC), 159–160

Prado, Nicolás de, 49, 294n41

preamble phase, in Jíkuri ritual, 141–143

Preciado, Carlos, 286n18

precipitation patterns, 22, 25

pregnancy. *See* childbirth and pregnancy

prevarication, 81, 302n60

Prionosciadium spp. (HRC), 208–209, 246–247, 331n542

Proboscidea parviflora (HRC), 180–181

processions, during *noríruachi* festivities, 105–109, 304n2

prodigiosa (HRC), 234

progesterone, herbal origins, 4

Prosopis spp. (HRC), 260–262, 352n1099

protective wards, herbal usage, 157, 208, 214, 238, 244, 257–259

Prunella vulgaris (HRC), 227

Prunus spp. (HRC), 255

Pseudotsuga spp., 28, 338n732

Psoralea spp. (HRC), 184, 322n295

psychological traits of Tarahumara, 66–68

Ptelea trifoliata, 160, 312n34

Pteridium aquilinum (HRC), 214–215, 333n587, 333n591

Pterocarpus podolitis (HRC), 345n920

pueblo buildings, 301n40

Punica granatum (HRC), 192

punishment, court trials, 76–77

Purshia tridentata (HRC), 174–175

Pusztai, Arpad, 327n445

396 ▾ Index

Quercus spp. (HRC), 238–239, 342n848; in ecological habitat, 27–28

rabbits, in ecological habitat, 30
radio schools, 55
railroads, 20–21
rainbows, in disease explanations, 120
rainfall patterns, 22, 25
Ranchería of Chuvíscar, 41, 291n13
Randia spp. (HRC), 232–233
rarajípari. *See* kickball races
Rarámuri, etymology. *See* Tarahumar people, overview
rasó (HRC), 234–235, 341n793
rasping ceremony, in Jíkuri ritual, 145–149
Ratibida mexicana (HRC), 231
Ratkay, Johannes M., 59–60, 299n21, 303n9
rattlesnakes. *See* snakes
rayó (HRC), 235
rebellions, colonial era, 47–51, 296nn67–68, 296nn70–71, 297n72, 297n79
reduccion system, 42–43, 46–47, 48, 292n35
religious beliefs, 69–70, 80–81, 91–93, 102–110, 299n21, 300n23. *See also* disease, conceptual explanations
remolino (HRC), 235
repartimientos system, 46–47, 295n57
repogá (HRC), 235–236, 341nn808–809
reptiles, in ecological habitat, 33–34. *See also* snakes
re'ré betéame, in legend, 69
resagí. *See* worm-based diseases
research methodology, 7–11, 267–268
respect, as cultural trait, 70–71, 133
respiratory ailments: in disease conceptual explanation, 115, 121, 126; herbal remedies, 156, 159, 166, 168–169, 173, 175, 180, 183, 186, 188, 194, 197, 199, 204, 206, 210, 213, 223, 227, 229, 233–235, 238, 240, 247, 249, 255, 265, 320n248, 345n920, 348n985, 348n995; western medicine perspective, 272, 277–278
Retana, Fernández de, 50, 51
retardation, 270, 354n13

reté kajera (HRC), 236
retribution, court trials, 76–77
Rewegáchi, in religious beliefs, 69
Reyes Ochoa, Alfonso, 245–246, 247, 315n124
rezanderos, 81–82, 105, 106, 304n2
Rhynchosia pyramidalis (HRC), 226
Ribes neglectum (HRC), 165
Ricnus communis (HRC), 199, 327n445
rikówi (HRC), 236
Río Fuerte Transitional Culture, 37–38, 291n13
Río Sinaloa, route of, 20
ripichawi (HRC), 236, 342n818
ripura (HRC), 237
Rituchi, fossils, 288n9
Riva Palacio, Vicente, 293nn38–39
Rivera, Mauricio, 80, 286n18
rivers of region, overview, 18, 20–22
riwérame (HRC), 237–238, 342n827
Robalo, Cecilio, 316n136, 317n166, 336n666, 336n681
Robinson, on plant identification, 328n469
ro'chíwari (HRC), 238, 342n837
Rodrigo del Castillo (on colonial epidemics), 293n39
Rogers, Nancy C., 274, 279
rojá (HRC), 238–239, 342n848
Rojas, Alfonso Villa, 7
rojá sewara (HRC), 239–240, 343n868
rojásowa (HRC), 240
Romano Pacheco, Arturo, 291n17
Romichi area, 17, 38
romina, cultural explanation, 130–131
ronínowa (HRC), 240–241, 343n878
ropiri, functions of, 79, 302n57
rorogochi (HRC), 242, 344n884
ro'sábari (HRC), 242–243, 344nn892–893
rosákame sikui (HRC), 242
Rosenkranz, George, 4
roundhouses, 38, 291n15
ruda (HRC), 243–244, 345n900
Rulfo, Juan, 7
Rumex crispus, 315n117
running ability, 6, 48, 67, 89–91. *See also* kickball races

rurikuchi (HRC), 244
rurubuchi (HRC), 244
ru'síwari: in bird myth, 31; in disease explanations, 121, 154, 306n26
Ruta graveolens (HRC), 243–244, 345n900
rutúburi dances, 91–93, 120

sábila (HRC), 244–245
Sahagún, Bernardino, 328n468, 336n666, 348n984
Salix bonplandiana (HRC), 260
Salmón, Enrique, writings on herbal remedies, 186, 207, 250, 259, 312n34, 314n109, 315n117, 324n363, 333n587, 353n1122
Salvatierra, Juan María, 294n41
salvia (HRC), 244, 345n915
Samachique, medical clinic, 9–10
San Bartolome agriculture, 45
San Carlos clinic, service patterns, 282
Sánchez, Manuel, 50
sangregado (HRC), 245–246, 345n920
San Juanito, weather patterns, 22
San Miguel de las Bocas, 45, 295n55
sa'parí (HRC), 246, 345n923
Sapindus spp. (HRC), 336n666
Sapium biloculare (HRC), 194–195, 325n399
sapogenins, chemical conversion, 4
saráame (HRC), 246
sarabí (HRC), 246–247
sarépari (HRC), 247
sayawi sa'para (HRC), 247
Sceloporus jarrovii (HRC), 332n574
Schimmer, O., 198–199
Scirpus spp. *(bakámwa)*, 134, 154, 165–166
Sebastiana spp. (HRC), 194–195, 325n399
seizures, 133, 308n54
Sejourné, Laurette, 4
Selaginella spp. (HRC), 198
self-healing practices, 135
Senecio spp. (HRC), 181–183, 250, 321n273
Septentrión River, 20–21
*se*ré* (HRC), 221–224, 335n658, 335n664, 336n666, 336n676

Serjania mexicana (HRC), 248, 346n940
sewáchari (HRC), 247, 346n935
sexual activity: after childbirth, 99; during *bajíachi* celebrations, 7, 88; court trials about, 68, 74, 81; Jesuit perspectives, 47; and Jíkuri ritual, 121, 153; during pregnancy, 95; tolerances, 68
sexually-transmitted diseases: cultural explanations, 121, 123, 130, 132, 279; prevalence, 279; remedies, 172, 178
shamans, in uprisings, 50–51. *See also* healers
sheep, domestic, 29
Sheridan, Thomas E., 292n34, 297n76, 297n83, 299n11
Sida rhombifolia (HRC), 215–216
Sierra Madre Occidental, 17, 287nn7–8
sierra region, physiographic overview, 17
simonillo (HRC), 247
sinowi ramerá (HRC), 248, 346n940
si'páame, 69, 136–137, 143–152, 308n5, 308n11
sipabuma, as disease explanation, 118
si'panema, in Jíkuri ritual, 152
si'píraka, 141, 145–146, 150–151
siríame grande: at Jíkuri ritual, 143, 144–145, 309n9; at *nawésari* ceremony, 79–81; during *noríruachi* festivities, 105, 110; in political structure, 73–74, 301n43
Sisters of Charity of St. Charles Borromeo, 286n16
Sisyrinchium arizonicum (HRC), 246, 345n923
sitagapi (HRC), 248, 346n944
sitákame (HRC), 249
skin color of Tarahumara, 59–64
skin disorders: cultural explanations, 128–129; herbal remedies, 156, 158–159, 167, 171, 175, 178, 182, 188, 191–194, 197, 218, 226–227, 229–230, 233, 236, 240, 242, 248–250, 315n117; western medical services, 282
skunks, in ecological habitat, 29
slash-and-burn agriculture, 27, 70
slavery, 46, 295n59
smallpox, 43, 44, 54, 293nn38–39
Smilax spp. (HRC), 183–184, 321n287

snakebites, western medical perspective, 280
snakes: in belief system, 264, 299n21, 305n15; bite remedies, 184, 201–202, 218, 229, 241, 254, 264, 324n363, 326n413; in disease explanations, 118–119, 126, 306n37; in ecological habitat, 33–34, 305n16; and Jíkuri ritual, 154; as medicinal remedy, 235, 247, 305n14; protective wards against, 259; and rituals, 154, 157, 247
snowfall patterns, 22, 288n23
só (HRC), 221–224, 335n658, 335n664, 336n666, 336n676
soap from plants, 163, 167, 220, 222, 336n666
social activities, 88–93, 303n9
social behaviors, 59, 67–68, 133, 134
social organization, 71–72, 82–83
sogíwari (HRC), 249–250
Solanum spp. (HRC), 177–178, 215, 263
solda (HRC), 250
sonogori, structure characteristics, 38, 291n15
sontarosi, in political structure, 72, 79
sopépari (HRC), 250
soul, in disease explanations, 115–119, 122, 137, 305n7, 305n11
sparrows, in ecological habitat, 31
spell casting. See witchcraft
Spicer, Edward, 50–51, 52, 297n76
spiders, in ecological habitat, 35
squash, diffusion of, 37, 290n8
squirrels, in ecological habitat, 30–31
staff, siríame's, 78, 79
state authorities, role of, 84–86
Steffel, Father, 89
Stemmadenia palmeri (HRC), 220
Stevia spp. (HRC), 240–241, 343n878
stomach ailments. See gastrointestinal ailments
sucking techniques of waniame, 138, 308nn10–13
suicide, 117, 282, 305n7
sukurúame, 117, 118, 119, 121, 137, 138–139
sunú (HRC), 250–251
sun worship, 68

supernatural explanations. See disease, conceptual explanations; religious beliefs
superstitions, kickball racing, 91, 118, 122
surachí okorá, prevalence, 132
surgeries, mystical, 131
sweat-bath format, 157–158, 311n17

Tabebuia palmeri (HRC), 227
Tagetes spp. (HRC), 168–170, 266, 353n1138
tapeworms. See worm-based diseases
Tarahumar people, overview, 5–6, 38–40, 71, 285n13, 292n25. See also specific topics, e.g., disease, cultural explanations; Jíkuri ritual; Plantago spp. (HRC)
Tardá, José, 49
Tauschia tarahumara (HRC), 202–203
tavachín (HRC), 251
Tecoma stans (HRC), 209, 209n548
Teloxys ambrosioides (HRC), 187–189
temperament of Tarahumara, 66–67
temperature patterns, 22–27
Tepehuan people, 38–39, 42, 48
Tephrosia spp. (HRC), 228, 338n727, 338n730
Tepórame (Indian leader), 48
tesgüino. See batari
tesoras, symbolism, 78, 79
Thalictrum spp. (HRC), 204
theft trials, 75–76
Thuja spp., 311n9
thunder, in religious beliefs, 69
Thyrallis glauca (HRC), 183
tikúwari (HRC), 251–252, 348nn984–985
tila (HRC), 252, 348n995
Tilia spp. (HRC), 252, 348n995
Tillandsia benthamiana (HRC), 239–240, 343n868
Tithonia fruticosa (HRC), 247, 346n935
Toboso people, 51, 296n67
toothache: herbal remedies, 159, 161–162, 164, 168, 180–183, 186, 194, 203, 218, 228, 238, 241, 246, 250, 256, 258, 260, 343n878; lizard remedy, 332n574
toronjil (HRC), 253
towita, 126, 127–128, 138, 233, 282

Tragia spp. (HRC), 230, 339n744
transhumance patterns, 25, 299n11
trauma. *See* wounds
travel fears, 119, 124, 126–127, 258
trébol (HRC), 253, 253n999
tree species, in ecological habitat, 27–29
Tropaeolum majus (HRC), 217–218
tuberculosis, 277–278
tubocurarine, western medicine uses, 4
*tu*chí* (HRC), 253
Tunera ulmifolia (HRC), 246
turpentine (HRC), 229, 338n735
tutuguri dances, 91–93, 120, 141, 142, 144
twentieth-century history, overview, 54–56
"The Two Brothers" legend, 40–41
typhoid and typhus, 43, 274, 293n39, 294n45

umbilical cord, 99, 303n9
unto sin sal (HRC), 253–254
Uranga, Ernesto, 82, 84, 299n7
Urbina, Manuel, 53, 332n568
uré (HRC), 254
Urera caracasana (HRC), 230, 339n744
urí (HRC), 254
urinary ailments, herbal remedies, 163, 167, 171, 173–174, 206, 211, 213–214, 218, 223, 231, 240, 248, 250, 254–255, 262, 324n376, 347n968, 348n995
Urique River (and canyon), 17, 18, 20, 21, 34
Urique (town), 21, 25
Urtica spp. (HRC), 230, 339n744
usabi (HRC), 255
Usnea subjusca (HRC), 236

vaccinations, 267, 272
valeriana (HRC), 255, 349n1025, 349n1027
Vallejo, Carlos, 288n9, 290n3
Varijío people, 39, 41, 44
Varohío people, 292n28
Vega, Marcial, 309n17
Velarde, Luis Javier, 325n399
Velásco, Juan Bautista, 188, 293n39
verbena (HRC), 256, 350n1029
Verde River, 20, 289n49

Verplancken, Luis G., 288n20
Vesga, Mateo de, 42
Vigueria decurrens (HRC), 227, 337n712
Villacís, Luis, 352n1099
Villa de Aguilar, 48, 294n41
Villalobos, Manuel, 286n18
Viola umbraticola (HRC), 253
Vitex mollis (HRC), 199–200, 327n450
Vitis spp. (HRC), 254
Vivó, Jorge A., 59
Vogel, Virgil J., 53, 312n34, 333n591, 336n666, 337n711, 352n1099
vultures, in ecological habitat, 33

Wadding, Father, 294n46
Wagner, Federico, 309n17, 334n638, 348n995, 350n1057
walking diseases, 125
Walúla legend, 119
Wampler, Joseph, 18
waniame, healing practices, 121, 130, 138, 308nn10–12
waré (HRC), 256
warfare, 41–42, 44, 46–48, 55, 296nn67–68, 296nn70–71
wasárowa (HRC), 256–258
wasía (HRC), 258–259, 350n1057
Waterfall Cave artifacts, 291n13
waterfalls, 21, 288n20
watorí (HRC), 259–260, 351n1069
watosí (HRC), 260
wawachí (HRC), 260
wawana, 128–129, 130, 175, 182, 197
Wedleia sp. (HRC), 227
wejazo, in Jíkuri ritual, 152, 310n13
western medical practices, 3–5, 53–55, 282–283, 356n50. *See also* health conditions, western medical perspective
wetajúpachi (HRC), 260
whirlwinds, role in disease, 123, 306n31
white people, in Tarahumara religious beliefs, 92–93
Whitt, Brondo, 288n20
wichá (HRC), 260–262, 352n1099

wichásuwa (HRC), 262
wichíwari, 134, 308n59
wichuri (HRC), 262–263, 352n1107
wiígame (HRC), 263
wikaráame, during *tutuguri* dances, 92–93
wikubema (HRC), 157–159
wipá or *wipáka* (HRC), 264, 352n1117, 353n1122
wisaró (HRC), 265
witchcraft: court trials, 301n48; in disease explanations, 117–118, 133, 305n7; herbal protections against, 244, 315n124; kickball races, 91; regional reputations, 140; *siríame grande* responsibilities about, 74; *sukurúame*'s role in, 138–139
wítzora (HRC), 265
woodpeckers, in ecological habitat, 32, 289n44
Woodsia mexicana (HRC), 214–215, 333n587, 333n591
Woodwardia spinulosa (HRC), 214–215, 333n587, 333n591
worm-based diseases: herbal remedies, 165, 189, 191, 192, 214, 254, 307n46; types of, 129–130, 132, 307n47; *waníame*'s healing practice, 138
worms, in ecological habitat, 34–35
wounds: cultural explanations, 123, 125, 130, 132; herbal remedies, 162–164, 166–168, 173, 175, 177–178, 180, 183, 187–188, 190–191, 195–197, 201–202, 206, 218–219, 229, 236, 240–242, 256, 258–260, 319n222, 326n420, 334n638, 338n735, 344n884, 345n920; western medical perspective, 279–280
wrestling contests, 106, 110, 152–153

Yaqui River, 21–22
Yepómera Mission, 49

zacate limón (HRC), 265
Zanthoxylum pentanome (HRC), 231–232, 339n764
Zapata Ortiz, Vicente, 45
Zea mays (HRC), 250–251. *See also* maize
Zeiner, Arthur R., 281
zempoal (HRC), 266, 353n1138
zempoalillo (HRC), 266, 353n1141
Zepeda, Nicolás de, 46
Zexmenia podocephala (HRC), 205, 330n512
Zingg, Robert M., writings on: cultural history evidence, 37–38, 291n13; disease concepts, 121; herbal remedies, 159, 185, 195, 210, 226, 240, 251, 255, 257, 264, 320n248, 324n363; lizard remedies, 332n574; witchcraft, 301n48
Zornia spp. (HRC), 195–196